T0390692

One Health

One Health

Integrated Approach to 21st Century Challenges to Health

Edited by

Joana C. Prata
University of Aveiro, Portugal

Ana Isabel Ribeiro
Epidemiologist and Health Geographer, Public Health Institute of University of Porto, Portugal

Teresa Rocha-Santos
University of Aveiro, Portugal

Academic Press is an imprint of Elsevier
125 London Wall, London EC2Y 5AS, United Kingdom
525 B Street, Suite 1650, San Diego, CA 92101, United States
50 Hampshire Street, 5th Floor, Cambridge, MA 02139, United States
The Boulevard, Langford Lane, Kidlington, Oxford OX5 1GB, United Kingdom

Notices

Knowledge and best practice in this field are constantly changing. As new research and experience broaden our understanding, changes in research methods, professional practices, or medical treatment may become necessary.

Practitioners and researchers must always rely on their own experience and knowledge in evaluating and using any information, methods, compounds, or experiments described herein. In using such information or methods they should be mindful of their own safety and the safety of others, including parties for whom they have a professional responsibility.

To the fullest extent of the law, neither the Publisher nor the authors, contributors, or editors, assume any liability for any injury and/or damage to persons or property as a matter of products liability, negligence or otherwise, or from any use or operation of any methods, products, instructions, or ideas contained in the material herein.

ISBN: 978-0-12-822794-7

For information on all Academic Press publications visit our website at https://www.elsevier.com/books-and-journals

Publisher: Andre G. Wolff
Acquisitions Editor: Elizabeth Brown
Editorial Project Manager: Sam W. Young
Production Project Manager: Sreejith Viswanathan
Cover Designer: Greg Harris

Typeset by TNQ Technologies

Contents

Contributors

Kahrić Adla, Department for Genetics and Biomedical Engineering, Center for Marine and Freshwater Biology Sharklab ADRIA, Sarajevo, Bosnia and Herzegovina

Luís Alves de Sousa, Public Health Doctor

Kulijer Dejan, Natural History Department, National Museum of Bosnia and Herzegovina, Sarajevo, Bosnia and Herzegovina

Šnjegota Dragana, Faculty of Natural Sciences and Mathematics, University of Banja Luka, Banja Luka, Bosnia and Herzegovina

Maria Pia Ferraz, Departamento de Engenharia Metalúrgica e de Materiais, Faculdade de Engenharia da Universidade do Porto, Porto, Portugal; i3S – Instituto de Investigação e Inovação em Saúde, Universidade do Porto, Porto, Portugal; INEB – Instituto de Engenharia Biomédica, Universidade do Porto, Porto, Portugal

Bernardo Mateiro Gomes, Porto Public Health Institute, Oporto Medical Faculty, North Health Region Administration, Porto, Portugal

Najla Haddaji, Department of Biology, Faculty of Sciences, University of Ha'il, Ha'il, Kingdom of Arabia Saudi; Laboratory of Analysis, Treatment and Valorization of Pollutants of the Environment and Products, Faculty of Pharmacy, Monastir, Tunisia

Balogun A. Halimah, National Veterinary Research Institute, Vom, Plateau State, Nigeria

Rui Leandro Maia, UFP Energy, Environment and Health Research Unit (FP-ENAS), University Fernando Pessoa, Porto, Portugal

Mariana Marrana, Preparedness and Resilience Department, World Organisation for Animal Health, Paris, France

Bolajoko Muhammad-Bashir, National Veterinary Research Institute, Vom, Plateau State, Nigeria

Dedić Neira, Department of Botany and Zoology, Masaryk University, Brno, Czechia

João Niza-Ribeiro, Departamento de Estudo de Populações, Vet-OncoNet, ICBAS, Instituto de Ciências Biomédicas Abel Salazar, Universidade do Porto, Porto, Portugal; EPIUnit - Instituto de Saúde Pública, Universidade do Porto, Porto, Portugal; Laboratório para a Investigação Integrativa e Translacional em Saúde Populacional (ITR), Porto, Portugal

Gisela Marta Oliveira, UFP Energy, Environment and Health Research Unit (FP-ENAS), University Fernando Pessoa, Porto, Portugal

Sandra Oliveira, Centro de Estudos Geográficos, Instituto de Geografia e Ordenamento do Território, Universidade de Lisboa, Rua Branca Edmée Marques, Cidade Universitária, Lisboa, Portugal

Chiara Palmieri, The University of Queensland, School of Veterinary Science, QLD, Australia

Katia C. Pinello, Departamento de Estudo de Populações, ICBAS, Instituto de Ciências Biomédicas Abel Salazar, Universidade do Porto, Porto, Portugal; Laboratório para a Investigação Integrativa e Translacional em Saúde Populacional (ITR), Porto, Portugal; EPIUnit - Instituto de Saúde Pública, Universidade do Porto, Porto, Portugal

Manuela Pontes, UFP Energy, Environment and Health Research Unit (FP-ENAS), University Fernando Pessoa, Porto, Portugal

Joana C. Prata, University of Aveiro, Portugal

Carlos Branquinho Rebelo, Department of Pathobiology and Population Sciences, Royal Veterinary College, Hertfordshire, United Kingdom

Ana Isabel Ribeiro, Epidemiologist and Health Geographer, Public Health Institute of University of Porto, Portugal

Jorge Rocha, Centro de Estudos Geográficos, Instituto de Geografia e Ordenamento do Território, Universidade de Lisboa, Rua Branca Edmée Marques, Cidade Universitária, Lisboa, Portugal

Teresa Rocha-Santos, University of Aveiro, Portugal

Joelma Ruiz, Joelma Ruiz Psychology Center, São Paulo, Brazil

Cláudia M. Viana, Centro de Estudos Geográficos, Instituto de Geografia e Ordenamento do Território, Universidade de Lisboa, Rua Branca Edmée Marques, Cidade Universitária, Lisboa, Portugal

Diogo Guedes Vidal, UFP Energy, Environment and Health Research Unit (FP-ENAS), University Fernando Pessoa, Porto, Portugal

Maria Lúcia Zaidan Dagli, Laboratory of Experimental and Comparative Oncology, School of Veterinary Medicine and Animal Science, University of São Paulo, São Paulo, Brazil

Biographies

Joana C. Prata, DVM, PhD, conducts research on environmental challenges under the One Health and on sustainable approaches to mitigate them under the Circular Economy. She holds a master's degree in Veterinary Medicine (2016) from the University of Porto, Portugal, being a certified veterinarian, and a PhD degree in Biology and Ecology of Global Change (2021) from the University of Aveiro, Portugal. She has published 40 scientific papers and has an h-index of 15 (2021).

Ana Isabel Ribeiro (MPH, PhD) is an epidemiologist and health geographer. She is a Researcher at the Public Health Institute of the University of Porto and Principal Investigator of two projects about neighborhoods and health. Ana is the leader/founder of the Research Lab "Health and Territory," a research group of the IRT—Laboratory for Integrative and Translational Research in Population Health. She is particularly interested in understanding how the social (e.g., deprivation) and biogeophysical context (e.g., pollution, green space) where individuals live influence their health and well-being and in emerging health risks such as climate change and environmental degradation. At the same time, she is an Invited Assistant Professor at the Faculty of Medicine (University of Porto), teaching Epidemiology, Demography, and Geographic Information Systems. Since 2019, she is an Associate Editor of the Journal Public Health Reviews.

Teresa Rocha-Santos has graduated in Analytical Chemistry (1996), obtained a PhD in Chemistry (2000) and an Aggregation in Chemistry (2018), both at the University of Aveiro, Portugal. Presently, she is a Principal Researcher at Centre for Environmental and Marine Studies (CESAM) & Department of Chemistry of University of Aveiro (since 2014) and Vice-coordinator of CESAM research centre (from 2021). Her research concentrates on the development of new analytical methodologies fit for purpose and on the study of emerging contaminants (such as microplastics) fate and behavior in the environment and during wastewater treatment. She published 170 scientific papers (October 2021) and has an h-index of 43 (October 2021). She is the editor of seven Books. She is a member of the editorial board of Current Opinion in Environmental Science and Health, Elsevier (since 2017), Data in brief, Elsevier (since 2018), Science of the Total Environment,

Elsevier (since 2018), Sensors, MDPI (since 2018), Molecules, MDPI (since 2018), Associate Editor of Euro-Mediterranean Journal for Environmental Integration, Springer (since 2016), and Associate Editor of Journal of Hazardous Materials (since 2019) and Co-Editor in Chief of Journal of Hazardous Materials Advances (since 2021).

Preface

The One Health approach provides an integrative view on health, encompassing the three interconnected pillars: human, animal, and environment. While humans and animals can share diseases (e.g., zoonosis), the environment provides a support system that modulates exposure and susceptibility. The complexity and interconnectedness of the modern world create challenges beyond the understanding of traditional linear thinking. Effective and sustainable solutions to modern health challenges can only be achieved by comprehensive approaches to complex systems involving multiple actors across disciplines. One Health is necessarily transdisciplinary, requiring greater collaboration between different professionals and stakeholders, from medical doctors, public health professionals and veterinarians to urban planners, sociologists, geographers, policymakers, engineers, farmers, and so on. Only through collaboration and greater understanding will humanity be able to endure the greatest challenges to health of the 21st century, including emerging diseases, climate change, food and water security and safety, and the degradation of ecosystems. As a discipline mainly founded in public health, One Health has so far been focused on infectious diseases. There is an urgent need to expand the One Health approach to other challenges threatening humanity, leveraging its collaborative approach to find sustainable solutions. Environmental challenges, such as climate change, have the potential to have devastating effects on human health and should be addressed under the One Health approach. This book offers a reference material by providing a framework and short examples to understand One Health in practice while balancing the presentation of traditional concepts on infectious diseases with environmental challenges. The book also attempts to highlight the role of different actors, beyond health workers, in preserving public health through the interpretation of complex systems. The book is organized into ten chapters, introducing core concepts and presenting key challenges to health that will shape the 21st century. The editors are grateful to the authors and Elsevier staff who helped assemble the chapters and provide an interdisciplinary overview of the various issues under One Health. The editors also welcome readers interested in this topic, from researchers to undergraduate and graduate students to professionals in multiple areas. Hopefully, this work can provide the foundation for more inclusive collaboration and transdisciplinarity, culminating in a complete One Health approach necessary to tackle the great challenges of the coming decades.

Joana C. Prata
Ana Isabel Ribeiro
Teresa Rocha-Santos

Chapter 1

An introduction to the concept of One Health

Joana C. Prata[a], **Ana Isabel Ribeiro**[b] **and Teresa Rocha-Santos**[a]
[a]*University of Aveiro, Portugal;* [b]*Epidemiologist and Health Geographer, Public Health Institute of University of Porto, Portugal*

1. Introduction

The new challenges faced by humankind raise awareness about the connection between human, animal, and environmental health. Working in separate disciplines at a local level is no longer sufficient to tackle the growing complexity of the modern world. The One Health approach offers a collaborative and transdisciplinary strategy to attain optimal health and wellbeing at the interface between humans, animals, and the environment, working on prevention and mitigation of diseases. This concept transcends anthropocentrism, attempting to simultaneously provide optimal health for humans, animals, and the environment, following a sustainable development (Bordier, Uea-Anuwong, Binot, Hendrikx, & Goutard, 2020). One aspect of this approach is the control of zoonoses, diseases shared between animals and humans (Zoonoses, 2021). Despite the most frequent application of the One Health approach being in the control of parasitic and infectious diseases shared between humans and animals, many other areas of convergence should be explored. For instance, environmental risk calls for a collaborative surveillance since humans and animals are frequently exposed by sharing the same resources (e.g., water), humans can be contaminated from the ingestion of animal products, and animals can be used as sentinels for human health (Bordier et al., 2020). Beyond contaminants, the environmental aspect of One Health should always be explored as an element that situates disease in a specific temporal, socioeconomic, and physical environment (Rock, Buntain, Hatfield, & Hallgrímsson, 2009). The application of One Health relies on the synergy between experts from many disciplines (e.g., veterinary, health, environment), stakeholders (e.g., politicians, businessman, farmers), and affected communities, creating a value-added and participatory approach that can give rise to innovative solutions. By providing tools for surveillance and

One Health. https://doi.org/10.1016/B978-0-12-822794-7.00004-6

improving the early detection of threats through a transdisciplinary interpretation of data, this approach follows the Precautionary Principle, which warrants precaution in face of uncertainty, recognizing the interconnectedness of systems (e.g., environmental, sociocultural, economic) and increasing their resilience (Akins, Lyver, Alrøe, & Moller, 2019). The benefits achieved by One Health go beyond the sum of its parts through the collaboration of multiple disciplines and stakeholders, applying a systems approach that conveys an operational overview of the problem across subjects. The need for such approach has been made clear in past outbreaks, such as Ebola, Avian Influenza, and severe acute respiratory syndrome (SARS) (Calistri et al., 2013). More recently, the COVID-19 pandemic, caused by SARS-CoV-2, likely originating from a spillover event (from a still unconfirmed animal source) in a wildlife market in Wuhan, China, has infected hundreds of millions of people, caused millions of deaths, and severe socioeconomic disruption (WHO Coronavirus, 2021). The response to the pandemic could have had benefited by a One Health approach by creating joint surveillance and testing, mobilizing resources, and involving communities in the fight against COVID-19 (Mushi, 2020).

2. The history of One Health

The concept of One Health has been present in the human subconscious since the beginning of the species. Many pastoral communities share healing practices between animals and humans, with myths supporting the interdependency of humans on the environment and animals (Zinsstag, Schelling, Waltner-Toews, & Tanner, 2011). Veneration and protection of the environment and animals, such as rational use of natural resources (e.g., hunting and foraging) on which they survived, was common in premodern societies but was later lost in the progression to a more anthropocentric and industrialized world. The first veterinary schools in Europe, founded in the late 18th century, benefited from the contribution of medical doctors to the veterinary profession (Woods & Bresalier, 2014). In the 19th century, Rudolf Virchow, known as the father of cellular pathology, contributed greatly to both medicine and veterinary medicine, recognizing the interconnecting and comparative aspects between these disciplines (Saunders, 2000). It is though that one of Virchow's students, Osler, introduced the term "One Medicine", but the concept was only revived in the 20th century by Calvin Schwabe (Zinsstag et al., 2011). In Schwabe's book, "One Medicine" is described as an interaction between human and animal health, which besides epidemiology of infectious and non-infectious diseases, addressed food security from animal sources (Schwabe, 1964).

Further development of the concept of One Health started in a set of conferences by the Wildlife Disease Association and the Society for Tropical

Veterinary Medicine in 1999 and 2001 addressing the need to encourage collaboration among specialists to promote health. The resulting Pilanesberg Resolution called for an international collaboration on animal health to improve the sustainable management of wildlife and livestock (Karesh, Osofsky, Rocke, & Barrows, 2002). In 2005, the scientific journals *The Veterinary Record* and *BMJ* published a joint issue on "Human and animal health: strengthening the link" to strengthen collaboration between the fields of veterinary and human health (Gibbs, 2014). In the same period, the concept of "One World, One Health" was first introduced in a conference of the Wildlife Conservation Society focused on interdisciplinary approach to the emergence of diseases, translated into 12 recommendations that became known as the Manhattan Principles (One World, 2021). Many entities (e.g., World Health Organization - WHO; World Organization for Animal Health - OIE) mostly view One Health as a contribution of other disciplines to human health, defining it as a collaboration of multiple sectors to achieve better public health outcomes, such as in zoonoses control, food safety, and antibiotic resistance (One Health, 2021a). This view limits the scope of One Health, only improving it slightly over the concept of One Medicine. A more general definition is provided by the One Health Initiative (One Health, 2021b), which defines One Health as an interdisciplinary collaboration and communication in all aspects of human, animal, and environmental health, including in research and education (Fig. 1.1). The concept can even be further expanded by

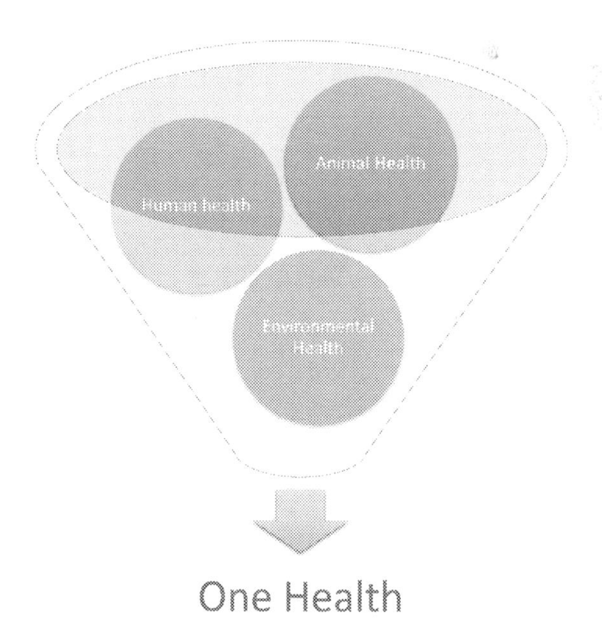

FIG. 1.1 One Health approach as transdisciplinary approach to human, animal, and environmental health.

considering the societal environment and by including different scales of systems biology (from molecules to populations) (Zinsstag et al., 2011). A One Health approach must consider all environmental aspects of disease (social, economic, cultural, and natural) that contribute to its occurrence, including non-communicable or non-infectious diseases, benefiting all disciplines through the sharing of information and resources across different scientific fields. Currently, the WHO, OIE and Food and Agriculture Organization (FAO) have a strategic collaboration to address challenges under the One Health approach (The FAO-OIE-WHO Collaboration, 2021).

Another concept that intersects with One Health is the ecosystems approach to health, or Ecohealth, due to the intention of bridging many interrelated disciplines. Ecohealth, which has been gaining momentum since the 90s, is the systematic and sustainable approach to human and animal health and wellbeing or welfare in the context of social and ecological interaction, promoting healthier ecosystems through the integration of different fields and producing real world solutions (Waltner-Toews, 2009). While One Health is founded on the management of zoonosis shared between human and animals, Ecohealth is founded on addressing environmental issues in the context of ecosystems (Zinsstag, 2012). Another difference is the emphasis of One Health on human and animal health equally, inevitably considering environmental factors, while Ecohealth is focused on improving human health through the optimization of ecosystems by combining public health and environmental management (Mi, Mi, & Jeggo, 2016). Both disciplines are evolving toward the same intersecting understanding of the role of socio-ecological systems in health (Mi et al., 2016). By sharing knowledge on animal health issues, in the case of One Health, and on an ecosystem and sustainable development, in the case of Ecohealth, both fields could benefit from increased collaboration, especially by involving a more multidisciplinary team (Zinsstag, 2012). One Health could also benefit from a practical approach applied by EcoHealth based on six principles: (i) systems thinking, by considering the relationship between the several dimensions requiring the expertise of social sciences; (ii) transdisciplinary research, involving scientists across disciplines and expert in soft skills (e.g., negotiation) which can create platform for the engagement of stakeholders and community representatives; (iii) participation, involving communities in the attempt to find locally rooted innovative solutions and resolving divergences impeding progress from a shared understanding of the problem and negotiation between actors; (iv) sustainability, protecting and improving ecosystems providing positive, ethical, and lasting dynamic changes addressing both short and long-term challenges to improve the livelihood of communities; (v) gender and social equality, addressing unequal conditions holding back disadvantaged groups by reducing inequities and improving ecosystems which can provide more livelihood opportunities;

(vi) knowledge to action, producing actionable but not necessarily exhaustive knowledge capable of evolving with the situation, which can be applied in practice to public health or influence policy, through strong communication, dissemination and networking strategies (Charron, 2012). However, cultural differences, difficulty in abandoning the original fields of work, and difficulty in engaging medical doctors can limit collaboration (Mi et al., 2016).

3. The benefits of a One Health approach

The World Bank has estimated that the direct costs of outbreak of zoonotic diseases in a decade (2000–10) as US$20 billion, with over US$200 billion in indirect economic losses (Bank, 2010). One Health approach to surveillance and response could have helped reduce outbreak events and/or severity, reducing economic losses and providing a more efficient approach to the use of resources. It has been estimated that an investment in One Health of US$25 billion over 10 years could generate benefits worth five times more (US$125 billion), in addition to improved ecosystems and saved DALYs (Disability Adjusted Life Years) (Grace, 2014). As an example, nomadic people in Chad benefited from a joint mobile vaccination program which vaccinated livestock, important for their livelihood, but also children and women, achieving a higher coverage rate, with the benefit of sharing resources between veterinarians and physicians and reducing the cost of the intervention (Schelling et al., 2007). Clear benefits are achieved from this approach when health resources are scarce, combined infrastructures and skillset are underutilized, and when there is a need to respond quickly and proportionally to a health emergency (Rushton, Häsler, De Haan, & Rushton, 2012). This approach creates an added value that goes beyond the individual disciplines, which would be ineffective in addressing the problem from all perspectives, at different time scales, and at a global level (Rüegg et al., 2017). One Health intends to provide durable benefits across disciplines (e.g., health, economy, environment), across species, and at different scales, from cells to ecosystems (Rüegg et al., 2017). Hasler, Cornelsen, Bennani, & Rushton (2014) identified the major benefits as improvements in human and animal health (including wellbeing or welfare, and better disease control), social and cultural improvement (e.g., poverty reduction), environmental protection (e.g., improving ecosystem resilience), economic benefits (e.g., reduction of costs), and higher quality and quantity of data (Hasler et al., 2014). However, many reports have no clear outcomes or only report broad qualitative metrics, which hinders the interpretation of results (Baum, Machalaba, Daszak, Salerno, & Karesh, 2017). Therefore, clear definition of outcomes and measurements of well-defined quantitative and qualitative data can support a more widespread application of the One Health approach.

4. Obstacles in the implementation of the One Health approach

Despite the clear benefits of One Health, many barriers still need to be transposed for an effective application of its principles. The first difficulty is in translating the concept of One Health into an actionable plan with a clear agenda. For many, One Health is still a broad and vague concept that is poorly recognized in society, with undetermined costs and benefits, often used for rebranding of work still conducted in the original disciplines (Gibbs, 2014). The resistance to the concept also originates from cultural differences and resistance in abandoning the original fields of research, especially difficult in engaging medical doctors which often see the concept as an intrusion of other disciplines into medicine (Mi et al., 2016). The integration of disciplines requires a high-level support with great institutionalization in education by providing necessary capacity building, but also at political level, by defining clear budgets and agendas and creating synergies (e.g., between Ministries of Health and Agriculture, for instance), hopefully reached at a global level (Lee & Brumme, 2013). However, this effort can be hindered by the lack of agreement in leadership to define priorities and the unclear translation of results of One Health into clear advantages or actionable policies.

The necessary restructuring of multiple strategies and disciplines is accompanied by a strong inertia, also resulting from different priorities, views, and concepts, that are difficult to articulate, and the traditional allocation of funds which favors competition instead of collaboration (HASLER et al., 2014). The lack of consensus is even felt in those already adopting a One Health approach. For instance, the definition of surveillance strategies considered under the One Health approach can vary from the collection of data from multiple domains to solely requiring intersectoral collaborations (Bordier et al., 2020). Many barriers exist in information sharing between fields, such as unharmonized collection, insufficient sharing within and between institutions, incomplete interpretation under the One Health scope, and legal constraints (e.g., confidentiality) (Bordier et al., 2020). For instance, a clear understanding of the disease is required by all participants in order to plan and apply a shared intervention (Johnson, Hansen, & Bi, 2018). Related to data collection is the insufficient application of health indicators to clearly translate the outcomes of the approach, contributing to its valorization (Hasler et al., 2014). Stakeholder engagement, across disciplines and sectors (e.g., private sectors), increases difficulty in the implementation effort by requiring management of multiple actors with different expectations and approaches, but contributes to a better overall outcome by promoting innovative solutions with long-term benefits with greater acceptability. Despite all these challenges, One Health is slowly gaining recognition, which will create a more receptive environment to overcome these barriers and limitations.

5. Practical principles in the application of One Health

Implementation of a One Health approach is complex, not facilitated by the current lack of a universal conceptual framework in which to plan and operationalize initiatives. Bordier and colleagues organized the One Health approach by working on three collaboration levels: (i) policy; (ii) institutional; and (iii) operational (Bordier et al., 2020). The policy level is responsible for defining the strategy regarding the objectives, areas of action, and resource allocation. The institutional level defines appropriate collaboration to achieve the targets set by the policy level, which requires definition of areas of implementation (e.g., sampling, laboratory testing, data analysis), the roles and responsibilities of each actor involved, technical mechanisms to support collaboration (e.g., shared database), and resources allocation (i.e., human, material, and financial resources). Finally, the operational level puts in practice collaborative actions by detailing procedures, creating technical mechanisms and tools to support collaboration (e.g., data sharing routine), and managing resource allocation.

At policy level, a cost-sharing scenario can help determine which allocation of resources across sectors can produce the most benefit. For instance, in Mongolia, Brucellosis vaccination program for livestock would produce benefits worth of US$ 18.3 million (3.2 times the amount invested, including health benefits). A contribution proportional to the benefit of each sector would only require 11% participation from the public health sector, as the private sectors (e.g., livestock breeders) would be the ones to benefit the most (Roth et al., 2003). Costs and complexity can be reduced by limiting collaboration to the minimum level possible to achieve the objective, with the aim of increasing the efficiency of resource use or added benefits (Babo Martins et al., 2017). Permanent joint-operational mechanisms need to be created at national levels, for instance by having individual ministries report to a common task force in direct contact with the executive level of government, and at international levels, following a similar approach such as by having a high-level United Nations mechanism which received information from WHO, OIE, FAO, and other relevant institutions (e.g., environmental health or wildlife organizations) (Bank, 2010). Sustainable Development Goals, a list of 17 goals defined by the United Nations which are necessary to reach prosperity for the people and the planet, can provide a foundation in which to develop specific One Health goals (Queenan, 2017). Highly relevant topics that can be addressed using a One Health approach include current public health issues (e.g., social acceptance of vaccination), management of natural resources (e.g., water), farming practices, and tropical diseases (e.g., food-borne parasitic zoonoses) (Binot et al., 2015). For instance, water management accounts for availability, water-borne diseases, use in food production, environmental contamination, wastewater management, flood control, and others.

At institution level, collaboration must involve multiple sectors (e.g., food safety, public health, animal health), across disciplines (e.g., social sciences, medicine), and involving both public and private partners (e.g., antibiotic resistance surveillance systems). This collaborative system must include an intersectoral framework (official or collaborative) and intersectoral strategy, supervision by the same authority, efficient communication channels (e.g., joint database, harmonization of data), and mechanisms to ensure commitment of all stakeholders (Bordier et al., 2020). Education and training are also essential to achieve the transdisciplinarity required for One Health implementation, which can be aided by the introduction in curricula or through accreditation (Mackenzie, McKinnon, & Jeggo, 2014). For instance, case-based classroom or workshop activities can be implemented to further disseminate the One Health approach, just as by using a "jigsaw" format for active learning (Bartlow & Vickers, 2020). At operation level, concrete implementation must allocate roles, responsibilities, and competencies, which requires strong connections and coordinated activities between actors (Rüegg et al., 2017). Operationalization will be specific for the problem and goals being pursued. In practice, a systems approach can be applied to better understand a complex system. Rüegg and colleagues suggest addressing the following concepts when trying to understand a One Health system: (i) aim and indicators of the system (not synonym to the aim of the initiative) which could respond to the purpose of the system (e.g., in food production) or indicators of selected attributes in the social-ecological system (e.g., productivity, health); (ii) stakeholders, who affect or are affected by the decisions, and actors, subgroups of stakeholders in specific activities; (iii) relationships, which should identify links between components, their nature (e.g., information share, goods transfer), their characteristics (e.g., slow/fast, weak/strong), and the most important links; (iv) dimensions, which are different levels of organization and which might have scales (e.g., the dimension of life ranges from molecules to populations) and might be restricted by boundaries (e.g., regulations); v) evolution of the system, which considers changes in dynamics and boundaries with time (Rüegg et al., 2018). The use of systems approach allows a smart evaluation of the system, making compromises on which to include, instead of being overly comprehensive.

A classical *top-down* approach is based on management by higher levels of organization following a broader understanding of the problem, as previously described. Due to the nature of One Health, an opposite *bottom-up* approach could also be adopted, with a possible co-existence of both approaches. A *bottom-up* approach is especially important to detect emerging health threats, lacking specific surveillance mechanisms. For instance, integrated with other risk management under One Health, environmental contamination can benefit from a *bottom-up* approach based on continuous surveillance and information sharing at local level, which could then be communicated to state or national levels (Eddy, Stull, & Balster, 2013). As contaminants often affect both humans and animals sharing the same environment, information sharing under the One Health approach allows early detection and intervention. Surveillance for environmental contaminants already exists in public health, expansion of the programs to integrate veterinary data and other fields would provide the tools for a premature and more sensitive threat identification. For instance, the Center for Disease Control and Prevention (CDC), in the US, has an Environmental Public Health Tracking Program to identify environmental hazards threatening human health (Environmental Public Health Tracking Program, 2021), which could be expanded to incorporate information from other areas (e.g., veterinary, ecological, environmental).

A One Health approach can provide the foundations to conduct risk assessment, supporting decision making. Risk assessment is a process which allows to identify harmful agents (hazard identification), determine how and what levels cause adverse effects (hazard characterization), determine who or what is exposed and to how much (exposure assessment), and determine the likelihood and the extent of the adverse effects (risk characterization). This information is then used to reduce or set acceptable levels of risk (risk management) and to determine how and to whom will the risks be communicated to (risk communication). The European research project COHESIVE is developing the One Health Risk Analysis System (OHRAS) focused on zoonosis, providing a simple implementation by: (1) setting goals and objectives; (2) determining involved stakeholders; (3) mapping the current organization of risk analysis; (4) designing activities such as signaling, risk assessment, and risk management; and (5) implementation. A One Health approach may also be implemented during health emergencies (e.g., during pandemics), accompanying the evolution of the outbreak. The first step is to identify and describe the outbreak with the available information, which can be obtained by contacting national and international agencies and consulting available literature on similar events. The second step is to evaluate potential sectors involved or impacted by the outbreak (e.g., socioeconomic impacts), which may comprise stakeholders. The third step is to create a One Health task force by inviting the necessary elements to address the problem, organize them in teams to conduct risk assessment, defining tasks for the group, gathering available scientific evidence, planning new studies, and defining shared terminology and methods. The final step is to define a strategy of risk communication, which is usually provided by the One Health task force to the authorities, who are responsible for policymaking. Clarity of information is especially important since misinformation can have severe socioeconomic and health costs. In all cases, the role of the One Health approach is to provide advice at a transdisciplinary level to address a problem with the best available information, which may not be necessarily exhaustive.

6. Global challenges which can benefit from One Health approach

The Anthropocene, the human-dominated geological epoch we are living in, is characterized by the occurrence of global changes, which are planetary scale changes to the environment resulting from human activity. These changes give rise to multiple global challenges that will define the XXI century and which would benefit from the transdisciplinary approach proposed by One Health (Fig. 1.2). Thus, it is important that One Health approach expands from local to worldwide problems, materializing into improvements in human, animal, and environmental health at a global scale. These key challenges, such as emerging infectious diseases and climate change, are shortly addressed here and will be discussed in more detail in the following chapters of this book.

6.1 Outbreaks of emerging infectious diseases

Emerging infectious diseases are defined by the WHO as serious public health treats at a global scale that suddenly appear or reemerge (WHO World Health Organization, 2014). About 60% of emerging infectious diseases are zoonoses, which globally cause one billion cases of illness and millions of deaths (Zoonotic disease, 2021). The emergence of infectious diseases is related to the dynamics of natural communities and environments, potentiating an adaptative emergence when a microorganism is capable of infecting a new host (e.g., humans) or a geographical emergence when pathogens or parasites expand their geographical ranges (Ogden et al., 2019). Factors involved in disease emergence and globalization include ecological changes (many resulting from anthropogenic interventions), climate change, changes in

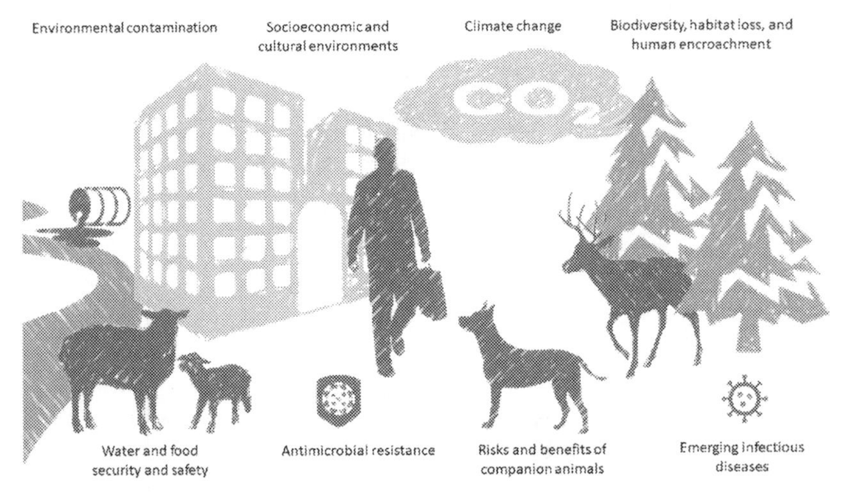

FIG. 1.2 Eight global challenges for the XXI century that should be addressed following a One Health approach.

human demography and behavior, human activities (e.g., travel, commerce, industry), insufficient public health measures, and microbial changes and adaptation (Zessin, 2006). Many of these factors are related to human cultural and socioeconomic issues, such as the increasing deforestation in the establishment of new agricultural and pastoral lands to feed a growing human population. Geographical emergence also requires the crossing of geographical barriers (e.g., invasive species transported during human activities), capability of surviving and completing the life cycle in the new environment, and propitious environmental conditions that create a suitable niche for the establishment and spreading of species (Ogden et al., 2019). Human encroachment in wildlife habitats, through increasing proximity and increasing the density of wildlife populations in fragmented habitats, facilitates spillover events, when the pathogens is transmitted from a reservoir population to a new host (Daszak, 2000). Conversely, humans and domesticated animals can introduce the disease to wild populations. For instance, wild reservoirs of tuberculosis can initially originate in the transmission from livestock, then creating a transmission dynamic between wildlife, livestock, and humans (Miller & Olea-Popelka, 2013). Influenza A virus also has complex ecologies involving multiple host species and frequent molecular evolution, mostly originating in avian hosts but evolving in pigs through the simultaneous infection with avian-like and mammalian-like virus ("mixing vessel") giving rise to strains potentially capable of infecting humans (Kahn, Ma, & Richt, 2014). Emerging infectious diseases originate from a closer contact between anthropogenic and wild systems, originating from multiple environmental factors, and requiring a joint surveillance which allows for early detection and control of spillover events.

6.2 Risks and benefits of companion animals

Companion animals, or pets, are present in many households worldwide. Based on a study conducted in 22 countries, 33% of households have a dog, 23% a cat, and 12% fish (Man's best friend, 2021). Pet ownership has been related to multiple benefits to human health. The human-animal bond benefits psychological and social health, decreasing anxiety, loneliness, and depression by reducing the perception of situations as negative or threatening (Friedmann & Son, 2009). For instance, dogs facilitate human companionship such as by encouraging strangers to meet and talk (Friedmann & Son, 2009). It is no surprise that pets can be used to facilitate therapy or as emotional support animals. Companion animals also encourage exercise and lower blood pressure, decreasing the risk of cardiovascular disease (Arhant-Sudhir, Arhant-Sudhir, & Sudhir, 2011). Pet ownership can also have negative effects on health, such as through animal bites, allergies to pets, infections, and infestations, which are especially relevant to risk groups such as pregnant women, children, and the immunocompromized. For instance, the prevalence of parasites (e.g., *Giardia, Cryptosporidium*) can increase the risk of parasitic infestation in human without proper prophylaxis and hygiene, such as routine parasite treatments (Robertson, Irwin, Lymbery, & Thompson, 2000). In many developing countries, rabies is still mostly transmitted to humans by dogs, posing a serious health risk that could be prevented by canine vaccination (Davlin & VonVille, 2012). Exotic pets, rising in popularity in the last years, increase the risk of zoonosis transmission, aggravated by the lack of public awareness of the problem (Chomel, Belotto, & Meslin, 2007). The Centers for Disease Control and Prevention, in the US, estimates that handling of reptiles is responsible for 7% of salmonella infections in humans (Chomel et al., 2007). Multiple salmonellosis cases were also identified from turtle handling, giving rise to a regulatory size restriction in the US to avoid fitting in small children's mouths (i.e., turtles can only be legally sold with shell sizes over 10 cm) (Chomel et al., 2007). Indeed, multiple varieties of *Salmonella* can be found in the intestine of reptiles (mostly asymptomatic infections) which can be transmitted to humans when lacking proper hygiene procedures, such as hand washing (Corrente et al., 2017). Many exotic pets also pose risks by being a driver of diversity loss when removed from their natural ecosystem (Bush, Baker, & MacDonald, 2014) or by being improperly released into the environment competing with autochthone species. Many domestic species can also contribute to the introduction of contagious disease among wildlife or the decline of wild populations due to their predatory behaviors (Voith, 2009). Therefore, companion animals can have both a positive and negative influence on health, mostly resulting from husbandry practices.

6.3 Water and food security and safety

The access to nutritious (security) and safe (safety) food is being threatened by multiple factors (e.g., climate change, infectious diseases, conflict), with 690 million (8.9% of the world population) suffering from hunger and 600 million cases of food-borne illnesses per year (Food Security, 2021). The growing human population and the rising urban populations increase the pressure on food production, transportation, and distribution systems, which are often underdeveloped in emerging economies, already strained from environmental changes (e.g., climate change) and elevated pressures of industrial agricultural production (e.g., soil erosion) posing a great threat to food security (Kelly, Ferguson, Galligan, Salman, & Osburn, 2013). Food safety can also be influenced by environmental changes due to the changing ecology of both plants and animals for human consumption, but also of their related pathogens, as well as the emergence of new zoonotic diseases and the chemical contamination of food (e.g., agrochemicals) (Flynn et al., 2019). For instance, in Europe the main zoonosis in humans are campylobacteriosis and salmonellosis, food-borne diseases (The European, 2019). The perception of risk or risky behaviors may also have cultural or sociocultural roots

(Dosman, Adamowicz, & Hrudey, 2001), for instance, how food is handled and processed. As an example, the growing popularity of consumption of raw milk in the US following disputable health claims is accompanied by numerous outbreaks, when compared to pasteurized milk (Lucey, 2015). Similarly, the access to clean water is also dependent on many environmental and socioeconomic factors. One billion people worldwide do not have access to safe and easily accessible water (Hunter, MacDonald, & Carter, 2010). Water is not only important to directly support human survival, but also to support farming and industrial systems which provide essential resources. Water scarcity originates from droughts (e.g., from climate change), overuse (e.g., from population growth, poor management), distance to water resources, or conflicts due to political interests (Gude, 2017). For instance, increased use of water in the irrigation of agricultural land (with an increasing water footprint) in China resulted in water scarcity (Xu et al., 2019). In addition to scarcity, water safety, the access to clean and healthy water with no microbiological nor chemical contamination, is necessary to preserve human health (e.g., avoiding water-borne infectious diarrheas) and requires sanitation and limited discharges of contaminants in natural systems (e.g., anthropogenic pollution from untreated sewage or agricultural leachates), translated into an increased sustainability in its use (Hunter et al., 2010). The availability of clean water is also important to multiple wildlife species, without which they can succumb to diseases or dehydration. In addition, many highly populated cities are dependent on distant water sources. Therefore, supply and demand of water must be considered in the management of scarce water resources (Veettil & Mishra, 2020). This holistic system must also consider the importance of water, sanitation, and hygiene (WASH) in the health of communities as a part of an integrated water resource management (Hadwen et al., 2015). Both food and water security and safety are intrinsically connected to the environment, human, and animal health, which should be addressed under the transdisciplinary approach of One Health.

6.4 Socioeconomic and cultural environments

Socioeconomic drivers, such as human population density, have been positively correlated to the distribution of emerging infectious diseases, in addition to ecological and environmental determinants (Jones et al., 2008). Effects of economic factors on health can be direct or indirect, such as by providing access to healthcare services or improving nutrition (Lenhart, 2019). Health outcomes are also influenced by social determinants, defined as the conditions in which people conduct their lives, including factors related to income, wealth, education, and access to spaces and resources (e.g., recreational areas, healthful foods) (Braveman & Gottlieb, 2014). A low socioeconomic status has been associated with premature mortality comparable to known risk factors, such as smoking or alcohol consumption (Stringhini et al., 2017).

Similarly, low-income neighborhoods often have lower accessibility to green spaces, or are damaged or present safety concerns, compromising the access to health benefits provided by these spaces (e.g., physical activity) (Hoffimann, Barros, & Ribeiro, 2017). Cultural factors also play an important role in One Health as they influence the interactions with other systems. For instance, cooking practices can modulate the access to contaminants in foodstuff, with the use of herbs in cooking decreasing mycotoxin content of chicken breasts by 60% but increasing their bioaccessibility (Sobral, Cunha, Faria, Martins, & Ferreira, 2019). Transmission of Ebola viral disease in West Africa, a likely zoonotic disease (suspected reservoirs in non-human primates, fruit bats, or duikers) coursing with bleeding from orifices and high mortality rates, was facilitated by cultural factors such as the consumption of bushmeat (meat of wild animals) and traditional funerary rituals (MaríSaéz et al., 2015; Phua, 2015). Similarly, the consumption of bushmeat while providing an important source of nutrients also exposes human populations to zoonotic pathogens and contaminants, such as metals and polycyclic aromatic hydrocarbons (Van Vliet et al., 2017). Spillover events originating disease outbreaks (e.g., HIV, SARS) have been related to bushmeat activities, such as consumption of wild animals but also exposure to their body fluids and feces (Kurpiers, Schulte-Herbrüggen, Ejotre, & Reeder, 2016). Bushmeat and illicit wildlife trade often occurs in wet markets which gather in close contact multiple species, dead or alive, with little regard to sanitation, enabling the spread of diseases in stressed animals kept in poor hygienic conditions (Aguirre, Catherina, Frye, & Shelley, 2020). The COVID-19 pandemic, coronavirus disease causing severe acute respiratory syndrome, likely originated in a wet market in Wuhan, China. Beside local traditions, wet markets provide nutrition and livelihood to multiple poor communities, being also influenced by socioeconomic factors, requiring a more delicate intervention than full prohibition (Traditional markets, 2021). Practices originating outbreaks are not limited to low-income communities. In the 80s, the bovine spongiform encephalopathy, a fatal neurological disease ("mad cow disease"), spread over 170,000 bovines across countries possibly due to changes in the industrial processing (lack of high temperature and/or solvent extraction) of sheep meat and bone meal included in ruminant feed that was no longer capable of inactivating prions (proteinaceous infectious particles) (Brewer, 1999). Thus, socioeconomic, cultural factors and common practices may strongly influence health outcomes.

6.5 Antimicrobial resistance

Antimicrobial resistance results from changes in bacteria, virus, fungi, or parasites that leads to resistance to medicines, increasing the severity of the disease due to the inability of treating it. This is a natural response to an environmental selective pressure (i.e., antimicrobial) which translates into mechanisms conferring resistance (e.g., production of β-lactamase which inactivates β-lactam antimicrobial drugs) and that can be transferred to other microorganisms (e.g., by transferring plasmids containing the resistance gene) (Holmes et al., 2016). To understand the scale of the problem, 10 million deaths per year will be associated with drug resistance by 2050 (O'Neil, 2016). Antimicrobial resistance originates from the widespread use in animal production (livestock and aquaculture), agriculture, human medicine, but also from environmental contamination with antimicrobials, being facilitated by the transmission of antimicrobial resistance genes between microorganisms and by globalization, which facilitates the rapid spread and mixing of these genes (Robinson et al., 2016). Antimicrobial resistance depends highly on cultural factors, such as the use of antibiotics as prophylactic or growth promoters in livestock, but also on socioeconomic factors, such as access to clean water and sanitation which can prevent indirect person to person transmission or the lack of access to healthcare or veterinary services (Collignon & McEwen, 2019). Hospitals present an ecological niche for antibiotic-resistant bacteria due to the high selective pressure from the use of antibiotics and spreading in the hands of healthcare workers, finally being released in effluents, contaminating the environment (Hocquet, Muller, & Bertrand, 2016). In animal production, there is a long-term selective pressure by antibiotics due to their use in prophylaxis and treatment of disease in large scale, facilitating the appearance of antibiotic resistance genes directly on animals or in the environment contaminated with antibiotic residues in feces or urine (Duan et al., 2019). Similarly, the stressful condition of fish and shellfish farming leads to infections with increased use of antimicrobials, which modify the normal flora or organisms increasing their susceptibility to antimicrobial-resistant pathogens, increase the selective pressure on antimicrobial-resistant pathogens with increased virulence, and contaminates aquaculture products with antimicrobial-resistant pathogens (Cabello, Godfrey, Buschmann, & Dölz, 2016). Wildlife populations can also be impacted through environmental contamination, comprising an important reservoir (White & Hughes, 2019). For instance, in Azores, Portugal, fecal samples from wild rabbits often contained antibiotic resistant *Escherichia coli* (e.g., 42.9% to streptomycin), presenting an important role in the spread of antimicrobial resistance and creating both an ecological and a public health problem (Marinho et al., 2014). One of the areas of collaboration between the WHO, OIE, FAO under the scope of One Health is antimicrobial resistance establishing a number of actions (e.g., supporting harmonized surveillance, promoting preventive actions to reduce the use of antimicrobial agents) with the aim of ensuring the effectivity of antimicrobial agents in the future, promoting the responsible use of antimicrobial agents, and ensuring global access to effective medicines (FAO/OIE/WHO, 2021). The complexity of the antimicrobial resistance would indeed benefit from a One Health approach, by addressing the intricate interaction between all systems (e.g., animal production, human health, wildlife, socioeconomic factors).

6.6 Environmental contamination

Environmental contamination is another area which can benefit from a One Health approach. Environmental contaminants can result from unsustainable practices in multiple areas in society, such as farming practices (e.g., use of chemicals) or sanitation (e.g., sewage treatment), potentially having adverse effects on all living organisms or threatening environmental processes (Musoke, Ndejjo, Atusingwize, & Halage, 2016). Outbreaks of disease in both animals and humans in proximity may suggest an environmental origin as they share common environments and food or water sources, generally with animals being considered sentinels due to the premature onset of disease (Buttke, 2011). By having biologically compressed lifespans, often being more sensitive to exposure, and not sharing confounding human behaviors (e.g., smoking), animals are ideal in surveillance for the early identification of threats (Neo & Tan, 2017). Caged canaries in coal mines acted as sentinels for carbon monoxide poisoning because of their increase sensitivity and well recognizable response (i.e., falling from perches when ill) that allowed miners to protect themselves (Rabinowitz, Scotch, & Conti, 2010). Determining canine blood lead concentration has been suggested also a surveillance tool for children, since similar environmental exposure is expected (Ostrowski, 1990). In the marine environment, marine mammals also act as sentinels for human health, with some cancers resulting from a combined effect of immunosuppressive contaminants and infectious agents (Reif, 2011). In addition to premature communication, a One Health approach could contribute to a unified knowledge on environmental contaminants. Microplastics, ubiquitous and persistent plastics <5 mm, originate from multiple anthropogenic products and activities and threaten both human, animal, and environmental health (e.g., including habitat properties and biogeochemical processes), benefiting from a transdisciplinary One Health approach (Prata et al., 2021). Besides the impact of contaminants in health, human and animal health practices can also have an impact on ecosystems. For instance, the use of non-steroidal anti-inflammatory drugs (e.g., diclofenac) in livestock leads to the decline of sensitive vulture populations (i.e., exposed through scavenging of carcasses), allowing dogs to feed on carcasses, increasing their populations and rabies transmission to humans (Markandya et al., 2008). Therefore, the One Health approach could contribute to a better understanding of environmental contaminants and with novel solutions applicable to both animal, human, and environmental health.

6.7 Climate change

Climate change is one of the greatest challenges of the current century that will influence all other fields, with no exception of health. Direct adverse effects of climate change can be secondary to extreme temperatures contributing to higher morbidity and mortality from cardiovascular diseases, in addition to heat stroke and hypothermia (Giorgini et al., 2017). Other impacts of climate change will also result in relevant health impacts to human and animal health. Agricultural production and pastures distribution and productivity will be highly influenced from changes in precipitation, forest fires, floods, and droughts (Parajuli, Thoma, & Matlock, 2019). Food security provided by animal production will also be affected by these factors. For instance, small ruminants, which are essential to food security and livelihood of many communities across the globe, will be negatively affected from heat stroke, low availability of pastures, and emerging infectious diseases (Joy et al., 2020). The marine environment will be equally affected from warming, acidification from CO_2 dissolution, and lower oxygen concentrations with a predicted negative impact on global primary production that determines the overall food availability of the ecosystem (Henson et al., 2017). In general, ecosystems will be highly influenced by changes in habitat properties and species redistribution (Bonebrake et al., 2018). Vector-borne diseases are also sensitive to weather and climate as arthropod vectors depend on favorable environmental conditions (e.g., temperature, refuge, reproduction sites) and properties of host communities for completing their life cycles, determining the rate of their development, and/or incubation periods of pathogens (Ogden, 2017). Other emerging infectious diseases may also be aggravated by climate change, such as by contributing to the arrival, establishment and dispersal of pathogens or reservoir species in an already vulnerable socioeconomic setting (Lindgren, Andersson, Suk, Sudre, & Semenza, 2012). Similarly, climate change will modulate environmental contaminants distribution (e.g., transport from altered precipitation patterns), organism's response (e.g., tolerance range), changes in abiotic properties of habitats that influence bioavailability (e.g., salinity), and synergy between stressors (Noyes et al., 2009). For instance, mycotoxins, toxins secondary to fungal metabolism important in occupational exposure, animal health, and food safety, can change in geographical distribution due to climate change as it will affect agricultural practices, cereals used in animal feeding, and ecological niches for mycotoxigenic fungi (Viegas et al., 2020). Thus, climate change will be a forcing factor modulating all other aspects of One Health.

6.8 Biodiversity, habitat loss, and encroachment into wildlife

Emergence of novel infectious agents is related to habitat loss and fragmentation, reduced species diversity, and human encroachment in species-rich areas (Wilkinson, Marshall, French, & Hayman, 2018). Anthropogenic activities with loss of wildlife, loss of habitat, and increase in density of species adapted to human-dominated landscapes (especially mammals) are correlated to a higher risk of spillover events (Johnson et al., 2020). For instance, in Australia, the distribution of species of flying foxes (*Pteropus* species), reservoirs for Hendra virus, were related to spillover events possibly due to changes in the environment, such as attraction to increased food sources in urban settings, fragmentation of landscapes and populations, and proximity to woody savanna susceptible to human intervention (e.g., farming, pasture) (Walsh, Wiethoelter, & Haseeb, 2017). Parasitic diseases are also affected by human encroachment in habits, for instance, by increasing the contact with wildlife reservoirs or creating opportunities for changes in their distribution (Thompson & Polley, 2014). For example, tick distribution is highly dependent on the presence of suitable habitats, which can be influence by deforestation, climate change, determining regions which they can complete their life cycle, and populations of hosts and predators, being responsible for changes in risk areas of tick-borne diseases (e.g. Lyme disease) (Dantas-Torres, 2015). Changes in species abundance can also have wider impacts due to the underlying interactions between all components in the ecosystem, termed trophic cascade. Although still debatable, a trophic cascade may have resulted from the introduction of wolves in the Yellowstone National Park, USA. It has been hypothesized that the preference of wolves for aspen stands shifted elk population's habitat selection, leading to the recovery of aspen populations (Fortin et al., 2005). Another indirect effect of biodiversity loss is the permanent loss of scientific development which may be relevant to health, for instance, with the discovery of novel anti-cancer drugs from marine organisms (Erwin, López-Legentil, & Schuhmann, 2010). Ecosystem services provide the conditions for human health and survival (e.g., by providing clean water or climate regulation) (Coutts & Hahn, 2015) and can be compromised by anthropogenic activity, such as the decrease in ecosystem's productivity by anthropogenic stressors (e.g., the amount of chlorophyll *a* in marine systems) (Johnston, Mayer-Pinto, & Crowe, 2015). Forests, for example, provide multiple ecosystem services such has pollination, biomass production, carbon sequestration, and pest regulation (Brockerhoff et al., 2017). Thus, environmental changes caused by human development, encroachment on wildlife habitats, and changes in habitat characteristics can contribute with threats to human and animal life.

7. Real world use of the One Health approach

Examples of the One Health approach can help clarify how it can be implemented in practice and which benefits were achieved in those cases. The following case studies exemplify in more detail the application of the One Health approach to real world problems. Additional cases can be found on the Veterinarians Without Borders website (Veterinarians Without Borders, 2021).

7.1 Rabies in Tanzania

Rabies is an acute and incurable viral encephalitis transmitted through the bites of infected animals, often by dogs which live in closer proximity to communities. An Integrated Bite Case Management, a surveillance system involving both public health and veterinary sectors, was attempted in 20 districts of Tanzania (Lushasi et al., 2020). The framework implicated health workers in the detection of suspected human bites, pharmacists in the application of post-exposure prophylaxis to exposed people, Livestock Field Officers which received reports from communities and exchanges information with health workers, and the laboratories which received samples from animals collected by Livestock Field Officers for rapid diagnostic testing. The final information was reported to district offices which reported directly to the Ministry of Health and to the Ministry of Livestock. A common reporting database was created and filled through a mobile phone app by health workers, which reported the incident, and by Livestock Field Officers when conducting epidemiological investigation and sample collection. Alerts were generated when health officials categorized bites as a high-risk incident (i.e., animal displayed suggestive signs, the animal died or disappeared afterward, bite from a wild animal, the person presented symptoms of rabies). Post-exposure prophylaxis was administered, and Livestock Field Officers received the notification to conduct an epidemiological investigation in the area through phone consultation or in person. The collaborative approach to rabies management resulted in increased reporting of bite incidents from 55.7 to 92.2 per month, and suspected risk of rabies from 26.9% to 64.9%, with most being caused by bites of domestic dogs (93.0%). As a result of reporting, 823 investigations by Livestock Field Officers were conducted, with the collection of 13 samples, with 49.1% of animals showing signs of rabies or being positive after sample collection. Most animals (79.1%) disappeared or were already dead (naturally or not) at the time of the investigation. An intersectoral program improving communication between healthcare and veterinary services, facilitated by technology (e.g., app), provided grounds for the prevention and control between sectors, generated more accurate data to support policy decisions, provided a rapid response to outbreaks and thus prevented human deaths, and allowed to evaluate the impact of interventions (e.g., mass dog vaccination programs).

7.2 Ciguatera fish poisoning in Cuba

Ciguatoxin is a group of natural marine toxins produced by the dinoflagellate microalgae (*Gambierdiscus*) which typically lives at the bottom of shallow tropical and subtropical waters (Friedman et al., 2017; Morrison, Prieto, Castro Dominguez, & Waltner-Toews, 2005; One Health for One World, 2010). Fish from these shallow coastal areas are susceptible to the accumulation of ciguatoxins, with bioaccumulation across the food web (i.e., generally higher in carnivore fish, such as barracuda, jack, and snappers), and which cannot be identified in food (i.e., odorless, colorless, tasteless) nor removed by any kind of storage, preparation, or cooking methods. Humans are exposed to ciguatoxins through the ingestion of fish, resulting in an illness called Ciguatera Fish Poisoning, characterized by gastrointestinal, neurological, and cardiac symptoms. The geographic distribution of outbreaks of Ciguatera Fish Poisoning appears to be expanding, either from increased public health reporting and increased international tourism and trade or from anthropogenic environmental disturbances. These environmental factors might include coastal eutrophication or rising ocean temperatures from climate change. Moreover, dinoflagellate monitoring might not be helpful since toxin production is either strain-dependent or triggered by an unknown factor. Additionally, no simple tests exist to determine if the fish is affected. In Cuba, outbreaks have been increasing after the fall of the Soviet Block in 1989 due to an increased exploitation of coastal resources, especially for food. The increase in outbreaks is thought to be related to socio-economic conditions but also from the resulting degradation of near-reef ecosystems. The impacts of the disease go beyond the affected individuals, with large economic costs and malnutrition in already disadvantaged communities. Management is hindered by the number of affected areas, the lack of information (e.g., lack of reliable tests), and intrinsic variation in factors (e.g., not all fish of the same species are affected, no clinical signs are always present, there are many types of ciguatoxins). An interdisciplinary team was gathered to collect epidemiological data and understand the social-ecological system related to event in each community. Interventions were conducted at community and national level, improving the communication between stakeholders (e.g., workshops, interviews, school interventions) and involving experts of relevant disciplines (e.g., coral reef ecology). The systematic approach was conducted on three coastal communities included semi-structured community interviews (i.e., ≥ 5 fishers; ≥ 5 people experienced in the disease, ≥ 5 men and ≥ 5 women not fishers nor experienced in the disease), interviews on informants at different levels (local, provincial, national) and different disciplines (fisheries, environment, health), survey on population of local domestic camping areas, analysis of epidemiological records, community workshops, and information gathering from journal articles and government reports. Thus, local stakeholders included local communities, healthcare, the Ministry of Public Health, the Ministry of Environment, the Ministry of the Fishing Industry, the National Toxicology Center, the Institute of Oceanology, the Coast Guard, the Sport Fishing Federation, among others. This approach resulted in the recognition of many interacting factors such as the lack of sufficient information to make informed choices (e.g., women cannot recognize fish species despite being the family cook; many individuals are unwilling to change their eating habits), unwillingness of fishers to sell ciguatoxic fish in their communities which does not apply if selling to intermediaries who then sell them to city dwellers, and frequency and severity of outbreaks related to environmental degradation (e.g., degradation of coral reefs). Thus, the effective management of Ciguatera Fish Poisoning can be achieved through education and awareness, effective protection of coral reefs from degradation, and better management of onshore developments and exploitation.

7.3 Mercury from fish consumption in the Amazon

The lower Tapajós River region is highly impacted by mercury contamination (Guimarães & Mergler, 2012). A project started in 1994 to explore the sources, transmission, and effects of mercury on communities, by involving professionals of different fields, such as researchers from social, health, and natural sciences. Campaigns and workshops involved local communities, health agents, and authorities. Many tests were conducted on water, fish, soil, and humans. The initial assumption that mercury originated from intense artisanal gold rush in the 80s was discredited when no gradient of concentrations was found. Instead, soils were highly enriched with high levels of natural mercury that was mobilized due to agricultural practices. Deforestation based on "slash and burn" produced ash which introduced reactive cations in soils, dislodging mercury, phosphorus, and nitrogen, which could then be washed away. Soil erosion further contributed to losses of carbon and nitrogen, which enriched rivers and lakes and increased methylmercury production (with highest bioavailability). Human exposure was related to environmental concentrations, following a seasonal variation, and with dose-related motor and visual deficits. Fish were contaminated with methylmercury, with multiple bioaccumulation patterns, varying between individuals and seasons. As a rule, carnivore fish species had the highest mercury levels. Ingestion of fish could not be avoided as it constituted an important source of protein for the community. Instead, villagers were encouraged to reduce carnivore fish ingestion, generally more contaminated, which resulted in a decrease in mercury exposure. Follow up assessments identified fruit consumption and high selenium levels (e.g., from Brazil nuts and certain fish) as factors which decreased mercury levels in inhabitants, which may be related to mechanisms of absorption or offsetting of toxic effects. The investigation also identified high exposure to lead likely originating from low-quality metal plates used and heated in the production of manioc flour. Complex socio-cultural factors (e.g., agricultural practices, diet preference), coupled with environment variables (i.e., high concentration of mercury in soils), originated environmental contamination which affected freshwater ecosystems and communities that depended on fish as a major protein source. The transdisciplinary effort allowed to identify that sustainable agricultural practices preventing soil erosion (e.g., alternative techniques and crops) are required to reduce mercury in fish, and in turn human exposure.

7.4 Anthrax in Western Uganda

Anthrax (*Bacillus anthracis*) is a zoonotic disease naturally present in many countries and regions, which has been controlled through vaccination campaigns and movement controls (Coffin, Monje, Asiimwe-Karimu, Amuguni, & Odoch, 2015). The ecology of anthrax is complex, involving soil characteristics (quality, flora, fauna), food and drought cycles, transport by insects and other, leading to dormant periods and outbreaks. Herbivorous mammals are exposed through ingestion or inhalation of spores in soil or forage, progressing rapidly to septicemia and death. Human infection can occur during the ingestion of contaminated meat, inhalation during butchering of carcasses, or through cutaneous infection via insects or open wounds. While inhalation and ingestion progress rapidly to septicemia and death, cutaneous infection progresses more slowly and can be treated with antibiotics if diagnosed early. West Uganda has reported many anthrax outbreaks involving both livestock, wild animals, and humans, especially in the vicinity of Queen Elizabeth National Park. A One Health approach was conducted in two districts in this area, Kasese and Sheema, where previous outbreaks had occurred. A team was gathered including experts from veterinary medicine, public health, ecology, wildlife management, and qualitative methodologies. These teams first conducted conventional public health surveys and then applied participatory methodologies from social science. Stakeholders in these districts included the communities, but also health, trade, and conservation specialists. Data collection was based on focus groups and participatory exercises, structured questionnaires, and observation. Communities near the national park (<10 km) had the greatest knowledge on anthrax, but also reported more wildlife conflicts (e.g., wildlife destruction of crops, losses of livestock to wild carnivores). Participants ranked anthrax as less important economic and social than other health threats, despite reporting illnesses and deaths in livestock and humans, and did not give it much importance due to its rarity compared to other diseases (e.g., brucellosis). Depreciation of animal products, costs of treatment, imposed quarantine, and market closures imposed during anthrax outbreaks had great impacts on the livelihood of communities. Communities intentionally underreported cases due to the imposed sanctions or did not report them due to the lack of knowledge, indicating the anthrax events were likely more frequent than officially reported. Community solutions to outbreaks were to quickly sell suspected animals, overcook suspected meat, or bury carcasses injected in petrol to discourage consumption. However, these solutions still hold risks as infection can occur through inhalation during butchering and burying carcasses can contaminate soil with spores. Moreover, veterinary care was unavailable due to costs and distance, as often were reporting structures. Proximity to the national park had strong influence in the vicinity due to increased governmental presence and economic incentives. Poverty, lack of education, and lack of incentives for reporting generally resulted in underreporting of outbreaks and strong barriers to changing risky behaviors. In the future, strategies must engage communities when planning interventions, engage health teams in communicating animal diseases, investigate and vaccinate areas at risk of anthrax outbreaks, minimize wildlife conflicts, and address poverty in these communities.

7.5 Methicillin resistant *Staphylococcus aureus* in Portugal

Methicillin-resistant *S. aureus* (MRSA) is multidrug-resistant pathogen. Portugal has one of the highest rates of MRSA and the second highest in Europe, with its presence spreading to many animal communities, environment, and wild animals (Igrejas et al., 2018). Despite its global relevance in antimicrobial resistance, MRSA is not well known, and its movement between humans, domestic animals, wild animals, and the environment are not yet understood. This knowledge could contribute to the implementation of measures to control zoonoses, as well as benefit individuals and society, due to the burden of disease (e.g., health outcomes, economic losses). Thus, MRSA is a good candidate for a One Health approach. The operation was implemented in the region of Trás-os-Montes e Alto Douro, Portugal, by collecting samples from hospitals, farms (soils, animals, animal handlers, farmers), and wildlife animals (collected by hunters). MRSA and specific strains (including biofilm forming) were identified in these samples. Additionally, questionnaires about antibiotic use were sent to relevant stakeholders. Data was collected in a web-based application and georeferenced. The effort to clarify how MRSA circulates in the environment (especially outside healthcare facilities), including occupational risk from exposure to livestock and the existence of reservoirs in wildlife animals, is still ongoing and involves multiple stakeholders. The research will identify which MRSA strains are associated with each species or habitats, how to reduce the impact of antibiotic use in the spread of resistance, how to improve sustainability and efficacy of antibiotic use, and estimate economic cost, proposing steps to avoid the spread and improve healthcare of this agent. Community engagement will occur through lectures in high school and university students and by using social and traditional media outlets. This One Health approach will help understand antimicrobial resistance in a complex system.

References

Aguirre, A. A., Catherina, R., Frye, H., & Shelley, L. (2020). Illicit wildlife trade, wet markets, and COVID-19: Preventing future pandemics. *World Medical & Health Policy, 12*(3), 256–265. https://doi.org/10.1002/wmh3.348

Akins, Lyver, Alrøe, & Moller. (2019). The universal precautionary principle: New pillars and pathways for environmental, sociocultural, and economic resilience. *Sustainability, 11*(8), 2357. https://doi.org/10.3390/su11082357

Arhant-Sudhir, K., Arhant-Sudhir, R., & Sudhir, K. (2011). Pet ownership and cardiovascular risk reduction: Supporting evidence, conflicting data and underlying mechanisms. *Clinical and Experimental Pharmacology and Physiology, 38*(11), 734–738. https://doi.org/10.1111/j.1440-1681.2011.05583.x

Babo Martins, S., Rushton, J., & Stärk, K. D. C. (2017). Economics of zoonoses surveillance in a 'one health' context: An assessment of *Campylobacter* surveillance in Switzerland. *Epidemiology and Infection, 145*(6), 1148–1158. https://doi.org/10.1017/S0950268816003320

Bank, T. W. (2010). *People, pathogens and our planet; Washington DC, USA*.

Bartlow, A. W., & Vickers, T. (2020). Solving the mystery of an outbreak using the one health concept. *The American Biology Teacher, 82*(1), 30–36. https://doi.org/10.1525/abt.2020.82.1.30

Baum, S. E., Machalaba, C., Daszak, P., Salerno, R. H., & Karesh, W. B. (2017). Evaluating One Health: Are we demonstrating effectiveness? *One Heal, 3*, 5–10. https://doi.org/10.1016/j.onehlt.2016.10.004

Binot, A., Duboz, R., Promburom, P., Phimpraphai, W., Cappelle, J., Lajaunie, C., ... Roger, F. L. (2015). A framework to promote collective action within the one health community of practice: Using participatory modelling to enable interdisciplinary, cross-sectoral and multi-level integration. *One Heal, 1*, 44–48. https://doi.org/10.1016/j.onehlt.2015.09.001

Bonebrake, T. C., Brown, C. J., Bell, J. D., Blanchard, J. L., Chauvenet, A., Champion, C., ... Danielsen, F. (2018). Managing consequences of climate-driven species redistribution requires integration of ecology, conservation and social science. *Biological Reviews, 93*(1), 284–305. https://doi.org/10.1111/brv.12344

Bordier, M., Uea-Anuwong, T., Binot, A., Hendrikx, P., & Goutard, F. L. (2020). Characteristics of one health surveillance systems: A systematic literature review. *Preventive Veterinary Medicine, 181*, 104560. https://doi.org/10.1016/j.prevetmed.2018.10.005

Braveman, P., & Gottlieb, L. (2014). The social determinants of health: It's time to consider the causes of the causes. *Public Health Reports, 129*(1_Suppl. 2), 19–31. https://doi.org/10.1177/00333549141291S206

Brewer, M. S. (1999). Current status of bovine spongiform encephalopathy-a review. *Journal of Muscle Foods, 10*(1), 97–117. https://doi.org/10.1111/j.1745-4573.1999.tb00389.x

Brockerhoff, E. G., Barbaro, L., Castagneyrol, B., Forrester, D. I., Gardiner, B., González-Olabarria, J. R., ... Taki, H. (2017). Forest biodiversity, ecosystem functioning and the provision of ecosystem services. *Biodiversity & Conservation, 26*(13), 3005–3035. https://doi.org/10.1007/s10531-017-1453-2

Bush, E. R., Baker, S. E., & MacDonald, D. W. (2014). Global trade in exotic pets 2006-2012. *Conservation Biology, 28*(3), 663–676. https://doi.org/10.1111/cobi.12240

Buttke, D. E. (2011). Toxicology, environmental health, and the "one health" concept. *Journal of Medical Toxicology, 7*(4), 329–332. https://doi.org/10.1007/s13181-011-0172-4

Cabello, F. C., Godfrey, H. P., Buschmann, A. H., & Dölz, H. J. (2016). Aquaculture as yet another environmental gateway to the development and globalisation of antimicrobial resistance. *The Lancet Infectious Diseases, 16*(7), e127–e133. https://doi.org/10.1016/S1473-3099(16)00100-6

Calistri, P., Iannetti, S., Danzetta, L. M., Narcisi, V., Cito, F., Di Sabatino, D., ... Carvelli, A. (2013). The components of 'one world - one health' approach. *Transboundary and Emerging Diseases, 60*, 4–13. https://doi.org/10.1111/tbed.12145

Charron, D. F. (2012). Ecohealth: Origins and approach. In *Ecohealth research in practice* (pp. 1–30). New York, NY: Springer New York. https://doi.org/10.1007/978-1-4614-0517-7_1

Chomel, B. B., Belotto, A., & Meslin, F.-X. (2007). Wildlife, exotic pets, and emerging zoonoses. *Emerging Infectious Diseases, 13*(1), 6–11. https://doi.org/10.3201/eid1301.060480

Coffin, J. L., Monje, F., Asiimwe-Karimu, G., Amuguni, H. J., & Odoch, T. (2015). A one health, participatory epidemiology assessment of anthrax (*Bacillus anthracis*) management in Western Uganda. *Social Science & Medicine, 129*, 44–50. https://doi.org/10.1016/j.socscimed.2014.07.037

Collignon, P., & McEwen, S. (2019). One Health—its importance in helping to better control antimicrobial resistance. *Travel Medicine and Infectious Disease, 4*(1), 22. https://doi.org/10.3390/tropicalmed4010022

Corrente, M., Sangiorgio, G., Grandolfo, E., Bodnar, L., Catella, C., Trotta, A., ... Buonavoglia, D. (2017). Risk for zoonotic *Salmonella* transmission from pet reptiles: A survey on knowledge, attitudes and practices of reptile-owners related to reptile husbandry. *Preventive Veterinary Medicine, 146*, 73–78. https://doi.org/10.1016/j.prevetmed.2017.07.014

Coutts, C., & Hahn, M. (2015). Green infrastructure, ecosystem services, and human health. *International Journal of Environmental Research and Public Health, 12*(8), 9768–9798. https://doi.org/10.3390/ijerph120809768

Dantas-Torres, F. (2015). Climate change, biodiversity, ticks and tick-borne diseases: The butterfly effect. *International Journal of Parasitology Parasites and Wildlife, 4*(3), 452–461. https://doi.org/10.1016/j.ijppaw.2015.07.001

Daszak, P. (2000). Emerging infectious diseases of wildlife– threats to biodiversity and human health. *Science, 287*(5452), 443–449. https://doi.org/10.1126/science.287.5452.443

Davlin, S. L., & VonVille, H. M. (2012). Canine rabies vaccination and domestic dog population characteristics in the developing world: A systematic review. *Vaccine, 30*(24), 3492–3502. https://doi.org/10.1016/j.vaccine.2012.03.069

Dosman, D. M., Adamowicz, W. L., & Hrudey, S. E. (2001). Socioeconomic determinants of health- and food safety-related risk perceptions. *Risk Analysis, 21*(2), 307–318. https://doi.org/10.1111/0272-4332.212113

Duan, M., Gu, J., Wang, X., Li, Y., Zhang, R., Hu, T., & Zhou, B. (2019). Factors that affect the occurrence and distribution of antibiotic resistance genes in soils from livestock and poultry farms. *Ecotoxicology and Environmental Safety, 180*, 114–122. https://doi.org/10.1016/j.ecoenv.2019.05.005

Eddy, C., Stull, P. A., & Balster, E. (2013). Environmental health-champions of one health. *Journal of Environmental Health, 76*(1), 46–48.

WHO (World Health Organization). (2014). A brief guide to emerging infectious diseases and zoonoses. https://apps.who.int/iris/handle/10665/204722. (Accessed 25 February 2021).

Environmental Public Health Tracking Program: Closing America's Environmental Public Health Gap https://www.cdc.gov/nceh/tracking/pdfs/aag04.pdf. (Accessed 25 February 2021).

Erwin, P. M., López-Legentil, S., & Schuhmann, P. W. (2010). The pharmaceutical value of marine biodiversity for anti-cancer drug discovery. *Ecological Economics, 70*(2), 445–451. https://doi.org/10.1016/j.ecolecon.2010.09.030

FAO/OIE/WHO Tripartite Collaboration on AMR. https://www.who.int/foodsafety/areas_work/antimicrobial-resistance/tripartite/en/. (Accessed 25 February 2021).

Flynn, K., Villarreal, B. P., Barranco, A., Belc, N., Björnsdóttir, B., Fusco, V., ... Teixeira, P. (2019). An introduction to current food safety needs. *Trends in Food Science & Technology, 84*, 1–3. https://doi.org/10.1016/j.tifs.2018.09.012

Food Security. https://www.worldbank.org/en/topic/food-security. (Accessed 25 February 2021).

Fortin, D., Beyer, H. L., Boyce, M. S., Smith, D. W., Duchesne, T., & Mao, J. S. (2005). Wolves influence elk movements: Behavior shapes a trophic cascade in Yellowstone National Park. *Ecology, 86*(5), 1320–1330. https://doi.org/10.1890/04-0953

Friedman, M., Fernandez, M., Backer, L., Dickey, R., Bernstein, J., Schrank, K., … Bienfang, P. (2017). An updated review of Ciguatera fish poisoning: Clinical, epidemiological, environmental, and public health management. *Marine Drugs, 15*(3), 72. https://doi.org/10.3390/md15030072

Friedmann, E., & Son, H. (2009). The human–companion animal bond: How humans benefit. *Veterinary Clinics of North America: Small Animal Practice, 39*(2), 293–326. https://doi.org/10.1016/j.cvsm.2008.10.015

Gibbs, E. P. J. (2014). The evolution of one health: A decade of progress and challenges for the future. *The Veterinary Record, 174*(4), 85–91. https://doi.org/10.1136/vr.g143

Giorgini, P., Di Giosia, P., Petrarca, M., Lattanzio, F., Stamerra, C. A., & Ferri, C. (2017). Climate changes and human health: A review of the effect of environmental stressors on cardiovascular diseases across epidemiology and biological mechanisms. *Current Pharmaceutical Design, 23*(22). https://doi.org/10.2174/1381612823666170317143248

Grace, D. (2014). The business case for one health. *Onderstepoort Journal of Veterinary Research, 81*(2). https://doi.org/10.4102/ojvr.v81i2.725

Gude, V. G. (2017). Desalination and water reuse to address global water scarcity. *Reviews in Environmental Science and Bio/Technology, 16*(4), 591–609. https://doi.org/10.1007/s11157-017-9449-7

Guimarães, J. R. D., & Mergler, D. (2012). A virtuous cycle in the Amazon: Reducing mercury exposure from fish consumption requires sustainable agriculture. In *Ecohealth research in practice* (pp. 109–118). New York, NY: Springer New York. https://doi.org/10.1007/978-1-4614-0517-7_10

Hadwen, W. L., Powell, B., MacDonald, M. C., Elliott, M., Chan, T., Gernjak, W., & Aalbersberg, W. G. L. (2015). Putting WASH in the water cycle: Climate change, water resources and the future of water, sanitation and hygiene challenges in Pacific Island countries. *Journal of Water, Sanitation and Hygiene for Development, 5*(2), 183–191. https://doi.org/10.2166/washdev.2015.133

Hasler, B., Cornelsen, L., Bennani, H., & Rushton, J. (2014). A review of the metrics for one health benefits. *Revue scientifique et technique (International Office of Epizootics), 33*(2), 453–464. https://doi.org/10.20506/rst.33.2.2294

Henson, S. A., Beaulieu, C., Ilyina, T., John, J. G., Long, M., Séférian, R., … Sarmiento, J. L. (2017). Rapid emergence of climate change in environmental drivers of marine ecosystems. *Nature Communications, 8*(1), 14682. https://doi.org/10.1038/ncomms14682

Hocquet, D., Muller, A., & Bertrand, X. (2016). What happens in hospitals does not stay in hospitals: Antibiotic-resistant bacteria in hospital wastewater systems. *Journal of Hospital Infection, 93*(4), 395–402. https://doi.org/10.1016/j.jhin.2016.01.010

Hoffimann, E., Barros, H., & Ribeiro, A. (2017). Socioeconomic inequalities in green space quality and accessibility—evidence from a Southern European city. *International Journal of Environmental Research and Public Health, 14*(8), 916. https://doi.org/10.3390/ijerph14080916

Holmes, A. H., Moore, L. S. P., Sundsfjord, A., Steinbakk, M., Regmi, S., Karkey, A., … Piddock, L. J. V. (2016). Understanding the mechanisms and drivers of antimicrobial resistance. *Lancet, 387*(10014), 176–187. https://doi.org/10.1016/S0140-6736(15)00473-0

Veterinarians Without Borders. https://www.vetswithoutborders.ca/. (Accessed 25 February 2021).

Hunter, P. R., MacDonald, A. M., & Carter, R. C. (2010). Water supply and health. *PLoS Medicine, 7*(11), e1000361. https://doi.org/10.1371/journal.pmed.1000361

Igrejas, G., Correia, S., Silva, V., Hébraud, M., Caniça, M., Torres, C., … Poeta, P. (2018). Planning a one health case study to evaluate methicillin resistant *Staphylococcus aureus* and its economic burden in Portugal. *Frontiers in Microbiology, 9*. https://doi.org/10.3389/fmicb.2018.02964

Johnson, I., Hansen, A., & Bi, P. (2018). The challenges of implementing an integrated one health surveillance system in Australia. *Zoonoses Public Health, 65*(1), e229–e236. https://doi.org/10.1111/zph.12433

Johnson, C. K., Hitchens, P. L., Pandit, P. S., Rushmore, J., Evans, T. S., Young, C. C. W., & Doyle, M. M. (2020). Global shifts in mammalian population trends reveal key predictors of virus spillover risk. *Proceedings of the Royal Society B: Biological Sciences, 287*(1924), 20192736. https://doi.org/10.1098/rspb.2019.2736

Johnston, E. L., Mayer-Pinto, M., & Crowe, T. P. (2015). Chemical contaminant effects on marine ecosystem functioning. *Journal of Applied Ecology, 52*(1), 140–149. https://doi.org/10.1111/1365-2664.12355

Jones, K. E., Patel, N. G., Levy, M. A., Storeygard, A., Balk, D., Gittleman, J. L., & Daszak, P. (2008). Global trends in emerging infectious diseases. *Nature, 451*(7181), 990–993. https://doi.org/10.1038/nature06536

Joy, A., Dunshea, F. R., Leury, B. J., Clarke, I. J., DiGiacomo, K., & Chauhan, S. S. (2020). Resilience of small ruminants to climate change and increased environmental temperature: A review. *Animals, 10*(5), 867. https://doi.org/10.3390/ani10050867

Kahn, R. E., Ma, W., & Richt, J. A. (2014). *Swine and Influenza: A challenge to one health research*. https://doi.org/10.1007/82_2014_392

Karesh, W. B., Osofsky, S. A., Rocke, T. E., & Barrows, P. L. (2002). Joining forces to improve our world. *Conservation Biology, 16*(5), 1432–1434.

Kelly, A. M., Ferguson, J. D., Galligan, D. T., Salman, M., & Osburn, B. I. (2013). One Health, food security, and veterinary medicine. *Journal of the American Veterinary Medical Association, 242*(6), 739–743. https://doi.org/10.2460/javma.242.6.739

Kurpiers, L. A., Schulte-Herbrüggen, B., Ejotre, I., & Reeder, D. M. (2016). Bushmeat and emerging infectious diseases: Lessons from Africa. In *Problematic wildlife* (pp. 507–551). Cham: Springer International Publishing. https://doi.org/10.1007/978-3-319-22246-2_24

Lee, K., & Brumme, Z. L. (2013). Operationalizing the one health approach: The global governance challenges. *Health Policy and Planning, 28*(7), 778–785. https://doi.org/10.1093/heapol/czs127

Lenhart, O. (2019). The effects of income on health: New evidence from the earned income tax credit. *Review of Economics of the Household, 17*, 377–410. https://doi.org/10.1007/s11150-018-9429-x

Lindgren, E., Andersson, Y., Suk, J. E., Sudre, B., & Semenza, J. C. (2012). Monitoring EU emerging infectious disease risk due to climate change. *Science, 336*(6080), 418–419. https://doi.org/10.1126/science.1215735

Lucey, J. A. (2015). Raw milk consumption. *Nutrition Today, 50*(4), 189–193. https://doi.org/10.1097/NT.0000000000000108

Lushasi, K., Steenson, R., Bernard, J., Changalucha, J. J., Govella, N. J., Haydon, D. T., … Mpolya, E. A. (2020). One health in practice: Using integrated bite case management to increase detection of rabid animals in Tanzania. *Frontiers in Public Health, 8*. https://doi.org/10.3389/fpubh.2020.00013

Mackenzie, J. S., McKinnon, M., & Jeggo, M. (2014). One health: From concept to practice. In *Confronting emerging zoonoses* (pp. 163–189). Tokyo: Springer Japan. https://doi.org/10.1007/978-4-431-55120-1_8

Man's best friend: Global pet ownership and feeding trends gfk.com/insights/mans-best-friend-global-pet-ownership-and-feeding-trends.(2021) (Accessed 25 February 2021).

Marinho, C., Igrejas, G., Gonçalves, A., Silva, N., Santos, T., Monteiro, R., … Poeta, P. (2014). Azorean wild rabbits as reservoirs of antimicrobial resistant *Escherichia coli*. *Anaerobe, 30*, 116–119. https://doi.org/10.1016/j.anaerobe.2014.09.009

Marí Saéz, A., Weiss, S., Nowak, K., Lapeyre, V., Zimmermann, F., Düx, A., … Merkel, K. (2015). Investigating the zoonotic origin of the west African Ebola epidemic. *EMBO Molecular Medicine, 7*(1), 17–23. https://doi.org/10.15252/emmm.201404792

Markandya, A., Taylor, T., Longo, A., Murty, M. N., Murty, S., & Dhavala, K. (2008). Counting the cost of vulture decline—an appraisal of the human health and other benefits of vultures in India. *Ecological Economics, 67*(2), 194–204. https://doi.org/10.1016/j.ecolecon.2008.04.020

Miller, M., & Olea-Popelka, F. (2013). One health in the shrinking world: Experiences with tuberculosis at the human–livestock–wildlife interface. *Comparative Immunology, Microbiology and Infectious Diseases, 36*(3), 263–268. https://doi.org/10.1016/j.cimid.2012.07.005

Mi, E., Mi, E., & Jeggo, M. (2016). Where to now for one health and Ecohealth? *EcoHealth, 13*(1), 12–17. https://doi.org/10.1007/s10393-016-1112-1

Morrison, K. E., Prieto, P. A., Castro Dominguez, A., & Waltner-Toews, D. (2005). An ecosystem approach to Ciguatera fish poisoning in Cuba: preliminary results. In *Proceedings of OCEANS 2005 MTS/IEEE* (pp. 1–6). IEEE. https://doi.org/10.1109/OCEANS.2005.1640021

Mushi, V. (2020). The holistic way of tackling the COVID-19 pandemic: The one health approach. *Tropical Medicine and Health, 48*(1), 69. https://doi.org/10.1186/s41182-020-00257-0

Musoke, D., Ndejjo, R., Atusingwize, E., & Halage, A. A. (2016). The role of environmental health in one health: A Uganda perspective. *One Heal, 2*, 157–160. https://doi.org/10.1016/j.onehlt.2016.10.003

Neo, J. P. S., & Tan, B. H. (2017). The use of animals as a surveillance tool for monitoring environmental health hazards, human health hazards and bioterrorism. *Veterinary Microbiology, 203*, 40–48. https://doi.org/10.1016/j.vetmic.2017.02.007

Noyes, P. D., McElwee, M. K., Miller, H. D., Clark, B. W., Van Tiem, L. A., Walcott, K. C., … Levin, E. D. (2009). The Toxicology of climate change: Environmental contaminants in a warming world. *Environment International, 35*(6), 971–986. https://doi.org/10.1016/j.envint.2009.02.006

Ogden, N. H. (2017). Climate change and vector-borne diseases of public health significance. *FEMS Microbiology Letters, 364*(19). https://doi.org/10.1093/femsle/fnx186

Ogden, N. H., Wilson, J. R. U., Richardson, D. M., Hui, C., Davies, S. J., Kumschick, S., … Pulliam, J. R. C. (2019). Emerging infectious diseases and biological invasions: A call for a one health collaboration in science and management. *Royal Society Open Science, 6*(3), 181577. https://doi.org/10.1098/rsos.181577

One Health.(2021). https://www.who.int/news-room/q-a-detail/one-health (Accessed 25 February 2021).

One Health.(2021). https://onehealthinitiative.com/about/ (Accessed 25 February 2021).

One Health for One World: A compendium of case studies.(2010).

One World – One Health.(2021). http://www.oneworldonehealth.org/ (Accessed 25 February 2021).

O'Neil, J. (2016). *Tackling drug-resistant infections globally: Final report and recommendations*.

Ostrowski, S. R. (1990). Sentinel animals (dogs) as predictors of childhood exposure to environmental lead contamination: Observations on preliminary results. In *Situ evaluation of biological hazards of environmental pollutants* (pp. 145–150). Boston, MA: Springer US. https://doi.org/10.1007/978-1-4684-5808-4_13

Parajuli, R., Thoma, G., & Matlock, M. D. (2019). Environmental sustainability of fruit and vegetable production supply chains in the face of climate change: A review. *The Science of the Total Environment, 650*, 2863–2879. https://doi.org/10.1016/j.scitotenv.2018.10.019

Phua, K.-L. (2015). Meeting the challenge of Ebola virus disease in a holistic manner by taking into account socioeconomic and cultural factors: The experience of West Africa. *Infectious Diseases: Research and Treatment, 8*. https://doi.org/10.4137/IDRT.S31568. IDRT.S31568.

Prata, J. C., da Costa, J. P., Lopes, I., Andrady, A. L., Duarte, A. C., & Rocha-Santos, T. (2021). A one health perspective of the impacts of microplastics on animal, human and environmental health. *The Science of the Total Environment, 777*, 146094. https://doi.org/10.1016/j.scitotenv.2021.146094

Queenan, K. (2017). Roadmap to a one health agenda 2030. *CAB Reviews: Perspectives in Agriculture, Veterinary Science, Nutrition and Natural Resources, 12*(014). https://doi.org/10.1079/PAVSNNR201712014

Rabinowitz, P. M., Scotch, M. L., & Conti, L. A. (2010). Animals as sentinels: Using comparative medicine to move beyond the laboratory. *ILAR Journal, 51*(3), 262–267. https://doi.org/10.1093/ilar.51.3.262

Reif, J. S. (2011). Animal sentinels for environmental and public health. *Public Health Reports, 126*(1_Suppl. 1), 50–57. https://doi.org/10.1177/00333549111260S108

Robertson, I. D., Irwin, P. J., Lymbery, A. J., & Thompson, R. C. A. (2000). The role of companion animals in the emergence of parasitic zoonoses. *International Journal for Parasitology, 30*(12–13), 1369–1377. https://doi.org/10.1016/S0020-7519(00)00134-X

Robinson, T. P., Bu, D. P., Carrique-Mas, J., Fèvre, E. M., Gilbert, M., Grace, D., ... Kariuki, D. (2016). Antibiotic resistance is the quintessential one health issue. *Transactions of the Royal Society of Tropical Medicine and Hygiene, 110*(7), 377–380. https://doi.org/10.1093/trstmh/trw048

Rock, M., Buntain, B. J., Hatfield, J. M., & Hallgrímsson, B. (2009). Animal-human connections, "one health," and the syndemic approach to prevention. *Social Science & Medicine, 68*(6), 991–995. https://doi.org/10.1016/j.socscimed.2008.12.047

Roth, F., Zinsstag, J., Orkhon, D., Chimed-Ochir, G., Hutton, G., Cosivi, O., ... Otte, J. (2003). Human health benefits from livestock vaccination for Brucellosis: Case study. *Bulletin of the World Health Organization, 81*(12), 867–876.

Rüegg, S. R., McMahon, B. J., Häsler, B., Esposito, R., Nielsen, L. R., Ifejika Speranza, C., ... Zinsstag, J. (2017). A blueprint to evaluate one health. *Frontiers in Public Health, 5*. https://doi.org/10.3389/fpubh.2017.00020

Rüegg, S. R., Nielsen, L. R., Buttigieg, S. C., Santa, M., Aragrande, M., Canali, M., ... Radeski, M. (2018). A systems approach to evaluate one health initiatives. *Frontiers in Veterinary Science, 5*. https://doi.org/10.3389/fvets.2018.00023

Rushton, J., Häsler, B., De Haan, N., & Rushton, R. (2012). Economic benefits or drivers of a 'one health' approach: Why should anyone invest? *Onderstepoort Journal of Veterinary Research, 79*(2). https://doi.org/10.4102/ojvr.v79i2.461

Saunders, L. Z. (2000). Virchow's contributions to veterinary medicine: Celebrated then, forgotten now. *Veterinary Pathology Online, 37*(3), 199–207. https://doi.org/10.1354/vp.37-3-199

Schelling, E., Bechir, M., Ahmed, M. A., Wyss, K., Randolph, T. F., & Zinsstag, J. (2007). Human and animal vaccination delivery to remote nomadic families, Chad. *Emerging Infectious Diseases, 13*(3), 373–379. https://doi.org/10.3201/eid1303.060391

Schwabe, C. W. (1964). *Veterinary medicine and human health*. Baltimore, Maryland, USA: The Williams & Wilkins Company.

Sobral, M. M. C., Cunha, S. C., Faria, M. A., Martins, Z. E., & Ferreira, I. M. P. L. V. O. (2019). Influence of oven and microwave cooking with the addition of herbs on the exposure to multi-mycotoxins from chicken breast muscle. *Food Chemistry, 276*, 274–284. https://doi.org/10.1016/j.foodchem.2018.10.021

Stringhini, S., Carmeli, C., Jokela, M., Avendaño, M., Muennig, P., Guida, F., … Bochud, M. (2017). Socioeconomic status and the 25 × 25 risk factors as determinants of premature mortality: A multicohort study and meta-analysis of 1·7 million men and women. *Lancet, 389*(10075), 1229–1237. https://doi.org/10.1016/S0140-6736(16)32380-7

The European Union one health 2018 zoonoses report. *EFSA Journal, 17*(12), (2019). https://doi.org/10.2903/j.efsa.2019.5926

The FAO-OIE-WHO Collaboration. (2021). *Tripartite Concept Note*. https://www.who.int/influenza/resources/documents/tripartite_concept_note_hanoi/en/ (Accessed 25 February 2021).

Thompson, R. C. A., & Polley, L. (2014). Parasitology and one health. *International Journal of Parasitology Parasites and Wildlife, 3*(3), A1–A2. https://doi.org/10.1016/j.ijppaw.2014.09.002

Traditional markets blamed for virus outbreak are lifeline for Asia's poor 2021.https://www.reuters.com/article/southeast-asia-health-markets/traditional-markets-blamed-for-virus-outbreak-are-lifeline-for-asias-poor-idUSL8N2A5201 (Accessed 25 February 2021).

Van Vliet, N., Moreno, J., Gómez, J., Zhou, W., Fa, J. E., Golden, C., … Nasi, R. (2017). Bushmeat and human health: Assessing the evidence in tropical and sub-tropical forests. *Ethnobiology and Conservation*. https://doi.org/10.15451/ec2017-04-6.3-1-45

Veettil, A. V., & Mishra, A. (2020). Water security assessment for the contiguous United States using water footprint concepts. *Geophysical Research Letters, 47*(7). https://doi.org/10.1029/2020GL087061

Viegas, S., Assunção, R., Twarużek, M., Kosicki, R., Grajewski, J., & Viegas, C. (2020). Mycotoxins feed contamination in a dairy farm – potential implications for milk contamination and workers' exposure in a one health approach. *Journal of the Science of Food and Agriculture, 100*(3), 1118–1123. https://doi.org/10.1002/jsfa.10120

Voith, V. L. (2009). The impact of companion animal problems on society and the role of veterinarians. *Veterinary Clinics of North America: Small Animal Practice, 39*(2), 327–345. https://doi.org/10.1016/j.cvsm.2008.10.014

Walsh, M. G., Wiethoelter, A., & Haseeb, M. A. (2017). The impact of human population pressure on flying Fox niches and the potential consequences for Hendra virus spillover. *Scientific Reports, 7*(1), 8226. https://doi.org/10.1038/s41598-017-08065-z

Waltner-Toews, D. (2009). Eco-health: A primer for veterinarians. *Canadian Veterinary Journal, 50*(5), 519–521.

White, A., & Hughes, J. M. (2019). Critical importance of a one health approach to antimicrobial resistance. *EcoHealth, 16*(3), 404–409. https://doi.org/10.1007/s10393-019-01415-5

WHO Coronavirus Disease (COVID-19) Dashboard. https://covid19.who.int/. (Accessed 25 February 2021).

Wilkinson, D. A., Marshall, J. C., French, N. P., & Hayman, D. T. S. (2018). Habitat fragmentation, biodiversity loss and the risk of novel infectious disease emergence. *Journal of The Royal Society Interface, 15*(149), 20180403. https://doi.org/10.1098/rsif.2018.0403

Woods, A., & Bresalier, M. (2014). One health, many histories. *The Veterinary Record, 174*(26), 650−654. https://doi.org/10.1136/vr.g3678

Xu, Z., Chen, X., Wu, S. R., Gong, M., Du, Y., Wang, J., … Liu, J. (2019). Spatial-temporal assessment of water footprint, water scarcity and crop water productivity in a major crop production region. *Journal of Cleaner Production, 224*, 375−383. https://doi.org/10.1016/j.jclepro.2019.03.108

Zessin, K.-H. (2006). Emerging diseases: A global and biological perspective. *Journal of Veterinary Medicine Series B, 53*(s1), 7−10. https://doi.org/10.1111/j.1439-0450.2006.01011.x

Zinsstag, J. (2012). Convergence of Ecohealth and one health. *EcoHealth, 9*(4), 371−373. https://doi.org/10.1007/s10393-013-0812-z

Zinsstag, J., Schelling, E., Waltner-Toews, D., & Tanner, M. (2011). From "one medicine" to "one health" and systemic approaches to health and well-being. *Preventive Veterinary Medicine, 101*(3−4), 148−156. https://doi.org/10.1016/j.prevetmed.2010.07.003

Zoonoses. https://www.who.int/news-room/fact-sheets/detail/zoonoses. (Accessed 25 February 2021).

Zoonotic disease: emerging public health threats in the region. http://www.emro.who.int/fr/about-who/rc61/zoonotic-diseases.html. (Accessed 25 February 2021).

Chapter 2

Public health, surveillance systems and preventive medicine in an interconnected world

Bernardo Mateiro Gomes[a], Carlos Branquinho Rebelo[b] and Luís Alves de Sousa[c]
[a]*Porto Public Health Institute, Oporto Medical Faculty, North Health Region Administration, Porto, Portugal;* [b]*Department of Pathobiology and Population Sciences, Royal Veterinary College, Hertfordshire, United Kingdom;* [c]*Public Health Doctor*

1. Public Health, One Health, surveillance: definitions

Public Health heavily relies on surveillance (Thacker et al., 2012) and monitoring of health and disease phenomena in space and time. The term Public Health Surveillance is classically defined as the "Public health surveillance is the ongoing systematic collection, analysis, and interpretation of data, closely integrated with the timely dissemination of these data to those responsible for preventing and controlling disease and injury" (Thacker & Berkelman, 1988). Essential Public Health Operations (EPHOs) comprise ten different items, whose two first ones are related to surveillance of population and health wellbeing (EPHO1) and monitoring and response to health hazards and emergencies (EPHO2) (Foldspang, 2015). Epidemic Intelligence usually refers to the mix between stable, indicator-based surveillance systems and more flexible, event-based surveillance systems. It is the foundation of a basic surveillance framework in which we must retain stability for monitoring and evaluation but simultaneously flexibility for embracing event sinalization that has to be filtered out but is an undeniable advantage for any population health surveillance system.

One Health has been defined as "*any added value in terms of health of humans and animals, financial savings or environmental services achievable by the cooperation of human and veterinary medicine when compared with the concepts of approaches of the two medicines working separately*" (Zinsstag et al., 2020). The COVID-19 pandemic has once again highlighted the

One Health. https://doi.org/10.1016/B978-0-12-822794-7.00006-X

importance of investments in dealing with zoonotic transmissions and the artificial gap between veterinary, environmental and human epidemiological surveillance. Coronaviruses are just one family of pathogens that need proper monitoring in bat populations (El-Sayed & Kamel, 2021) for safety purposes in the nearby population and beyond. Since there was more than enough evidence to justify a closer look at the possibility of a new zoonotic emergence, more actions should have been taken (Li et al., 2005; Tang et al., 2006) to avoid the COVID-19 pandemic. Concurrent and recurrent threats need proportional investment, like the frequently emerging and highly pathogenic avian influenza virus (Shi & Gao, 2021). Public health surveillance has been previously considered inadequate to deal with the rhythm of new emergent threats and the possibilities of bioterrorism as well (Choi, 2012). An economic case for investment in global surveillance systems and prevention of pandemic threats is clear (Dobson et al., 2020; Pike et al., 2014), being reinforced by the ravaging impact of COVID-19, estimated at more than 16 trillion dollars (Cutler & Summers, 2020). Cost-effectiveness of One Health related interventions and the clear demonstration of savings is also a powerful tool not only to select approaches but also to gather political will (Grace et al., 2016).

Surveillance, as a separate article of the International Health Regulations (IHR) in 2005, and with core capacity requirements (World Health Organization (WHO), 2008a,b), is in need of a review to address not only the known problems of data sharing (Edelstein, Lee, Herten-Crabb, Heymann, & Harper, 2018), but also to reflect the growing need of a more capacitated zoonotic surveillance in risk areas. In this instance, the Terrestrial Animal Health Code (TAHC) of the World Organisation for Animal Health (OIE) is further ahead (Thiermann, 2015), considering updates to respond to already known threats such as antimicrobial resistance (AMR). Even the definition of "risk" of a zoonosis should be carefully addressed and set apart from "threat" (Figuié, 2014), since its uncertainty and capacity of global impact are not instantly recognized, and response should be based on precaution, not classic prevention. The shift of a paradigm from an "*international management of threats*" to a "*global governance of risks*" (Figuié, 2014) or even "*deep prevention*" (Vinuales et al., 2021) has profound implications that are still unmet, despite international efforts to improve preparedness, prevention, detection and response.

Roughly more than half of the human infectious diseases are caused by pathogens shared with animals (Karesh et al., 2012), with increasing risks to be faced (Gibb et al., 2020; Lloyd-Smith et al., 2009) and there is a critical point to be surveyed: the crossing of species barriers, which is mainly associated with animal-based food systems (Kock et al., 2020). For example, Sars-CoV-2 emergence has been linked with not only bats but also pangolins as possible source (Xiao et al., 2020) and intermediate hosts, respectively, before the jump into human populations (Wacharapluesadee et al., 2021; Zhang et al., 2020). Simultaneously, the circulation of Sars-CoV-2 between other species has exposed new risks of transmission (van Aart et al., 2021). Recently, a canine coronavirus (an alphacoronavirus not typically associated with human disease) was put under suspicion of human infections after a retrospective study was performed (Vlasova et al., 2021). With the current surveillance systems, the probability of capture of a new phenomenon of species-barrier jump is higher in healthcare, dangerously too late to prevent secondary cases due not only to ascertainment bias but also lack of capacity of recognizing a new disease.

Climate change will open up new exposure scenarios for humans, modifying determinants in which population and infectious organisms coexist. American Geophysical Union has coined the term for the field that joins Humans, Health and Earth System (Almada, Golden, Osofsky, & Myers, 2017). Almada et al. argue the need of the term for encompassing the scale of anthropogenic changes, including the term Planetary Health, also underlined by others as the appropriate context for the COVID-19 pandemic (Jowell & Barry, 2020). Global Health is another term that is frequently used, although its proper scope is under debate (Salm et al., 2021). Climatic changes also underline the need for reinforcement of vector surveillance in an increasing number of locations (Elbers, Koenraadt, & Meiswinkel, 2015; Semenza & Zeller, 2014) and integrated vector surveillance is one of the studied examples of proven savings with a One Health Approach (Paternoster et al., 2017).

Despite a bigger focus being understandably allocated to pathogenic organisms to humans, One Health surveillance encompasses much more. Pathogenic organisms to animals are obviously to be considered, especially those whose survival feeds the overstretched human chains, but one has to consider biodiversity surveillance and habitat maintenance. These should be considered in broader terms, not only as an important determinant of zoonotic jumps, but also as indicators of changes in the environment that can have unpredictable consequences (Smith et al., 2015; Steffen et al., 2015). Plants are often overlooked when One Health is mentioned, despite, among other factors, deforestation being one of the main determinants of emergence of zoonotic disease (Chua, Chua, & Wang, 2002; Wolfe et al., 2005), and even establishing feedback loops involving higher transmission of disease (MacDonald & Mordecai, 2019).

Food consumption patterns need also to be monitored alongside the soil demands and habitat impacts it induces (Almada et al., 2017), namely on freshwater sources and forests. Anthropogenic changes to the food production context have avoided past widespread famines but represent a threat to the collective future, through several mechanisms (Pingali, 2012). Deepening the knowledge about infectious disease emergence mechanisms is crucial. Identifying areas at risk and associated factors such as land-use changes and wildlife biodiversity (Allen et al., 2017), reflecting those realities in metrics such as risk indexes can prioritize the proper preparation of sustainable initiatives of surveillance. Food and Agriculture Organization of the United

Nations (FAO) has long recognized the need for joint work with other organizations in surveillance within the framework of One Health (Lubroth, 2012).

Beyond the obvious interest in predicting extreme meteorological events, surveillance systems should also be prepared to measure consequences of unusual circumstances that can be used to show counterfactuals and impact for policy and public communication. One example: lockdowns and the decrease of human movement during the COVID-19 pandemic has provided an unique setting to measure anthropogenic effects on wildlife and environment (Corlett et al., 2020; Rutz et al., 2020).

2. Surveillance systems: between health, veterinary and environmental frameworks

Human public health surveillance is considered to represent the continuous collection, analysis, interpretation and dissemination of information concerning human diseases, injuries, and death, in order to promote the health of the public (Jekel, Katz, Elmore, & Wild, 2007). On the other hand, animal health surveillance encompasses an analogous systematic and continuing process of collecting, collating, and analyzing information related to animal health as well as the timely report of information to decision makers (Thiermann, 2015). Classically, animal health is more focused on domestic animals, especially livestock but with limited scope on wildlife, with notifiable animal diseases falling short of the need of detection of emerging pathogens.

Surveillance processes are foundationally supported on the existence of public health surveillance systems, incorporating a collection of practices and components that allow public health specialists to conduct surveillance (Groseclose & Buckeridge, 2017). The enabling elements of these systems usually include an array of components within clinical, laboratory, technological and legal dimensions. As an example, typical components of a surveillance system will include clinical consulting and reporting as well as public health and laboratory workers for conducting, respectively, epidemiological inquiries and laboratory diagnostics to detect or confirm infection or disease, information technologies to support data collection, analysis, and dissemination, and an adequate legal and regulatory framework (German, Horan, Lee, Milstein, & Pertowski, 2001; Groseclose & Buckeridge, 2017).

Inherently, the components of surveillance systems establish the frame upon which the system's objectives are attained (German et al., 2001). Concurrently, the objectives of surveillance systems across human and animal public health are largely similar, including monitoring of disease trends and health outcomes, identifying outbreaks or epidemics, as well as early detection of emerging diseases, evaluating and planning healthcare/disease control programs, supporting decision-making regarding research priorities, and providing evidence of absence of (or *'freedom from'*) disease (German et al., 2001; Groseclose & Buckeridge, 2017).

Fundamentally, public health surveillance systems should yield information to support and guide public health decision-making, particularly for disease prevention and health promotion, healthcare and disease control program planning and management, quality improvement, and healthcare provision and resource allocation (Groseclose & Buckeridge, 2017).

2.1 Concepts and definitions in One Health surveillance

Most central concepts on surveillance have an effective overlapping meaning and practical application across human and animal public health. In this section, we will discuss some of the most fundamental terms in public health surveillance and surveillance systems as well as their similarities and differences between human and veterinary fields.

A primary concept that must be separated from surveillance is the term monitoring which, according to OIE Animal Health Terrestrial Code, can be regarded as the intermittent process and analysis of routine measurements and observations, aimed at identifying variations in the environment or health status of a population (Thiermann, 2015).

Regarding the type of infection and disease occurrence, three main categories are relevant when characterizing human and veterinary public health surveillance. Firstly, a disease can be considered endemic when it is present in a particular population continuously through time and maintained at a given baseline level. Endemic diseases can be expressed by varying degrees of incidence and disease severity, whilst often a portion of the population will express some form of immunity toward this disease. Secondly, exotic diseases are usually not present in the population under surveillance, either due to previous elimination or sustained control measures. Hence, in these circumstances, a population remains fully susceptible unless vaccinated. Thirdly, emerging diseases, such as the infection by Nipah virus, Avian influenza or West Nile virus, comprise those previously existing diseases whose incidence has increased recently in time and/or in geographical range, or that have appeared in a population for the first time (Thiermann, 2015).

2.1.1 Active and passive surveillance

Concerning data sources and their corresponding data flows to public health surveillance systems, they can include a variety of fields, stakeholders and procedures essential to public health action. They might vary from a simple manual system collecting data from a single source, to electronic systems that manage data from multiple sources in several formats (German et al., 2001). Data collection and reporting procedures are traditionally used to classify the type of surveillance as either active or passive (Dufour & Hendrikx, 2005; European Center for Disease Prevention and Control (ECDC), 2014; Drewe et al., 2012).

Passive surveillance is the most common type of surveillance and relies on the clinicians, laboratory staff or other related sources to either take the initiative or comply with mandatory rules to report data to public health authorities or departments (European Center for Disease Prevention and Control (ECDC), 2014). Typically, this type of surveillance is able to more easily detect and report clinically sick individuals or outbreaks to be noticed. For example, within veterinary public health, passive surveillance is useful for diseases that almost always produce clear clinical signs, such as foot-and-mouth disease in cows. Concomitantly, passive surveillance can be affected by under-reporting, for reasons that span from unreported asymptomatic cases or untested samples from clinically suspicions events (European Center for Disease Prevention and Control (ECDC), 2014; Drewe et al., 2012).

According to ECDC, an active surveillance system is a system that relies on the public health authorities' initiative to contact physicians, laboratory or hospital staff or other relevant sources to report data. Within veterinary public health, active surveillance typically requires regular screening of animals/products at risk (including clinically healthy appearing test units), preferably based on random sampling or risk-based procedures. This type of surveillance is associated with high-quality data standards, particularly regarding completeness, validity and timeliness. The main disadvantages of active surveillance are the considerable resources needed, such as available diagnostic tests and infrastructure, and the ensuing need to focus on high-priority diseases that explain the additional expense (European Center for Disease Prevention and Control (ECDC), 2014).

The related term - *enhanced passive surveillance* - has been used to define passive surveillance systems that have been adapted and refined by public health professionals to standardize and improve the usage of the information collected (Ouagal et al., 2010). In practical terms, enhanced passive has been used to describe either surveillance systems or their components. According to Drewe et al., enhanced passive surveillance systems are used to identify trends arising from apparently isolated disease events or syndromes. Any activity favoring awareness of co-occurring disease syndromes either in multiple locations or data sources may be considered examples of an enhanced passive surveillance system (Drewe et al., 2012).

2.1.2 Compulsory and voluntary surveillance

Of the several approaches in which surveillance systems might prompt surveillance data from their sources, compulsory and voluntary reporting are the two main ways. Compulsory reporting is based on mandatory data submission, by either legal demand, professional statute, policy or guidance, whereas alternative reporting methods are to varying extent voluntary. Whilst a legal or professional edict to report can improve a surveillance system's completeness and timeliness, it does not necessarily translate into better representativeness or validity of the reported data (European Center for Disease Prevention and Control (ECDC), 2014).

2.1.3 Comprehensive and sentinel surveillance

According to ECDC, a surveillance system is considered comprehensive when cases occurring within the entire population of the catchment geographical area are reported. On the other hand, sentinel systems depend on reports from a selected group of physicians, hospitals, laboratories, and other institutions. The latter systems are based on a representative selection of reporting stakeholders established as part of the system-design process. Primarily, the choice between sentinel and comprehensive systems is based on the surveillance system's purpose and objectives, as well as the features of the disease under surveillance, and the availability of resources (European Center for Disease Prevention and Control (ECDC), 2014).

2.1.4 Syndromic surveillance

Syndromic surveillance uses health-related information such as combinations of clinical signs or other related data that often preclude a formal diagnosis. This may be used to indicate a sufficient likelihood of a change in the health of the population either to warrant additional investigation or to enable an opportune assessment of the impact of health threats that require public health action. Syndromic surveillance is used to detect a variety of diseases or pathogens, including emerging diseases, a feature that makes it particularly relevant for early-warning surveillance systems, alongside the capacity for collection of "real-time" data (Sosin, 2003; Drewe, Hoinville, Cook, Floyd, & Stark, 2012). In practical terms, syndromic surveillance has effectively been applied to early detection of increases in the incidence of endemic diseases, such as seasonal influenza, as well as for checking the health impact of environmental threats, like flooding (Mirhaji, 2009; Sosin, 2003). From its conceptual definition, syndromic surveillance systems could be used to detect undefined events for which tests are unavailable.

2.1.5 Participatory surveillance

Participatory surveillance is a type of surveillance used in veterinary public health that started in the 1980s along with rural development programs in low-income countries, serving as a means to allow communities to recognise problems with livestock and develop solutions. It takes advantage of traditional information networks using participatory methods to conduct risk-based surveillance. It relies upon qualitative health data collected from key community informants, focusing on their knowledge on health events, risks, and control opportunities (Drewe et al., 2012).

2.1.6 Risk-based surveillance

Another specific surveillance type typically used by veterinary public health is risk-based surveillance. It takes information about the probability of occurrence and the extent of the consequence of a given health hazard to plan, design, and/or interpret the results obtained from surveillance systems (Drewe et al., 2012). Risk-based surveillance can include one or several of the following methods (Drewe et al., 2012):

(i) risk-based prioritization, which focuses applying risk assessment to determine which health hazards should be selected for surveillance;
(ii) risk-based requirement, which uses risk assessment to revise the surveillance intensity required to achieve a pre-specified stated surveillance purpose;
(iii) risk-based sampling, which focuses on applying risk evaluation in order to design a sampling strategy to reduce the cost or enhance the accuracy and representativeness of surveillance; and
(iv) risk-based analysis, that intends to analyze and revise a given disease status using risk assessment methods.

One particularly important example of risk assessment and risk-based surveillance in veterinary public health is the *Proof of freedom survey* and *Import Risk analysis*. This type of analysis aims to provide evidence, with a degree of statistical certainty, that a given disease is absent in animal populations in a given country or zone. These types of surveys are an essential tool in veterinary epidemiology as they support trading partners with evidence that certain diseases are not likely to be present in animals from the exporting country (Drewe et al., 2012).

2.1.7 Pathogen surveillance

Contrary to the more common assertion of disease surveillance, the concept of pathogen surveillance is important in animal health surveillance, since infection with a particular pathogen might not produce visible signs of disease in all species under surveillance. In addition, this explains that certain diseases and infections caused by the same pathogen are notifiable. Nonetheless, occurrence of a given pathogen is important for understanding the potential for transmission between species, including to humans (Drewe et al., 2012).

2.1.8 Epidemic Intelligence

According to Paquet et al., epidemic intelligence covers all activities related to prompt identification of potential health hazards, their verification, assessment and investigation in order to recommend public health control measures (Paquet et al., 2006). It incorporates both an indicator-based and an event-based surveillance components. Indicator-based surveillance is closely related to traditional routine surveillance systems, where structured data is collected in a standardized way on specific diseases. On the other hand, event-based surveillance refers to data gathering from non-traditional sources, such as news items, reports, internet and media sources, that might contain relevant intelligence sources for potential health events (Paquet et al., 2006; Drewe et al., 2012).

2.2 Between human health, veterinary health and environmental frameworks

The One Health idea demands the deconstruction of the existing compartmentalized structures among human and veterinary health, as well as environment and ecosystems health, whilst promoting a more sustainable and adequate governance of health. This will imply the necessary transgression of current old-fashioned boundaries between disciplines and the creation of a new way of defining, thinking and managing health matters (Bordier, Uea-Anuwong, Binot, Hendrikx, & Goutard, 2020). According to several authors, albeit evidence on these proposals is sparse in some areas, the application of the OH concept to surveillance is expected to increase efficiency, cost-effectiveness and cost-benefits (Babo Martins, Rushton, & Stärk, 2017; Bordier et al., 2020; Stärk et al., 2015). Concurrently, international projects to strengthen collaboration across sectors and disciplines and OH surveillance have been increasingly encouraged at the global, national and local-level (Bordier et al., 2020).

Within Europe, several initiatives focusing on OH Surveillance have been implemented in the last decade. According to a review on these national projects by Ellis-Iversen et al., OH surveillance initiatives could be classified in four types, namely cross-sector laboratory analyses, multidisciplinary groups to discuss surveillance outcomes, surveillance outcome communications to the general public, and governance structures across sectors (Ellis-Iversen, Petersen, & Helwigh, 2019).

2.2.1 Dimensions and degrees of collaboration in One Health surveillance systems

According to a systematic review on the characteristics of One Health surveillance systems by Bordier et al., current systems can be grouped in any of four main dimensions, depending on the extent of cooperation across sectors and disciplines: (i) institutional collaboration across sectors aiming at improving governance and operation of a given surveillance system; (ii) cooperation along the hierarchical ladder of the decision-making process; (iii) cooperation across disciplines; and (iv) cooperation through public-private partnerships (Bordier et al., 2020).

Looking at the first operational dimension of institutional cooperation, several degrees of collaboration existed at all steps of the surveillance process, namely through planning, data collection, laboratory testing, data management, data sharing, data analysis and communication of results (Bordier et al., 2020).

Secondly, the commitment of a diverse set of disciplines, such as biosciences, social sciences and engineering was an important condition and characteristic of cooperative One Health surveillance initiatives. According to Bordier et al., disciplines referring to biosciences, namely medicine, microbiology, epidemiology, entomology, ornithology, and parasitology exhibited greater preponderance (Bordier et al., 2020).

Thirdly, cooperation along the hierarchical scales of the decision-making process included various administrative and jurisdictional structures both at different organisational levels of a given country, either locally, regionally or centrally, as well as at supra-national structures such as international bodies and institutions (Bordier et al., 2020).

Lastly, cooperation in One Health surveillance can be based on the development of public-private partnerships across sectors, primarily represented by organisations focusing on veterinary, medicine, private laboratory, farming, feed and food logistics and distribution, pharmaceuticals, or through professional associations (Bordier et al., 2020).

2.2.2 Environmental framework

A less spoken but rather exemplary matter on One Health is environmental surveillance, particularly represented by the impact of environmental contaminants such as dioxins, heavy metals, polychlorinated biphenyl (PCB) or phycotoxins (Bordier et al., 2020). Both humans and animals are potentially exposed to a set of common chemical elements as they share the same food and water sources, within the same overarching environmental structure. Humans can be further exposed through the ingestion of contaminated animal food products included into their diet (Buttke, 2011). Furthermore, due to their particular and earlier environmental exposure, animals can serve as sentinels for environmental health risks to humans (Pearce & Douwes, 2013; Reif, 2011). Necessarily, environmental contamination will demand an exceedingly interdisciplinary approach to properly react and respond to a given health risk (Bordier et al., 2020).

Considering a more encompassing framework of assessing environmental hazards and structuring surveillance of environmental exposures, Thacker et al. have proposed that surveillance of environmental hazards includes the following levels of investigation (Thacker et al., 1996):

1. *Hazard surveillance*, focusing on identifying and demonstrating the existence of a given hazard(s) within a specific environment (e.g. air pollution);
2. *Exposure surveillance*, aiming at determining the degree of exposure to a particular chemical in a specified at-risk population (e.g. lead poisoning)

3. *Outcome surveillance*, aiming at quantifying the extent of adverse effects occurring after exposure to a given chemical (e.g. birth defects).

Conceptually, all three aforementioned levels of investigation are useful to One Health surveillance initiatives and frameworks. For example, looking at hazard surveillance, pollutant chemicals in water, soil or air can potentially affect aquatic and terrestrial animals, and therefore, monitoring and alleviating their environmental release will likely be advantageous for all species and the general environment (Thacker et al., 1996).

2.3 Specific surveillance issues

Within the WHO depiction of One Health, as '*an approach to designing and implementing programmes, policies, legislation and research in which multiple sectors communicate and work together to achieve better public health outcomes*' (World Health Organization (WHO), 2018), there is a clear underlining of three particular surveillance areas, namely zoonotic and emerging diseases, food safety, and antimicrobial resistance.

Within the European Union, Decision 1082/2013/EU creates the legal foundation for the One Health domain into the *European Union Health Security framework*, pushing for a clear focus on animal health and food safety, preparedness and response to zoonotic threats, and antimicrobial resistance (European Commission (EUC), 2013). Additionally, in 2017 the European Commission adopted the *European One Health Action Plan* against antimicrobial resistance (European Commission, 2017), whilst also supporting capability enhancing activities with simulation exercises and workshops, and promoting coordination between sectors (European Centre for Disease Prevention and Control, 2018). The EU vision for One Health ambitions to build up cooperation and coordination across human and animal public health sectors, to develop, implement and integrate Early Warning and Response System and existing surveillance, alert and information systems, and to advance new regulations on veterinary medicines and feed (European Centre for Disease Prevention and Control, 2018).

In this section, we will discuss some of the most relevant elements on One Health surveillance concerning some of these areas of interest.

2.3.1 Zoonotic diseases surveillance and outbreak investigations

An exemplary case displaying the importance of the One Health approach, particularly within a low-income country setting, occurred with the 2014 Ebola outbreak in Sierra Leone. Aside from the several challenges and difficulties highlighted by the direct management of the human public health side of this outbreak, including the need to improve access to diagnostics, to strengthen disease surveillance, and to enhance capacity and capability among the healthcare workforce, specific One Health priority areas were also considered (Philips and Markham, 2014; European Centre for Disease Prevention and Control, 2018). In this regard, these priority proposals focused on the development and implementation of early warning systems integrating human, veterinary, wildlife, and ecological related data, the local and national public health systems reinforcement and strengthening (European Centre for Disease Prevention and Control, 2018). Notwithstanding, at low-income settings, initiatives to build up preparedness and response capacity often are curtailed by insufficient or unavailable resources. Therefore, there will be a paramount need for adequate and sustainable financing of One Health efforts in these settings (European Centre for Disease Prevention and Control, 2018).

Likewise, the United States Agency for International Development's (USAID) Emerging Pandemic Threats program has developed the PREDICT project, working on more than 20 countries in Central Africa, South and South-East Asia and Latin America, through a consortium of global and in-country partners (Kelly et al., 2017). This project aims to improve surveillance of potential zoonotic viruses using "risk-based" surveillance. As an example, the PREDICT project performs sampling of wildlife animals that are known to harbor zoonotic viruses with human and non-human spillover potential (Kelly, 2016; Levinson et al., 2013).

At an opposing side, within high-income settings, the problem posed by the recurrent occurrence and continuous spatial dissemination of West Nile virus in Europe has also raised the need for better surveillance and improved interaction between human and veterinary domains (European Centre for Disease Prevention and Control, 2018). West Nile Fever (WNF) is a vector-borne zoonotic viral disease that infects humans and horses and has birds as its primary reservoir, and can also be transmissible between humans through contaminated blood products. In the past decade, European public health authorities have studied and documented the epidemiologic evolution of West Nile fever based on human routine traditional passive surveillance systems, complemented by veterinary surveillance with sero-prevalence studies across several species, particularly in horses. Assuming a One Health approach might boost surveillance of this disease by improving cooperation and public action coordination among experts for human and veterinary public health, as well as generating a more comprehensive epidemiologic understanding. A proposal for the future preparedness of WNF within the One Health approach should hence include a joint surveillance system, encompassing humans, horses, mosquitos, and birds (European Centre for Disease Prevention and Control, 2018).

Globally, there are several zoonotic disease surveillance systems and projects primarily based on data collation and analysis. The Global Early Warning System for Major Animal Diseases (GLEWS) is a combined initiative by WHO, FAO and OIE that intends to collate information on disease events collected by each organization (Pinto et al., 2011). The Global Public Health Intelligence Network (GPHIN) is a web-based network performing website crawling and data scraping from websites, digital news and other internet-based outlets in order to track disease outbreaks in humans, animals and plants (Blench, 2008). The Global Outbreak Alert and Response Network (GOARN) is a network of institutions and organisations providing international public health resources and technical assistance for rapid identification, confirmation and response to public health emergencies and international outbreaks (World Health Organization (WHO), 2008a,b). Lastly, the Program for Monitoring Emerging Diseases (ProMED) is a web-based service to detect uncommon health events related to emerging infectious diseases and toxins affecting humans, animals and plants. It produces daily reports and commentary by global team of experts in many One-Health disciplines, and has been the first to report on numerous outbreaks including SARS-CoV, MERS-CoV and Ebola virus disease (Yu & Madoff, 2004).

2.3.2 Foodborne disease outbreak investigations

Food production and safety, due to their overlap on human and animal health, is an extremely important domain when considering a joint approach to the prevention, control and mitigation of foodborne diseases. This involves the application of proper guidelines and standards by animal feed manufacturers and farmers to safeguard food safety, as well as the responsibility of retailers and consumers in averting downstream contamination. Nonetheless, certain human pathogenic microorganisms can exist as normal flora agents in food animals. For example, *Campylobacter* and *Salmonella* bacteria often infect poultry, which can result in subsequent contamination of chicken meat and eggs. Similarly, human and animal microorganisms can also contaminate fruit and vegetables during the production chain, namely during washing, packaging and transportation. As an essential public health function, food-borne diseases surveillance will likely benefit from a One Health approach, in particular through the creation of epidemiologic and laboratory teams with and joint use and application of new molecular typing and genomic methods for rapid cluster detection and evaluation (European Centre for Disease Prevention and Control, 2018; Bordier et al., 2020).

2.3.3 Surveillance of emerging pathogens

The emergence of Middle East respiratory syndrome (MERS-CoV) has emphasized the preponderance and urgency of applying a One Health approach, particularly because there are various possible animal reservoirs for human diseases. While the first reported human cases were identified in 2012 in Saudi Arabia, dromedary camels were later identified as the primary animal reservoir (European Centre for Disease Prevention and Control, 2018). The identification and detection of this emerging pathogen was a multi-institutional cooperation effort, among WHO, the Qatar Ministry of Health, and Erasmus Medical Center. Within a One Health perspective, this joint initiative allowed for a swift gathering of data and knowledge improvement on MERS-CoV, and later advanced a more detailed understanding on the disease incidence and transmission. Nevertheless, this experience has also underlined several challenges in One Health preparedness, namely limited governmental cooperation, deficient support on the veterinary health sector, knowledge gaps on disease transmission between humans and camels, deficient screening of animal imports and exports, and the lack of coordination and guidance on laboratory procedures and microbiological sampling standards (European Centre for Disease Prevention and Control, 2018).

2.3.4 Antimicrobial resistance

Antimicrobial resistance can be outlined as primarily an environmental matter. Its impacts are manifold, from human health to livestock, domestic and wildlife animal health, ecosystems and plant health, as well as food safety and hygiene (Butaye, van Duijkeren, Prescott, & Schwarz, 2014; Queenan et al., 2016; Radhouani et al., 2014).

In recent years, several high-income countries have made consistent progress setting up integrated surveillance systems dedicated to antimicrobial resistance. For example, the Danish Integrated Antimicrobial Resistance Monitoring and Research Program (DANMAP), established in 1995, gathers and analyses data on antibiotic use and antimicrobial resistance for a set of predefined zoonotic and pathogenic bacteria. Sample isolates are gathered from an array of sources, such as healthy animals at slaughterhouses, sick animals under veterinary investigation, food items sampled from wholesale and retail outlets, humans under clinical evaluation at a given healthcare system, and healthy people (Hammerum et al., 2007). Consequently, this program was fundamental in guiding policy on the use of antibiotics in livestock, both at the national level in Denmark, as well as cross-nationally in Europe (Hammerum et al., 2007). Comparable initiatives were formed, such as the Canadian Integrated Program for Antimicrobial Resistance Surveillance (CIPARS) and the National Antimicrobial Resistance Monitoring System (NARMS) in the United States (Acar & Moulin, 2013).

Other examples of antimicrobial resistance surveillance are the European Food- and Waterborne Diseases and Zoonoses Network (FWD-Net) and the European Antimicrobial Resistance Surveillance Network (EARS-Net), both under central coordination by the European Center for Disease Control and Prevention (ECDC). The former network focuses on food- and waterborne diseases and zoonoses as well as antimicrobial resistance in human *Salmonella* spp. and *Campylobacter* spp. infections in Europe (Pol-Hofstad et al., 2012). The latter network incorporates reported data from European national and regional institutions, organizations and laboratories, responsible for antimicrobial resistance. They collect data from blood and cerebrospinal fluid isolates on specified bacterial microorganisms, such as *Escherichia coli*,

Klebsiella pneumonia or *Pseudomonas aeruginosa*. Importantly, the analysis of the data on the EARS-Net network supports the definition of antimicrobial agent combination panels for each bacterial species under surveillance as well as support the revision of The European Committee on Antimicrobial Susceptibility Testing (EUCAST) guidelines (Bronzwaer et al., 1999; Gagliotti et al., 2011).

Regarding the animal and food sectors in Europe, according to Directive 2003/99/EC and Decision 2013/652/EU, the European Food Safety Authority (EFSA) is responsible to coordinate the active surveillance of antimicrobial resistance on healthy livestock animals and food products thereof focusing on zoonotic bacteria, such as *Salmonella* spp. and *Campylobacter* spp., and *E. coli* as indicator bacteria. The European Antimicrobial Resistance Surveillance network in veterinary medicine (EARS-Vet) is the future proposal network for this legal framework for veterinary surveillance on antibiotic resistance (Mader et al., 2021).

A relevant point made by Queenan et al. highlights the fact that none of the aforementioned surveillance systems comprises wildlife and domestic or companion animals as well as ecosystems (Queenan et al., 2016). Antimicrobial resistant pathogens have been identified in a diverse set of environmental samples, ranging from hospital and farm effluents, to sewage and wastewater, demonstrating that these pathogens can spread through the environment. Additionally, wildlife animals, albeit seldom exposed, have also been found to acquire antimicrobial resistant microorganisms (Radhouani et al., 2014).

As shown above, most antibiotic resistance surveillance networks have been established in high-income countries, focusing on specific pathogens and with limited geographical representativeness. Simultaneously, the absence of international standards for antimicrobial resistance data collection and reporting on human and veterinary populations has hampered comparability and policy guidance (Dar et al., 2016). In 2012, even though OIE members pushed a formal agenda toward the adoption of new international standards on the harmonisation of national antimicrobial resistance surveillance (Orand, 2012), most low- and middle-income countries did not possess the basic structure and processes to collect data on antimicrobial usage in animals (Nisi et al., 2013), or the needed human and veterinary microbiology facilities (Dar et al., 2016).

On a global perspective, a WHO report pointed out key gaps in global capacity for antimicrobial resistance surveillance. Particularly, it highlighted the absence of consensus and guidance on methodology, data sharing and institutional coordination among global regions (World Health Organization (WHO), 2014).

2.3.5 Environmental hazards

As previously mentioned, chemical contamination within environmental domains can serve as a sentinel event for hazardous exposures both for humans and animals.

Mercury is an important example of valuable application of the One Health surveillance principle. Environmental contamination with mercury can generate serious adverse effects across multiple biological systems and ecosystems, something which has been clearly demonstrated with the Minimata outbreak in Japan in 1956 (Harada, 1995), with clear involvement of fishes through the food chain in the damage done to cats and humans throughout months (Tsuchiya, 1992).

Lead poisoning is also an important example of applied animal sentinel surveillance within One Health surveillance. In 2006, in Western Australia, an unexpected event with over 10,000 dead song birds alerted authorities to a potential environmental catastrophe. This sentinel event in animals triggered an investigation that ultimately showed lead poisoning in samples collected from humans, birds, and the environment, particularly in drinking water and soil (Gulson et al., 2009). The attributed source was the transportation of lead concentrate through the involved location, despite the involvement of other unidentified sources (Gulson et al., 2009).

2.4 Barriers to surveillance

The establishment of an enduring, sustainable, and robust cross-sectoral surveillance system in a One Health context is beneficial in the long term, as stakeholders from different fields who have learned to communicate, work and share responsibilities together will be able to carry out those tasks more efficiently in future collaborations. In turn, such experience will help to enhance the overall mitigation strategies to the emergence of new zoonoses.

Some specialists argue however, that this collaboration has not been given significant consideration from researchers nor from program evaluators (Babo Martins, Rushton, & Stärk, 2016). The advantages of cooperation amongst a multitude of different public health sectors are not disease or program specific, as it concerns the capacity of a whole system to address and take action to different challenges. Consequently, the generated benefits are commonly not reported in surveillance and program evaluations. Such advantages potentially include intellectual capital, a stronger knowledge base, social capital, or other feelings of reassurance and trust, which may translate into tangible long-term benefits (Babo Martins et al., 2016).

Different organizations are required to overcome disciplinary and organizational separation when implementing an integrated approach in a One Health context (Kruk et al., 2015). Cooperating stakeholders and decision makers from involved organizations develop new interdisciplinary skills and competencies, and in addition, intersectoral teams, processes, and tools for management of programmes often become established, for example by development of formal collaboration or communication agreements (Kruk et al., 2015). Collectively, an experienced workforce will be better adapted to solve upcoming challenges, increasing the system's sustainability.

There are, however, challenges to this collaboration.

The One Health principle is currently under a balancing act of past leveraged challenges and captivating opportunities in the future. Firstly, the interdisciplinary condition of OH conditions each sector to look and envision this concept a priori through its own lens. Secondly, both the absence of a shared understanding as well as existing structural and political barriers can curtail the implementation of the One Health (European Centre for Disease Prevention and Control, 2018). In the following points, we describe in more detail some of the challenges and opportunities facing the One Health idea.

2.4.1 Communication and coordination

Communication and coordination are an essential element in the implementation of OH (European Centre for Disease Prevention and Control, 2018). Necessarily, unsuccessful communication among working sectors will not only hamper the potential for response, but it will also curtail the capability to surpass the existing cultural and political conditions exacerbated during acute events like outbreaks (Bordier et al., 2020). Several authors have shown that efficient communication and consultation channels can provide adequate and collaborative stakeholder commitment with a OH surveillance system (Sleigh et al., 1998a,b; Talaska, 1994; Wielinga et al., 2014).

According to Bordier et al., the development of mechanisms for coordinated situational awareness dedicated to early warning systems across sectors, the creation and revision of coordinated response plan with standardized case definitions and laboratory tests, and the identification and training of focal points across sectors are some of the suggestions for improving communication in OH surveillance systems (Bordier et al., 2020). Hence, it is of vital importance to define, prepare and train adequate communication channels and frameworks during non-emergency periods.

2.4.2 Lack of conceptual framework

The absence of unified or agreed conceptual framework to define OH surveillance is currently undermining its operationalisation (Bordier et al., 2020). According to several authors, the existence of a suitable framework, either legal or institutional, serves as a lever for cooperation (Abbas, Venkataramanan, Pathak, & Kakkar, 2011; Adamson, Marich, & Roth, 2011; Bordier et al., 2020; Lapiz et al., 2012).

Looking at existing OH surveillance systems, it is important that a conceptual framework describes the different organisational levels and structures of cooperation, as well as the determinants controlling their governance and practical application, both in the medium- and the long-term. Regarding this proposal, Bordier et al. have proposed a framework that responds to a minimum of three different organisational levels, namely policy, institutional, and operational-wise (Bordier et al., 2020).

2.4.3 Integration and sharing of data and biological samples

Either at national, regional or even international levels, data and biological sample sharing among institutions is an obstacle to One Health and most commonly indicates an absence of in-country capacity and coordination between sectors (Bordier et al., 2020). A set of suggestions to improve this are primarily centered on the development of standard frameworks for datasets and biological sample sharing, as well as data harmonization and interoperability on data structures across sectors. These frameworks must address a priori the issues regarding data ownership, cross-border and cross-sector data access, processing and storage (Bordier et al., 2020).

The integration of data from different sectors and disciplines can be further enhanced by establishing and developing joint databases, using compatible information systems defined by design (Bordier et al., 2020; Morgan, 2006; Talaska, 1994; Vrbova et al., 2016; Witt et al., 2004). A fundamental role in the operationalization of this point will be the definition of harmonized data and communication standards for laboratories, particularly for national reference laboratories working in human, veterinary and environmental domains (Ammon & Makela, 2010; Bordier et al., 2020).

2.4.4 Capacity building

Limitations and constraints in overall capacity for One Health will be a crucial element preventing proper implementation of this concept. Controlling and dealing with outbreaks implies that local human resources are prepared and trained with a readily available epidemiological and laboratory infrastructure. Complementary actions that can reinforce sustainable capacity building include the breakage of formal compartments between public health positions and academia or multi-institutional collaboration. This latter point could be further reinforced by allowing funding streams usually attributed to academia to be used in joint research initiatives with human and veterinary public health officials. Lastly, overcoming capacity limitations in One Health will require proper funding, infrastructure and personnel assigned and retained at reference laboratories or networks (Bordier et al., 2020).

2.4.5 Risk perception

According to a systematic review on several initiatives in OH surveillance systems by Bordier et al., differential risk perception among sectors is a limiting element for One Health. The central principles in risk management are traditionally different between human, where the precautionary principle is applied, and veterinary health, where less strict consideration is used. Cross-sector and cross-border discussion and agreement along with better communication are needed to overcome this risk perception challenge.

2.4.6 *Associated costs*

Cross-sector and cross-border surveillance systems are not inexpensive. The expenses include labor, logistics and material costs, transportation, and other administrative and implementation costs. Although the benefits may occur predominantly in one sector, the costs may however be mostly associated with another, thus complicating the economical relationships in the One Health context. A good example concerns the surveillance costs for the bovine spongiform encephalopathy in the UK. Before any surveillance measures were implemented, 3 million infected cattle entered the human food chain (Smith & Bradley, 2003). The economic losses were estimated at several billion Euros, and a global effort for surveillance in bovines was established. The current feeding practices in the livestock sector have been altered to avoid disease and health-related costs for humans (Zinsstag et al., 2020).

3. Preparing for the future

3.1 Reorganizing internally and externally

There is a case for OH departments instead of compartimentation of human, veterinary and environmental sections. The elimination of some zoonotic diseases, the increasing challenges imposed by climate change and other environmental challenges will push to the limit not only activities in the field but also the capacity to process huge amounts of data and the need of specialized teams handling not only complex data but also gathering the expertise of different fields. National public health institutions should be reinforced in their funding and staff to deal with the demanding times of an interconnected and changing world (Frenk, 2010; Frieden & Koplan, 2010).

To change a silo culture of different fields, one should bulk up multidisciplinarity by starting joint training in faculties between medicine, veterinary and environmental sciences (Rabinowitz et al., 2017), with already known frameworks of interprofessional education and practice (World Health Organization (WHO), 2010). The perception of the interdependence and lines of common research and action can open up new opportunities and pave the way for positions within the One Health framework. Veterinarians have been the driving force of OH initiatives and health professionals should be sensitized to the importance of working in this subject, involving social sciences professionals as well (Marcotty et al., 2013).

Rethinking postgraduate training should consider the equity possibilities that online and distance training brought during the pandemic. It has become easier to give access to experts at a low fee or even for free. Although this can cause some issues to academic institutions, public health organizations can benefit not only from that but also consider teleworking as a way to reinforce staff without reallocation costs and deterrence.

One response during the COVID-19 pandemic has been the creation of multidisciplinary centers.

Oxford University has announced in May 2021 the Pandemic Sciences Center, with the purpose of increasing collaboration in global research in this field.

USAID has launched a project of 85 million dollars addressing the training of a One Health Workforce - Next Generation scoping the years of 2019–24, capaciting a network of universities to train multidisciplinary teams. AFROHUN (African One Health University Network) and SEAOHUN (Southeast One Health University Network) are two one other projects addressing the same issue. Amidst joint initiatives, the One Health European Joint Program (OHEJP) is one of the several initiatives that is aiming to reinforce transdisciplinary cooperation through sharing of projects and training, with focus on foodborne zoonoses, AMR and emerging threats. In fact, it is not realistic to expect that individual countries are able to mount efficient detection and response without deep collaboration within subregional, regional and worldwide frameworks, starting obviously in IHR. GLEWS (Global Early Warning System) is a joint initiative from whom, FAO and OIE that joins data from different platforms to respond to the needs of monitoring, early detection and prediction of biological and environmental threats (Kshirsagar et al., 2013). EcoHealth Alliance must also be quoted as an international actor dedicated to the One Health approach, resulting from the merging of Wildlife Trust and the Consortium for Conservation Medicine.

International standards start in a common language and definitions - one of the efforts within the OHEJP is the construction and availability of a One Health Glossary to allow easier communication between different sectors (Buschhardt et al., 2021). This is part of a broader project called ORION (One health suRveillance Initiative on harmOnization of data collection and interpretation) that focuses specifically on the *"semantic and technical interoperability between the sectors, with focus on surveillance information"* (ORION Project, 2019). The project involves institutions from seven European countries but can pave the way for a broader use of common terms.

A new One Health High-Level Expert Panel is to be setup to support FAO, OIE, the United Nations Environment Program and WHO and one of the first steps is exactly the development of risk assessment and surveillance frameworks. The World Bank is moving toward new ways of funding One Health initiatives as well (World Bank, 2021).

The European Commission (EC) recently proposed to create the European Health Emergency Preparedness and Response Authority (HERA) which function will also be improving risk assessment, besides management and communication (Villa et al., 2021). The proposed benchmarking of monitoring, surveillance and matching policies at a national level, together with the Sendai framework (Aitsi-Selmi, Egawa, Sasaki, Wannous, & Murray, 2015) should prove effective in promoting important changes. The need for a new pandemic cooperation framework (Forman, Atun, McKee, & Mossialos, 2020) has been made clear and the negotiation of an international pandemic treaty has been announced (European Union, 2021), with one of the priorities being surveillance improval and reporting. Nonetheless, one must retain skepticism toward these announcements, since proper criticism should be undertaken regarding what has failed in the previously set mechanisms (Baum et al., 2021; Burkle, 2020; Nature, 2021).

Gaps in infrastructure are also a caveat. In a review of 400 public health events of international concern, Bogich and colleagues (Bogich et al., 2012) have pointed out breakdown or absence of public health infrastructure as the driving factor in a relevant proportion of outbreaks - 40%, underlining also the tendency of vertical surveillance programs, with lack of flexibility. Low resource settings can offer innovation in One Health that richer countries can take advantage of in reorganizing their public health services (Nsubuga et al., 2006; Rattanaumpawan et al., 2018). Less crystallized public health structures are more flexible and offer practical examples of outcomes of better cross-sectoral disease surveillance (Thomas et al., 2021).

Access to healthcare is crucial to provide a comprehensive answer to challenges of One Health. With barriers to access, detection, treatment and ultimately control of new and old threats is harder. So, an extra argument for Universal Health Care comes not only from the ethical need of leaving no one behind but also the danger that represents for the whole population (Lal et al., 2020). This has been beautifully put by Basu and colleagues in the context of dynamics of tuberculosis transmission and institutional amplifiers, proving that approaches to high-risk settings of infectious diseases are more efficient and effective than using the same resources in the general population (Basu, Stuckler, & McKee, 2011).

Health Systems reforms have to take into account the specific needs of One Health, among others (Haldane et al., 2021; Lal et al., 2020; Shroff et al., 2021), including appropriate leadership, funding, training, supply chains, access. Specifically testing and contact tracing must be reinforced and properly integrated with a flexible and speedy framework of action - slow adaptive responses and highly bureaucratic processes will hamper the adequate response.

State sovereignty should be retained but the reinforcement of international cooperation to stop disease emergence through collaborative projects is key to a better global outcome (Figuié, 2014; King, 2002). At an international level, one of the main focuses is the difficulty of sharing international outbreak data (Edelstein et al., 2018; Stoffel et al., 2020) and cross-sector communication, collaboration and knowledge (Filter et al., 2021).

Taking Portugal's example, we could consider the following facts:

1. difficulties in data access between different organizations in a government;
2. the lack of and need for efficiency and taking advantage of analytical skills in different branches of Human, Veterinary and Environmental agencies;
3. the sinergies and speed increase in monitoring, preparedness and response when a One Health approach is present in epidemic surveillance;
4. the lack of connection between academia and public health institutions.

With these prerogatives, one can make a solid case for the creation of national One Health Epidemiological Surveillance Centers in which all of detection, filtering, analysis and signaling can take place. Formalization, funding and recognition of these institutions can be considered a founding step of reform in the public sector. None of these are possible without political will and a shared vision of the importance of integrated surveillance. Integrated surveillance systems must then prove cost-effectiveness and proper metrics and evaluation tools are needed to demonstrate the added value of this approach (Hoinville et al., 2013). For a more favorable "One Health" structure, Governments must shift funding sources that are used for surveillance and preparedness activities, from emergency and unstable "packages" to a stable paradigm (Frenk, 2010).

The initial focus on surveillance as a priority would allow that a diversity of functions non related to surveillance could still function in a more specific branch.

3.2 Speed is a must

Slow reaction of the health system to the zoonotic events and lack of connection to the veterinary sector are problems well known and of widespread importance. The number of events related to zoonotic threats is certain to increase (Machado et al., 2021). Michael Ryan, executive director of the Health Emergencies Program of the World Health Organization (WHO) stated the famous sentence in the beginnings of the COVID-19 pandemic: "*Speed trumps precision*". This can be applicable to surveillance, control measures but also the communication to the public. Communication of epidemic findings and/or warnings can be boosted by social media, traditional media but also developed, instant, warning systems like Cell Broadcast (Sid, 2020). Unfortunately, the *2018 Report on the implementation of the European emergency number 112* has shown that only 4 of 28 countries are currently taking advantage of that technology. It is no coincidence that one must look into Civil Protection to leverage existing resources and optimize surveillance and response to emergencies, biological or not. Lessons should be learned from other areas of concern like wildfires, with the quick detection and deployment of international help via, for example, the European Forest Fire Information System (EFFIS). Speed of detection, communication and action has been at the heart of criticism of the Independent Panel for Pandemic Preparedness and Response (Sirleaf, 2021). Mckinsey has put forward a value from 357 billion dollars during ten years (Craven, Sabow, Van der Veken, & Wilson, 2020) to develop stable response systems, strengthen mechanisms of detection, an integrated epidemic-prevention agenda, readying health systems for surges and scale R&D activities. This is a small fraction of the estimated cost of the COVID-19 pandemic (Cutler & Summers, 2020).

Contact tracing capabilities are also one of the limiting factors in the response to pandemic threats and emerging diseases, despite heterogeneous effectiveness depending on the characteristics of the disease (Eames & Keeling, 2003; Keeling, Hollingsworth, & Read, 2020; Klinkenberg, Fraser, & Heesterbeek, 2006). The investment on this response should also be considered depending upon the expected results (Armbruster & Brandeau, 2007) and considering prospective and retrospective tracing (Bradshaw, Alley, Huggins, Lloyd, & Esvelt, 2021). For an appropriate response, preparation must not only involve health professionals and community workers but also citizens in general. Education and mechanisms that allow their participation is essential to quickly detect and respond to biological threats, including the induction of receptivity to cooperate with control efforts. The use of apps for participative surveillance is one of the present-future challenges in One Health. For contact tracing purposes, there were high hopes that COVID-19 would serve as a model for future pandemic control in that sense and modeling supported its effect (Ferretti et al., 2020; Cencetti et al., 2021). There was some indication that contact tracing apps were useful (Lewis, 2021) but uptake was highly heterogeneous and short of initial expectations (Toussaert, 2021). There are doubts surrounding the application of these apps in different cultural, geographical and social contexts and the need for a clear evaluation process has been put forward (Anglemyer et al., 2020; Colizza et al., 2021). South Korea and Israel were more successful also probably due to the usage of apps that went beyond the limits allowed in Europe in data protection and privacy (Khosla, 2020; Park et al., 2020), despite allegations that such obstacles could be overcome (O'Connell & O'Keeffe, 2021; Reichert et al., 2020). In resource deprived areas, the use of participative surveillance has extra added value, improved by access to apps and smartphones (Karimuribo et al., 2017; Robertson et al., 2010; Smolinski et al., 2017; Wood et al., 2019). Other areas beyond infectious diseases, like mental health, are likely to benefit from the use of apps in surveillance (Wang et al., 2018). Mobile phone data has potentials that goes beyond participative surveillance, including mobility and proximity data that can feed on surveillance systems and predictions (Grantz et al., 2020; Leung et al., 2021; Venkatramanan et al., 2021).

There are several concerns regarding the use of technology and the conflicts between the public interest and individual privacy that will require a continuous effort to balance an effective response without breaching human rights (Amit et al., 2020; Keshet, 2020; Khosla, 2020). The inclusion and development of technological solutions must be properly integrated in a bigger scope, especially regarding the reform and framework of health systems as a whole (Lal et al., 2020) and continuous work on ethical guidelines for this technology must continue (Morley et al., 2020). Continuously testing the data for privacy breaches, using appropriate methods, is also a priority (Carvalho, Faria, Antunes, & Moniz, 2021).

Detection methods and laboratory adaptations are needed to quickly work on unknown pathogens, essential to the discovery of new pathogens such as Middle East Respiratory Syndrome - Virus (Zaki et al., 2012) or Sars-CoV-2 (Zhou et al., 2020). Metagenomic sequencing and surveillance will also gain new momentum with the use of new tools (Kalantar et al., 2020) to provide easier access to low resources countries (Filkins & Schlaberg, 2021), and improve detection of a vast number of microorganisms, including those of unknown relevance, difficult to culture or divergent from the usual genotype. Harnessing the total possibilities of metagenomics, with samples collected all over the globe and compared, will require the matching computing power and expertise toward the intent of a "*genomics-informed, real-time, global pathogen surveillance system*" (Gardy & Loman, 2018).

3.3 Beyond health

Biodiversity, ecosystems and nature in general must be placed at a higher priority in politics and society to allow sustainable funding of surveillance systems. Determinants of zoonotic risk should be faced and surveyed to offer broader perspective and data to persuade political and societal action.

Some of the corrections needed are beyond the traditional scope of Health, involving changes of economic activities whose characteristics promote the appearance of new zoonotic problems. One of the solutions found for prevention of Nipah spillover was the modification of pig farming practices in the affected areas. The perception of local and eventual global risk must lead not only to proper incentives in the changes of these practices but also the funding of proper surveillance systems, adjusted to each context and properly integrated in the existing infectious diseases surveillance framework. Surveillance of food habits and animal handling in certain areas is critical to prevention of diseases like Ebola or even MERS.

Education and empowerment of local communities is also key in reinforcing detection and response as part of early warning and response systems (Smolinski et al., 2017). Spiegel and colleagues (Spiegel et al., 2005) outlined the need for a socio-ecological perspective throughout the analysis of several approaches to dengue in different countries, stressing the importance of cultural, geographic and social characteristics. Dengue is one of the models for the participative surveillance framework due to the dire need of participation of the population in the control and detection of mosquitoes, crossing data from health workers and citizens using simple reports of mosquitoes presence through apps, and generating real time risk maps (Babu et al., 2019). Additional examples can be found with other diseases like Chagas disease (Abad-Franch, Vega, Rolón, Santos, & de Arias, 2011), avian influenza (Azhar et al., 2010) and others (Mariner et al., 2011).

The role of social structure, behavior and networks in transmission of infectious diseases (Salje et al., 2016) should not only reinforce the need of community participation but also the role of sociology and anthropology to assist surveillance of infectious diseases (Janes, Corbett, Jones, & Trostle, 2012). Studying, measuring and monitoring the effects of legal actions and frameworks should also be part of a One Health approach (Burris, Anderson, & Wagenaar, 2021).

Creating common, intersectorial objectives is a way forward to promote One Health. In the context of food security, Denmark has been successful in implementation of One Health approach to the *Salmonella* problem (Alban et al., 2002, 2012, 2013; Alban & Stärk, 2005), through several measures, including the implementation of financial incentives to the decrease of use of antibiotics. Prioritization will always depend on the political will of defining cross sectoral targets, indicators and funding the effort to achieve them. Canada has taken a step for a national multisectorial initiative with the creation of the One Health Network for the Global Governance of Infectious Diseases and Antimicrobial Resistance - Global 1HN (Ruckert, 2020). Economic and cultural traditions as wet markets in Asia are to be shaped and surveilled in a way to protect workers, local populations and ultimately global health as evidence builds up of their potential role in zoonotic jumps (Xiao, Newman, Buesching, Macdonald, & Zhou, 2021).

An initial investment in prioritization (Ihekweazu et al., 2021) and surveillance upgrades can help drive the political will by underling more precise estimates of burden of disease and justifying proportional resource allocation. On top of that, promoting change that is easier to implement could also facilitate further changes, especially in the public perception and recruitment for bigger challenges of modification of production and consumption patterns.

3.4 One Health prediction

Data mining and modeling approaches to the study of relationships between the empirical data involving animal population indicators and human outbreaks, connecting them with predicted transmission patterns also shed light for predictions of outbreaks (Han, O'Regan, Paul Schmidt, & Drake, 2020; Pandit et al., 2020).

Using social media to monitor and predict the evolution of outbreaks (Zhang et al., 2018) is not new but the use of artificial intelligence (Rahman et al., 2021), natural language processing, machine learning and anomaly detection algorithms has pushed its use to new limits (Budd et al., 2020; Calandra & Favareto, 2020; McCall, 2020). We can find one of those examples in BlueDot, a canadian company that developed a disease surveillance analytic program prepared to screen news, airline data and reports in 65 different languages and that has been the first to emit a warning signal regarding Wuhan outbreak (Allam, Dey, & Jones, 2020; McCall, 2020). Previously, it had been successful in helping predict Zika spread (Bogoch et al., 2016).

Social media and traditional media monitoring is also within the scope of infodemiology, a field of increasing interest and importance due to the "infodemic" phenomenon. The impact on vaccine uptake and engagement in preventive behaviors justifies the corresponding triad of real-time surveillance, accurate diagnosis, and rapid response (Scales et al., 2021) applied to the infodemic.

Use of heavy processing capacities will always be dependent on the reliability of the reported data, which is especially difficult to assure in remote settings. But even crowdsourced data show the potential to help monitor and respond in health emergencies (Leung and Leung, 2020; Sun et al., 2020), with the advantage of speed of data acquisition and early mitigation measures. Crowdourcing will not replace valuable data and the need of standardized procedures to share it alongside the need of creating validation pipelines for predictions based on either approach, especially in emergency situations. The standardized approach to data is even more important because of the difficult integration between animal, human and environmental data. Meaningful and proper data collection continues to be a limiting factor to be addressed, together with the acceptance of society to allow more or less sensitive information to be collected. Individual devices collecting data are one of the frontiers that can be useful with careful ethical considerations in humans (Michael et al., 2006; Sepai et al., 2008) but with already broadened use in animal populations. Regarding his last aspect, despite the fast deployment of new technologies such as unmanned aircraft systems (Linchant et al., 2015) and others in livestock (Shen et al., 2014), standardization is also a prominent issue (Campbell, Urbano, Davidson, Dettki, & Cagnacci, 2016; Doherr & Audigé, 2001), besides ethical and impact considerations (Bodey et al., 2018). Data sharing in Public Health for disease surveillance, given the possibilities of new technologies pose several other ethical challenges to address (Kostkova, 2018).

Modeling human and non-human populations alongside environmental factors requires extraordinarily complex skills which are only possible with international cooperation and the training of multidisciplinary teams (LaDeau et al., 2017). Further understanding of the complete ecological framework of diseases, particularly neglected tropical diseases (Garchitorena et al., 2017) is also necessary for the proper prediction. Shifts of vector populations is one of the targets of prediction (Caldwell et al., 2021) but also changing patterns of transmitted diseases by unchanging vector populations (Mordecai, Ryan, Caldwell, Shah, & LaBeaud, 2020). An added complexity is the actual use of whole genome sequencing and risk prediction based on specific changes in infectious agents (Morse et al., 2012). One can also expect the prediction of vaccine needs (Menachery et al., 2015) and acceleration of vaccine development in response to the detection of a new relevant pathogen, through several efforts, including initiatives like Coalition for Epidemic Preparedness Innovations (Gouglas, Christodoulou, Plotkin, & Hatchett, 2019) with the ambition of reducing the calendar to less than six months time (Krammer, 2020).

4. Conclusion

On a belief of the need of an unprecedented international effort to address the major challenges in one health, climate change and planet health research (Amuasi et al., 2020), there are efforts than can be addressed to improving One Health surveillance setting at a local, regional and especially national level, in response to the already existing challenges. Detection and prediction must be quick enough to make a difference and all of this must translate into useful decision tools to guide policy and action. Morgan et al. have summarized core principles for an integrated disease surveillance system: population-based, laboratory confirmation, digital data, data transparency and adequate financing (Morgan et al., 2021), dimensions that have been described throughout this chapter.

We hope that these pages not only allow a broad perspective of concepts and current endeavors in OH surveillance but also show ways to move forward in this field.

References

van Aart, A. E., Velkers, F. C., Fischer, E. A., Broens, E. M., Egberink, H., Zhao, S., ... Smit, L. A. (2021). SARS-CoV-2 infection in cats and dogs in infected mink farms. *Transboundary and Emerging Diseases*.

Abad-Franch, F., Vega, M. C., Rolón, M. S., Santos, W. S., & de Arias, A. R. (2011). Community participation in Chagas disease vector surveillance: Systematic review. *PLoS Neglected Tropical Diseases, 5*(6), e1207.

Abbas, S. S., Venkataramanan, V., Pathak, G., & Kakkar, M. (2011). Rabies control initiative in Tamil Nadu, India: A test case for the "One Health" approach. *International Health, 3*, 231–239.

Acar, J. F., & Moulin, G. (2013). Integrating animal health surveillance and food safety: The issue of antimicrobial resistance. *Revue scientifique et technique, 32*, 383–392.

Adamson, S., Marich, A., & Roth, I. (2011). One health in NSW: Coordination of human and animal health sector management of zoonoses of public health significance. *NSW Public Health Bulletin, 22*, 105.

Aitsi-Selmi, A., Egawa, S., Sasaki, H., Wannous, C., & Murray, V. (2015). The Sendai framework for disaster risk reduction: Renewing the global commitment to people's resilience, health, and well-being. *International Journal of Disaster Risk Science, 6*(2), 164–176.

Alban, L., Baptista, F. M., Møgelmose, V., Sørensen, L. L., Christensen, H., Aabo, S., & Dahl, J. (2012). *Salmonella* surveillance and control for finisher pigs and pork in Denmark—a case study. *Food Research International, 45*(2), 656–665.

Alban, L., Dahl, J., Andreasen, M., Petersen, J. V., & Sandberg, M. (2013). Possible impact of the "yellow card" antimicrobial scheme on meat inspection lesions in Danish finisher pigs. *Preventive Veterinary Medicine, 108*(4), 334–341.

Alban, L., & Stärk, K. D. C. (2005). Where should the effort be put to reduce the Salmonella prevalence in the slaughtered swine carcass effectively? *Preventive Veterinary Medicine, 68*(1), 63–79.

Alban, L., Stege, H., & Dahl, J. (2002). The new classification system for slaughter-pig herds in the Danish *Salmonella* surveillance-and-control program. *Preventive Veterinary Medicine, 53*(1–2), 133–146.

Allam, Z., Dey, G., & Jones, D. S. (2020). Artificial intelligence (AI) provided early detection of the coronavirus (COVID-19) in China and will influence future Urban health policy internationally. *Artificial Intelligence, 1*(2), 156–165.

Allen, T., Murray, K. A., Zambrana-Torrelio, C., Morse, S. S., Rondinini, C., Di Marco, M., … Daszak, P. (2017). Global hotspots and correlates of emerging zoonotic diseases. *Nature Communications, 8*(1), 1–10.

Almada, A. A., Golden, C. D., Osofsky, S. A., & Myers, S. S. (2017). A case for planetary health/geohealth. *GeoHealth, 1*(2), 75–78.

Amit, M., Kimhi, H., Bader, T., Chen, J., Glassberg, E., & Benov, A. (2020). Mass-surveillance technologies to fight coronavirus spread: The case of Israel. *Nature Medicine*, 1–3.

Ammon, A., & Makela, P. (2010). Integrated data collection on zoonoses in the European Union, from animals to humans, and the analyses of the data. *International Journal of Food Microbiology, 139*, S43–S47.

Amuasi, J. H., Walzer, C., Heymann, D., Carabin, H., Huong, L. T., Haines, A., & Winkler, A. S. (2020). Calling for a COVID-19 one health research coalition. *Lancet, 395*(10236), 1543–1544.

Anglemyer, A., Moore, T. H., Parker, L., Chambers, T., Grady, A., Chiu, K., … Bero, L. (2020). Digital contact tracing technologies in epidemics: A rapid review. *Cochrane Database of Systematic Reviews, 8*.

Armbruster, B., & Brandeau, M. L. (2007). Contact tracing to control infectious disease: When enough is enough. *Health Care Management Science, 10*(4), 341–355.

Azhar, M., Lubis, A. S., Siregar, E. S., Alders, R. G., Brum, E., McGrane, J., … Roeder, P. (2010). Participatory disease surveillance and response in Indonesia: Strengthening veterinary services and empowering communities to prevent and control highly pathogenic avian influenza. *Avian Diseases, 54*(s1), 749–753.

Babo Martins, S., Rushton, J., & Stärk, K. D. C. (2016). Economic assessment of zoonoses surveillance in a 'One Health' context: A conceptual framework. *Zoonoses and Public Health, 63*(5), 386–395.

Babo Martins, S., Rushton, J., & Stärk, K. D. (2017). Economics of zoonoses surveillance in a "one health" context: An assessment of *Campylobacter* surveillance in Switzerland. *Epidemiology and Infection, 145*, 1148–1158.

Babu, A. N., Niehaus, E., Shah, S., Unnithan, C., Ramkumar, P. S., Shah, J., … Jose, C. P. (2019). Smartphone geospatial apps for dengue control, prevention, prediction, and education: MOSapp, DISapp, and the mosquito perception index (MPI). *Environmental Monitoring and Assessment, 191*(2), 1–17.

Basu, S., Stuckler, D., & McKee, M. (2011). Addressing institutional amplifiers in the dynamics and control of tuberculosis epidemics. *The American Journal of Tropical Medicine and Hygiene, 84*(1), 30–37.

Baum, F., Freeman, T., Musolino, C., Abramovitz, M., De Ceukelaire, W., Flavel, J., ... Villar, E. (2021). Explaining covid-19 performance: What factors might predict national responses? *BMJ*, 372.

Blench, M. (2008). Global public health intelligence network (GPHIN). In *Proceedings of the 8th conference of the Association for machine translation in the Americas: Government and commercial uses of MT* (pp. 299–303).

Bodey, T. W., Cleasby, I. R., Bell, F., Parr, N., Schultz, A., Votier, S. C., & Bearhop, S. (2018). A phylogenetically controlled meta-analysis of biologging device effects on birds: Deleterious effects and a call for more standardized reporting of study data. *Methods in Ecology and Evolution, 9*(4), 946–955.

Bogich, T. L., Chunara, R., Scales, D., Chan, E., Pinheiro, L. C., Chmura, A. A., ... Brownstein, J. S. (2012). Preventing pandemics via international development: A systems approach. *PLoS Medicine, 9*(12), e1001354.

Bogoch, I. I., Brady, O. J., Kraemer, M. U., German, M., Creatore, M. I., Kulkarni, M. A., ... Khan, K. (2016). Anticipating the international spread of Zika virus from Brazil. *The Lancet, 387*(10016), 335–336.

Bordier, M., Uea-Anuwong, T., Binot, A., Hendrikx, P., & Goutard, F. L. (2020). Characteristics of one health surveillance systems: A systematic literature review. *Preventive Veterinary Medicine, 181*, 104560.

Bradshaw, W. J., Alley, E. C., Huggins, J. H., Lloyd, A. L., & Esvelt, K. M. (2021). Bidirectional contact tracing could dramatically improve COVID-19 control. *Nature Communications, 12*(1), 1–9.

Bronzwaer, S. L. A. M., Goettsch, W., Olsson-Liljequist, B., Wale, M. C. J., Vatopoulos, A., & Sprenger, M. J. W. (1999). European antimicrobial resistance surveillance system (EARSS): Objectives and organisation. *Euro Surveillance, 4*(4), 41–44.

Budd, J., Miller, B. S., Manning, E. M., Lampos, V., Zhuang, M., Edelstein, M., ... McKendry, R. A. (2020). Digital technologies in the public-health response to COVID-19. *Nature Medicine, 26*(8), 1183–1192.

Burkle, F. M. (2020). Political intrusions into the international health regulations treaty and its impact on management of rapidly emerging zoonotic pandemics: What history tells us. *Prehospital and Disaster Medicine, 35*(4), 426–430. https://doi.org/10.1017/S1049023X20000515

Burris, S., Anderson, E. D., & Wagenaar, A. C. (2021). The "legal epidemiology" of pandemic control. *New England Journal of Medicine*.

Buschhardt, T., Günther, T., Skjerdal, T., Torpdahl, M., Gehtmann, J., Filippitzi, M. E., ... Team, T. O. G. (2021). A one health glossary to support communication and information exchange between the human health, animal health and food safety sectors. *One Health*, 100263.

Butaye, P., van Duijkeren, E., Prescott, J. F., & Schwarz, S. (2014). Antimicrobial resistance in bacteria from animals and the environment. *Veterinary Microbiology, 171*(3–4), 269–272.

Buttke, D. E. (2011). Toxicology, environmental health, and the "One Health" concept. *Journal of Medical Toxicology, 7*(4), 329–332.

Calandra, D., & Favareto, M. (2020). Artificial intelligence to fight COVID-19 outbreak impact: An overview. *European Journal of Social Impact and Circular Economy, 1*(3), 84–104.

Caldwell, J. M., LaBeaud, A. D., Lambin, E. F., Stewart-Ibarra, A. M., Ndenga, B. A., Mutuku, F. M., ... Mordecai, E. A. (2021). Climate predicts geographic and temporal variation in mosquito-borne disease dynamics on two continents. *Nature Communications, 12*(1), 1–13.

Campbell, H. A., Urbano, F., Davidson, S., Dettki, H., & Cagnacci, F. (2016). A plea for standards in reporting data collected by animal-borne electronic devices. *Animal Biotelemetry, 4*(1), 1–4.

Carvalho, T., Faria, P., Antunes, L., & Moniz, N. (2021). Fundamental privacy rights in a pandemic state. *PLoS One, 16*(6), e0252169. https://doi.org/10.1371/journal.pone.0252169. PMID: 34077454.

Cencetti, G., Santin, G., Longa, A., Pigani, E., Barrat, A., Cattuto, C., ... Lepri, B. (2021). Digital proximity tracing on empirical contact networks for pandemic control. *Nature Communications, 12*(1), 1–12.

Choi, B. C. (2012). The past, present, and future of public health surveillance. *Scientifica*, 2012.

Chua, K. B., Chua, B. H., & Wang, C. W. (2002). Anthropogenic deforestation, El Niiio and the emergence of Nipah virus in Malaysia. *Malaysian Journal of Pathology, 24*(1), 15–21.

Colizza, V., Grill, E., Mikolajczyk, R., Cattuto, C., Kucharski, A., Riley, S., ... Fraser, C. (2021). Time to evaluate COVID-19 contact-tracing apps. *Nature Medicine, 27*(3), 361–362.

Corlett, R. T., Primack, R. B., Devictor, V., Maas, B., Goswami, V. R., Bates, A. E., ... Roth, R. (2020). Impacts of the coronavirus pandemic on biodiversity conservation. *Biological Conservation, 246*, 108571.

Craven, M., Sabow, A., Van der Veken, L., & Wilson, M. (2020). *Not the last pandemic: Investing now to reimagine public health systems*. McKinsey Report.

Cutler, D. M., & Summers, L. H. (2020). The COVID-19 pandemic and the $16 trillion virus. *JAMA, 324*(15), 1495–1496.

Dar, O. A., Hasan, R., Schlundt, J., Harbarth, S., Caleo, G., Dar, F. K., & Heymann, D. L. (2016). Exploring the evidence base for national and regional policy interventions to combat resistance. *The Lancet, 387*(10015), 285–295.

Dobson, A. P., Pimm, S. L., Hannah, L., Kaufman, L., Ahumada, J. A., Ando, A. W., ... Vale, M. M. (2020). Ecology and economics for pandemic prevention. *Science, 369*(6502), 379–381.

Doherr, M. G., & Audigé, L. (2001). Monitoring and surveillance for rare health-related events: A review from the veterinary perspective. *Philosophical Transactions of the Royal Society of London. Series B: Biological Sciences, 356*(1411), 1097–1106.

Drewe, J. A., Hoinville, L. J., Cook, A. J. C., Floyd, T., & Stark, K. D. C. (2012). Evaluation of animal and public health surveillance systems: a systematic review. *Epidemiology & Infection, 140*(4), 575–590. In press.

Dufour, B., & Hendrikx, P. (2005). *Epidemiological surveillance in animal health*. Association pour l'Étude de l'Épidémiologie des Maladies Animales.

Eames, K. T., & Keeling, M. J. (2003). Contact tracing and disease control. *Proceedings of the Royal Society of London. Series B: Biological Sciences, 270*(1533), 2565–2571.

Edelstein, M., Lee, L. M., Herten-Crabb, A., Heymann, D. L., & Harper, D. R. (2018). Strengthening global public health surveillance through data and benefit sharing. *Emerging Infectious Diseases, 24*(7), 1324.

El-Sayed, A., & Kamel, M. (2021). Coronaviruses in humans and animals: The role of bats in viral evolution. *Environmental Science and Pollution Research*, 1–12.

Elbers, A. R. W., Koenraadt, C. J., & Meiswinkel, R. (2015). Mosquitoes and culicoides biting midges: Vector range and the influence of climate change. *Revue scientifique et technique, 34*(1), 123–137.

Ellis-Iversen, J., Petersen, C. K., & Helwigh, B. (2019). *Inspiration and ideas - one health integration in surveillance*. Technical University of Denmark.

European Centre for Disease Prevention and Control. (2018). *Towards one Health preparedness*. Stockholm: ECDC.

European Centre for Disease Prevention and Control (ECDC). (2014). *Data quality monitoring and surveillance system evaluation – a handbook of methods and applications*. Stockholm: ECDC.

European Commission. (2017). *A European one Health action plan against antimicrobial resistance AMR*. Brussels: European Commission. (Accessed 2 June 2021). Available from https://ec.europa.eu/health/amr/sites/amr/files/amr_action_plan_2017_en.pdf.

European Commission (EUC). (2013). *Decision No 1082/2013/EU of the European Parliament and of the Council of 22 October 2013 on serious cross-border threats to health and repealing decision No 2119/98/EC. Brussels: European Commission*. (Accessed 2 June 2021). Available from https://ec.europa.eu/health/sites/health/files/preparedness_response/docs/decision_serious_crossborder_threats_22102013_en.pdf.

European Union, The General Secretariat of the Council. (2021). *Towards an international treaty on pandemics*. Consilium. https://www.consilium.europa.eu/pt/infographics/towards-an-international-treaty-on-pandemics/.

Ferretti, L., Wymant, C., Kendall, M., Zhao, L., Nurtay, A., Abeler-Dörner, L., … Fraser, C. (2020). Quantifying SARS-CoV-2 transmission suggests epidemic control with digital contact tracing. *Science, 368*(6491).

Figuié, M. (2014). Towards a global governance of risks: International Health Organisations and the surveillance of emerging infectious diseases. *Journal of Risk Research, 17*(4), 469–483.

Filkins, L., & Schlaberg, R. (2021). Metagenomic applications for infectious disease testing in clinical laboratories. In *Application and integration of omics-powered diagnostics in clinical and public health microbiology* (pp. 111–131). Cham: Springer.

Filter, M., Buschhardt, T., Dórea, F., de Abechuco, E. L., Günther, T., Sundermann, E. M., … Ellis-Iversen, J. (2021). One Health surveillance Codex: Promoting the adoption of One Health solutions within and across European countries. *One Health*, 100233.

Foldspang, A. (2015). *Towards a public health profession: The roles of essential public health operations and lists of competences*.

Forman, R., Atun, R., McKee, M., & Mossialos, E. (2020). *12 Lessons learned from the management of the coronavirus pandemic*. Health Policy.

Frenk, J. (2010). The global health system: strengthening national health systems as the next step for global progress. *PLoS medicine, 7*(1). https://doi.org/10.1371/journal.pmed.1000089

Frieden, T. R., & Koplan, J. P. (2010). Stronger national public health institutes for global health. *The Lancet, 376*(9754), 1721–1722.

Gagliotti, C., Balode, A., Baquero, F., Degener, J., Grundmann, H., Gür, D., … Heuer, O. (2011). *Escherichia coli* and *Staphylococcus aureus*: Bad news and good news from the European antimicrobial resistance surveillance network (EARS-Net, formerly EARSS), 2002 to 2009. *Euro Surveillance, 16*(11), 19819.

Garchitorena, A., Sokolow, S. H., Roche, B., Ngonghala, C. N., Jocque, M., Lund, A., … De Leo, G. A. (2017). Disease ecology, health and the environment: A framework to account for ecological and socio-economic drivers in the control of neglected tropical diseases. *Philosophical Transactions of the Royal Society B: Biological Sciences, 372*(1722), 20160128.

Gardy, J. L., & Loman, N. J. (2018). Towards a genomics-informed, real-time, global pathogen surveillance system. *Nature Reviews Genetics, 19*(1), 9.

German, R. R., Horan, J. M., Lee, L. M., Milstein, B., & Pertowski, C. A. (2001). Updated guidelines for evaluating public health surveillance systems; recommendations from the Guidelines Working Group. *Morbidity and Mortality Weekly Report Recommendations and Reports, 50*(RR-13), 1–35.

Gibb, R., Redding, D. W., Chin, K. Q., Donnelly, C. A., Blackburn, T. M., Newbold, T., & Jones, K. E. (2020). Zoonotic host diversity increases in human-dominated ecosystems. *Nature, 584*(7821), 398–402.

Gouglas, D., Christodoulou, M., Plotkin, S. A., & Hatchett, R. (2019). CEPI: Driving progress toward epidemic preparedness and response. *Epidemiologic Reviews, 41*(1), 28–33.

Grace, D., Bett, B. K., Rich, K. M., Wanyoike, F. N., Lindahl, J. F., & Randolph, T. F. (2016). *Economics of One Health.*

Grantz, K. H., Meredith, H. R., Cummings, D. A., Metcalf, C. J. E., Grenfell, B. T., Giles, J. R., … Wesolowski, A. (2020). The use of mobile phone data to inform analysis of COVID-19 pandemic epidemiology. *Nature Communications, 11*(1), 1–8.

Groseclose, S. L., & Buckeridge, D. L. (2017). Public health surveillance systems: Recent advances in their use and evaluation. *Annual Review of Public Health, 38*, 57–79.

Gulson, B., Korsch, M., Matisons, M., Douglas, C., Gillam, L., & McLaughlin, V. (2009). Windblown lead carbonate as the main source of lead in blood of children from a seaside community: An example of local birds as "canaries in the mine". *Environmental Health Perspectives, 117*(1), 148–154.

Haldane, V., De Foo, C., Abdalla, S. M., Jung, A. S., Tan, M., Wu, S., … Legido-Quigley, H. (2021). Health systems resilience in managing the COVID-19 pandemic: Lessons from 28 countries. *Nature Medicine*, 1–17.

Hammerum, A. M., Heuer, O. E., Emborg, H. D., Bagger-Skjøt, L., Jensen, V. F., Rogues, A. M., & Monnet, D. L. (2007). Danish integrated antimicrobial resistance monitoring and research program. *Emerging Infectious Diseases, 13*(11), 1633.

Han, B. A., O'Regan, S. M., Paul Schmidt, J., & Drake, J. M. (2020). Integrating data mining and transmission theory in the ecology of infectious diseases. *Ecology Letters, 23*(8), 1178–1188.

Harada, M. (1995). Minamata disease: Methylmercury poisoning in Japan caused by environmental pollution. *Critical Reviews in Toxicology, 25*(1), 1–24.

Hoinville, L. J., Alban, L., Drewe, J. A., Gibbens, J. C., Gustafson, L., Häsler, B., … Stark, K. D. C. (2013). Proposed terms and concepts for describing and evaluating animal-health surveillance systems. *Preventive Veterinary Medicine, 112*(1–2), 1–12.

Ihekweazu, C., Michael, C. A., Nguku, P. M., Waziri, N. E., Habib, A. G., Muturi, M., … Gloria, O. (2021). Prioritization of zoonotic diseases of public health significance in Nigeria using the one-health approach. *One Health*, 100257.

Janes, C. R., Corbett, K. K., Jones, J. H., & Trostle, J. (2012). Emerging infectious diseases: The role of social sciences. *The Lancet, 380*(9857), 1884–1886.

Jekel, J. F., Katz, D. L., Elmore, J., & Wild, D. (2007). *Epidemiology, Biostatistics and Preventive Medicine*. Elsevier Health Sciences.

Jowell, A., & Barry, M. (2020). COVID-19: A matter of planetary, not only national health. *The American Journal of Tropical Medicine and Hygiene, 103*(1), 31–32.

Kalantar, K. L., Carvalho, T., de Bourcy, C. F., Dimitrov, B., Dingle, G., Egger, R., … DeRisi, J. L. (2020). IDseq—an open source cloud-based pipeline and analysis service for metagenomic pathogen detection and monitoring. *GigaScience, 9*(10), giaa111.

Karesh, W. B., Dobson, A., Lloyd-Smith, J. O., Lubroth, J., Dixon, M. A., Bennett, M., … Heymann, D. L. (2012). Ecology of zoonoses: Natural and unnatural histories. *The Lancet, 380*(9857), 1936–1945.

Karimuribo, E. D., Mutagahywa, E., Sindato, C., Mboera, L., Mwabukusi, M., Njenga, M. K., … Rweyemamu, M. (2017). A smartphone app (AfyaData) for innovative one health disease surveillance from community to national levels in Africa: Intervention in disease surveillance. *JMIR Public Health and Surveillance, 3*(4), e94.

Keeling, M. J., Hollingsworth, T. D., & Read, J. M. (2020). Efficacy of contact tracing for the containment of the 2019 novel coronavirus (COVID-19). *Journal of Epidemiology & Community Health, 74*(10), 861−866.

Kelly, T. R., Karesh, W. B., Johnson, C. K., Gilardi, K. V., Anthony, S. J., Goldstein, T., … Mazet, J. A. K. (2017). One Health proof of concept: Bringing a transdisciplinary approach to surveillance for zoonotic viruses at the human-wild animal interface. *Preventive Veterinary Medicine, 137*, 112−118.

Keshet, Y. (2020). Fear of panoptic surveillance: Using digital technology to control the COVID-19 epidemic. *Israel Journal of Health Policy Research, 9*(1), 1−8.

Khosla, R. (2020). Technology, health, and human rights: A cautionary tale for the post-pandemic world. *Health and Human Rights, 22*(2), 6.

King, N. B. (2002). Security, disease, commerce: Ideologies of postcolonial global health. *Social Studies of Science, 32*(5−6), 763−789.

Klinkenberg, D., Fraser, C., & Heesterbeek, H. (2006). The effectiveness of contact tracing in emerging epidemics. *PLoS One, 1*(1), e12.

Kock, R. A., Karesh, W. B., Veas, F., Velavan, T. P., Simons, D., Mboera, L. E. G., … Zumla, A. (2020). 2019-nCoV in context: Lessons learned? *The Lancet Planetary Health, 4*(3), e87−e88.

Kostkova, P. (2018). Disease surveillance data sharing for public health: The next ethical frontiers. *Life Sciences, Society and Policy, 14*(1), 1−5.

Krammer, F. (2020). Pandemic vaccines: How are we going to Be better prepared next time? *Med, 1*(1), 28−32.

Kruk, M. E., Myers, M. S., Varpilah, T. S., & Dahn, B. T. (2015). What is a resilient health system? Lessons from Ebola. *The Lancet, 385*(9980), 1910−1912.

Kshirsagar, D. P., Savalia, C. V., Kalyani, I. H., Kumar, R., & Nayak, D. N. (2013). Disease alerts and forecasting of zoonotic diseases: An overview. *Veterinary World, 6*(11), 889.

LaDeau, S. L., Han, B. A., Rosi-Marshall, E. J., & Weathers, K. C. (2017). The next decade of big data in ecosystem science. *Ecosystems, 20*(2), 274−283.

Lal, A., Erondu, N. A., Heymann, D. L., Gitahi, G., & Yates, R. (2020). Fragmented health systems in COVID-19: Rectifying the misalignment between global health security and universal health coverage. *The Lancet*.

Lapiz, S. M. D., Miranda, M. E. G., Garcia, R. G., Daguro, L. I., Paman, M. D., Madrinan, F. P., … Briggs, D. J. (2012). Implementation of an intersectoral program to eliminate human and canine rabies: The bohol rabies prevention and elimination project. *PLoS Neglected Tropical Diseases, 6*, 1891.

Leung, G. M., & Leung, K. (2020). Crowdsourcing data to mitigate epidemics. *The Lancet Digital Health, 2*(4), e156−e157.

Leung, K., Wu, J. T., & Leung, G. M. (2021). Real-time tracking and prediction of COVID-19 infection using digital proxies of population mobility and mixing. *Nature Communications, 12*(1), 1−8.

Levinson, J., Bogich, T. L., Olival, K. J., Epstein, J. H., Johnson, C. K., Karesh, W., & Daszak, P. (2013). Targeting surveillance for zoonotic virus discovery. *Emerging Infectious Diseases, 19*(5), 743.

Lewis, D. (2021). Contact-tracing apps help reduce COVID infections, data suggest. *Nature*, 18−19.

Linchant, J., Lisein, J., Semeki, J., Lejeune, P., & Vermeulen, C. (2015). Are unmanned aircraft systems (UAS s) the future of wildlife monitoring? A review of accomplishments and challenges. *Mammal Review, 45*(4), 239–252.

Li, W., Shi, Z., Yu, M., Ren, W., Smith, C., Epstein, J. H., ... Wang, L. F. (2005). Bats are natural reservoirs of SARS-like coronaviruses. *Science, 310*(5748), 676–679.4.

Lloyd-Smith, J. O., George, D., Pepin, K. M., Pitzer, V. E., Pulliam, J. R., Dobson, A. P., ... Grenfell, B. T. (2009). Epidemic dynamics at the human-animal interface. *Science, 326*(5958), 1362–1367.

Lubroth, J. (2012). FAO and the One Health approach. In *One Health: The Human-animal-environment interfaces in emerging infectious diseases* (pp. 65–72).

MacDonald, A. J., & Mordecai, E. A. (2019). Amazon deforestation drives malaria transmission, and malaria burden reduces forest clearing. *Proceedings of the National Academy of Sciences, 116*(44), 22212–22218.

Machado, D. J., Scott, R., Guirales, S., & Janies, D. A. (2021). *Fundamental evolution of all Orthocoronavirinae including three deadly lineages descendent from Chiroptera-hosted coronaviruses: SARS-CoV, MERS-CoV and SARS-CoV-2*. Cladistics.

Mader, R., Damborg, P., Amat, J. P., Bengtsson, B., Bourély, C., Broens, E. M., ... Madec, J. Y. (2021). Building the European antimicrobial resistance surveillance network in veterinary medicine (EARS-Vet). *Eurosurveillance, 26*(4), 2001359.

Marcotty, T., Thys, E., Conrad, P., Godfroid, J., Craig, P., Zinsstag, J., ... Boelaert, M. (2013). Intersectoral collaboration between the medical and veterinary professions in low-resource societies: The role of research and training institutions. *Comparative Immunology, Microbiology and Infectious Diseases, 36*(3), 233–239.

Mariner, J. C., Hendrickx, S., Pfeiffer, D. U., Costard, S., Knopf, L., Okuthe, S., ... Jost, C. C. (2011). Integração de abordagens participativas em sistemas de vigilância. *Revue scientifique et technique, 30*(3), 653–659.

McCall, B. (2020). COVID-19 and artificial intelligence: Protecting health-care workers and curbing the spread. *The Lancet Digital Health, 2*(4), e166–e167.

Menachery, V. D., Yount, B. L., Debbink, K., Agnihothram, S., Gralinski, L. E., Plante, J. A., ... Baric, R. S. (2015). A SARS-like cluster of circulating bat coronaviruses shows potential for human emergence. *Nature Medicine, 21*(12), 1508–1513.

Michael, K., McNamee, A., & Michael, M. G. (2006). *The emerging ethics of humancentric GPS tracking and monitoring*.

Mirhaji, P. (2009). Public health surveillance meets translational informatics: A desiderata. *Journal of the Association for Laboratory Automation, 14*, 157–170.

Mordecai, E. A., Ryan, S. J., Caldwell, J. M., Shah, M. M., & LaBeaud, A. D. (2020). Climate change could shift disease burden from malaria to arboviruses in Africa. *The Lancet Planetary Health, 4*(9), e416–e423.

Morgan, D. (2006). Control of arbovirus infections by a coordinated response: West nile virus in England and Wales. *Immunology and Medical Microbiology, 48*, 305–312.

Morgan, O. W., Aguilera, X., Ammon, A., Amuasi, J., Fall, I. S., Frieden, T., ... Dowell, S. F. (2021). Disease surveillance for the COVID-19 era: Time for bold changes. *The Lancet*.

Morley, J., Cowls, J., Taddeo, M., & Floridi, L. (2020). *Ethical guidelines for COVID-19 tracing apps*.

Morse, S. S., Mazet, J. A., Woolhouse, M., Parrish, C. R., Carroll, D., Karesh, W. B., ... Daszak, P. (2012). Prediction and prevention of the next pandemic zoonosis. *The Lancet, 380*(9857), 1956–1965.

Nature. (2021). The world must learn from COVID before diving into a pandemic treaty. *Nature, 592*(7853), 165–166. https://doi.org/10.1038/d41586-021-00866-7. PMID: 33824523.

Nisi, R., Brink, N., Diaz, F., & Moulin, G. (2013). Antimicrobial use in animals: Analysis of the OIE survey on monitoring of the quantities of antimicrobial agents used in animals. In *Proceeding of the OIE global conference on the responsible and prudent use of antimicrobial agents for animals held in March.*

Nsubuga, P., White, M. E., Thacker, S. B., Anderson, M. A., Blount, S. B., Broome, C. V., … Trostle, M. (2006). Public health surveillance: A tool for targeting and monitoring interventions. In , 2. *Disease control priorities in developing countries* (pp. 997–1018).

O'Connell, J., & O'Keeffe, D. T. (2021). Contact tracing for covid-19—a digital inoculation against future pandemics. *New England Journal of Medicine.*

Orand, J. P. (2012). Antimicrobial resistance and the standards of the World Organisation for Animal Health. *Revue scientifique et technique (International Office of Epizootics), 31*(1), 335–342.

ORION Project. (2019). *ORION: One health suRveillance Initiative on harmOnization of data collection and interpretatioN. One Health EJP.* https://onehealthejp.eu/jip-orion/.

Ouagal, M., Hendrikx, P., Saegerman, C., & Berkvens, D. (2010). Comparison between active and passive surveillance within the network of epidemiological surveillance of animal diseases in Chad. *Acta Tropica, 116*(2), 147–151.

Pandit, P., & Han, B., A. (2020). Rise of machines in disease ecology. *The Bulletin of the Ecological Society of America, 101*(1), e01625.

Paquet, C., Coulombier, D., Kaiser, R., & Ciotti, M. (2006). Epidemic intelligence: A new framework for strengthening disease surveillance in Europe. *Euro Surveillance, 11*(12), 5–6.

Park, S., Choi, G. J., & Ko, H. (2020). Information technology–based tracing strategy in response to COVID-19 in South Korea—privacy controversies. *JAMA, 323*(21), 2129–2130.

Paternoster, G., Babo Martins, S., Mattivi, A., Cagarelli, R., Angelini, P., Bellini, R., … Tamba, M. (2017). Economics of one health: Costs and benefits of integrated West Nile virus surveillance in Emilia-Romagna. *PLoS One, 12*(11), e0188156.

Pearce, N., & Douwes, J. (2013). Research at the interface between human and veterinary health. *Preventive Veterinary Medicine, 111*, 187–193.

Philips, M., & Markham, Á. (2014). Ebola: A failure of international collective action. *Lancet, 384*(9949), 1181.

Pike, J., Bogich, T., Elwood, S., Finnoff, D. C., & Daszak, P. (2014). Economic optimization of a global strategy to address the pandemic threat. *Proceedings of the National Academy of Sciences, 111*(52), 18519–18523.

Pingali, P. L. (2012). Green revolution: Impacts, limits, and the path ahead. *Proceedings of the National Academy of Sciences, 109*(31), 12302–12308.

Pinto, J., Ben Jebara, K., Chaisemartin, D., De La Rocque, S., & Abela, B. (2011). *The FAO/OIE/WHO global early warning system.*

Pol-Hofstad, I. E., Jacobs-Reitsma, W. F., Maas, H., Pinna, E. D., Mevius, D., & Mooijman, K. A. (2012). *Third external quality assurance scheme for Salmonella typing: European food-and waterborne diseases and zoonoses network. Third external quality assurance scheme for Salmonella typing: European food-and waterborne diseases and zoonoses network.*

Queenan, K., Häsler, B., & Rushton, J. (2016). A one health approach to antimicrobial resistance surveillance: Is there a business case for it? *International Journal of Antimicrobial Agents, 48*(4), 422–427.

Rabinowitz, P. M., Natterson-Horowitz, B. J., Kahn, L. H., Kock, R., & Pappaioanou, M. (2017). Incorporating one health into medical education. *BMC Medical Education, 17*(1), 1–7.

Radhouani, H., Silva, N., Poeta, P., Torres, C., Correia, S., & Igrejas, G. (2014). Potential impact of antimicrobial resistance in wildlife, environment and human health. *Frontiers in Microbiology, 5*, 23.

Rahman, M. M., Khatun, F., Uzzaman, A., Sami, S. I., Bhuiyan, M. A. A., & Kiong, T. S. (2021). A comprehensive study of artificial intelligence and machine learning approaches in confronting the coronavirus (COVID-19) pandemic. *International Journal of Health Services*, 00207314211017469.

Rattanaumpawan, P., Boonyasiri, A., Vong, S., & Thamlikitkul, V. (2018). Systematic review of electronic surveillance of infectious diseases with emphasis on antimicrobial resistance surveillance in resource-limited settings. *American Journal of Infection Control, 46*(2), 139–146.

Reichert, L., Brack, S., & Scheuermann, B. (2020). Privacy-preserving contact tracing of covid-19 patients. *IACR Cryptology ePrint Arch., 2020*, 375.

Reif, J. S. (2011). Animal sentinels for environmental and public health. *Public Health Reports, 126*, 50–57.

Robertson, C., Sawford, K., Daniel, S. L., Nelson, T. A., & Stephen, C. (2010). Mobile phone–based infectious disease surveillance system, Sri Lanka. *Emerging Infectious Diseases, 16*(10), 1524.

Ruckert, A., et al. (2020). Governing antimicrobial resistance: a narrative review of global governance mechanisms. *Journal of public health policy, 41*, 515–528. https://doi.org/10.1057/s41271-020-00248-9

Rutz, C., Loretto, M. C., Bates, A. E., Davidson, S. C., Duarte, C. M., Jetz, W., … Cagnacci, F. (2020). COVID-19 lockdown allows researchers to quantify the effects of human activity on wildlife. *Nature Ecology & Evolution, 4*(9), 1156–1159.

Salje, H., Lessler, J., Paul, K. K., Azmar, A., Rahman, M. W., Rahman, M., … Cauchemez, S. (2016). How social structures, space, and behaviors shape the spread of infectious diseases using chikungunya as a case study. *Proceedings of the National Academy of Sciences, 113*(47), 13420–13425.

Salm, M., Ali, M., Minihane, M., & Conrad, P. (2021). Defining global health: Findings from a systematic review and thematic analysis of the literature. *BMJ Global Health, 6*(6), e005292. https://doi.org/10.1136/bmjgh-2021-005292

Scales, D., Gorman, J., & Jamieson, K. H. (2021). The covid-19 infodemic—applying the epidemiologic model to counter misinformation. *New England Journal of Medicine*.

Semenza, J. C., & Zeller, H. (2014). Integrated surveillance for prevention and control of emerging vector-borne diseases in Europe. *Euro Surveillance, 19*(13), 20757.

Sepai, O., Collier, C., Van Tongelen, B., & Casteleyn, L. (2008). Human biomonitoring data interpretation and ethics; obstacles or surmountable challenges? *Environmental Health, 7*(1), 1–5.

Shen, M., Liu, L., Yan, L., Lu, M., Yao, W., & Yang, X. (2014). Review of monitoring technology for animal individuals in animal husbandry. *Nongye Jixie Xuebao= Transactions of the Chinese Society for Agricultural Machinery, 45*(10), 245–251.

Shi, W., & Gao, G. F. (2021). Emerging H5N8 avian influenza viruses. *Science, 372*(6544), 784–786.

Shroff, Z. C., Marten, R., Vega, J., Peters, D. H., Patcharanarumol, W., & Ghaffar, A. (2021). Time to reconceptualise health systems. *The Lancet*.

Sid, A. (2020). *Cell Broadcast alerts as tools for community outreach in Ebola virus disease Outbreaks in Central and Western Africa (Doctoral dissertation)*. Available at https://nrs.harvard.edu/URN-3:HUL.INSTREPOS:37365384.

Sleigh, A., Jackson, S., Li, X., & Huang, K. (1998b). Eradication of schistosomiasis in Guangxi, China. Part 2: Political economy, management strategy and costs, 1953-92. *Bulletin of the World Health Organization, 76*(5), 497–508.

Sleigh, A., Li, X., Jackson, S., & Huang, K. (1998a). Eradication of schistosomiasis in Guangxi, China. Part 1: Setting, strategies, operations, and outcomes, 1953-92. *Bulletin of the World Health Organization, 76*(4), 361–372.

Smith, P. G., & Bradley, R. (2003). Bovine spongiform encephalopathy (BSE) and its epidemiology. *British Medical Bulletin, 66*(1), 185–198.

Smith, M. R., Singh, G. M., Mozaffarian, D., & Myers, S. S. (2015). Effects of decreases of animal pollinators on human nutrition and global health: A modelling analysis. *The Lancet, 386*(10007), 1964–1972.

Smolinski, M. S., Crawley, A. W., Olsen, J. M., Jayaraman, T., & Libel, M. (2017). Participatory disease surveillance: Engaging communities directly in reporting, monitoring, and responding to health threats. *JMIR Public Health and Surveillance, 3*(4), e62.

Sosin, D. M. (2003). Syndromic surveillance: The case for skillful investment. *Biosecurity and Bioterrorism: Biodefense Strategy, Practice, and Science, 1*(4), 247–253.

Spiegel, J., Bennett, S., Hattersley, L., Hayden, M. H., Kittayapong, P., Nalim, S., ... Gubler, D. (2005). Barriers and bridges to prevention and control of dengue: The need for a social–ecological approach. *EcoHealth, 2*(4), 273–290.

Stärk, K. D. C., Arroyo Kuribreña, M., Dauphin, G., Vokaty, S., Ward, M. P., Wieland, B., & Lindberg, A. (2015). One Health surveillance – more than a buzz word? *Preventive Veterinary Medicine, 120*(1), 124–130.

Steffen, W., Broadgate, W., Deutsch, L., Gaffney, O., & Ludwig, C. (2015). The trajectory of the anthropocene: The great acceleration. *The Anthropocene Review, 2*(1), 81–98.

Stoffel, C., Schuppers, M., Buholzer, P., Muñoz, V., Lechner, I., Sperling, U., ... De Nardi, M. (2020). The ongoing crises in China illustrate that the assessment of epidemics in isolation is no longer sufficient. *Transboundary and Emerging Diseases, 67*(3), 1043–1044.

Sun, K., Chen, J., & Viboud, C. (2020). Early epidemiological analysis of the coronavirus disease 2019 outbreak based on crowdsourced data: A population-level observational study. *The Lancet Digital Health, 2*(4), e201–e208.

Talaska, T. (1994). A *Salmonella* data bank for routine surveillance and research. *Bulletin of the World Health Organization, 72*(1), 69–72.

Tang, X. C., Zhang, J. X., Zhang, S. Y., Wang, P., Fan, X. H., Li, L. F., ... Guan, Y. (2006). Prevalence and genetic diversity of coronaviruses in bats from China. *Journal of Virology, 80*(15), 7481.

Thacker, S. B., & Berkelman, R. L. (1988). Public health surveillance in the United States. *Epidemiologic Reviews, 10*(1), 164–190.

Thacker, S. B., Qualters, J. R., Lee, L. M., & Centers for Disease Control and Prevention. (2012). Public health surveillance in the United States: Evolution and challenges. *Morbidity and Mortality Weekly Report, 61*(Suppl. 1), 3–9.

Thacker, S. B., Stroup, D. F., Parrish, R. G., & Anderson, H. A. (1996). Surveillance in environmental public health: Issues, systems, and sources. *American Journal of Public Health, 86*(5), 633–638.

Thiermann, A. B. (2015). International standards: The world organisation for animal health terrestrial animal health Code. *Revue scientifique et technique (International Office of Epizootics), 34*(1), 277–281.

Thomas, L. F., Rushton, J., Bukachi, S., Falzon, L. C., Howland, O., & Fevre, E. (2021). Cross-sectoral zoonotic disease surveillance in western Kenya: Identifying drivers & barriers within a resource constrained setting. *Frontiers in Veterinary Science, 8*, 572.

Toussaert, S. (2021). Upping uptake of COVID contact tracing apps. *Nature Human Behaviour, 5*(2), 183–184.

Tsuchiya, K. (1992). Historical perspectives in occupational medicine. The discovery of the causal agent of minamata disease. *American Journal of Industrial Medicine, 21*(2), 275–280.

Venkatramanan, S., Sadilek, A., Fadikar, A., Barrett, C. L., Biggerstaff, M., Chen, J., … Marathe, M. (2021). Forecasting influenza activity using machine-learned mobility map. *Nature Communications, 12*(1), 1–12.

Villa, S., van Leeuwen, R., Gray, C. C., van der Sande, M., Konradsen, F., Fröschl, G., … Raviglione, M. (2021). HERA: A new era for health emergency preparedness in Europe? *The Lancet.*

Vinuales, J., Moon, S., Le Moli, G., & Burci, G. L. (2021). A global pandemic treaty should aim for deep prevention. *The Lancet, 397*(10287), 1791–1792.

Vlasova, A. N., Diaz, A., Damtie, D., Xiu, L., Toh, T. H., Lee, J. S. Y., … Gray, G. C. (2021). Novel canine coronavirus isolated from a hospitalized pneumonia patient, East Malaysia. *Clinical Infectious Diseases.*

Vrbova, L., Patrick, D. M., Stephen, C., Robertson, C., Koehoorn, M., Parmley, E. J., … Galanis, E. (2016). Utility of algorithms for the analysis of integrated Salmonella surveillance data. *Epidemiology and Infection, 144*(10), 2165–2175.

Wacharapluesadee, S., Tan, C. W., Maneeorn, P., Duengkae, P., Zhu, F., Joyjinda, Y., … Wang, L. F. (2021). Evidence for SARS-CoV-2 related coronaviruses circulating in bats and pangolins in Southeast Asia. *Nature Communications, 12*(1), 1–9.

Wang, K., Varma, D. S., & Prosperi, M. (2018). A systematic review of the effectiveness of mobile apps for monitoring and management of mental health symptoms or disorders. *Journal of Psychiatric Research, 107*, 73–78.

Wielinga, P. R., Jensen, V. F., Aarestrup, F. M., & Schlundt, J. (2014). Evidence-based policy for controlling antimicrobial resistance in the food chain in Denmark. *Food Control, 40*, 185–192.

Witt, C. J., Brundage, M., Cannon, C., & Cox, K. (2004). Department of defense West Nile virus surveillance in 2002. *Military Medicine, 169*(6), 421–428.

Wolfe, N. D., Daszak, P., Kilpatrick, A. M., & Burke, D. S. (2005). Bushmeat hunting, deforestation, and prediction of zoonotic disease. *Emerging Infectious Diseases, 11*(12), 1822.

Wood, C. S., Thomas, M. R., Budd, J., Mashamba-Thompson, T. P., Herbst, K., Pillay, D., … Stevens, M. M. (2019). Taking connected mobile-health diagnostics of infectious diseases to the field. *Nature, 566*(7745), 467–474.

World Bank. (2021). *Safeguarding animal, human and ecosystem health: One Health at the World Bank.* World Bank. https://www.worldbank.org/en/topic/agriculture/brief/safeguarding-animal-human-and-ecosystem-health-one-health-at-the-world-bank.

World Health Organization (WHO). (2008). *International health regulations (2005).* World Health Organization.

World Health Organization (WHO). (2008). *Global Outbreak alert & response network (GOARN)* (Accessed 4 June 2021). Available from: http://www.who.int/csr/outbreaknetwork/en/.

World Health Organization (WHO). (2010). *Framework for action on interprofessional education and collaborative practice (No. WHO/HRH/HPN/10.3).* World Health Organization.

World Health Organization (WHO). (2014). *Antimicrobial resistance global report on surveillance: 2014 summary (No. WHO/HSE/PED/AIP/2014.2).* World Health Organization.

World Health Organization (WHO). (2018). *One Health*. Geneva: WHO (Accessed 3 June 2021). Available from: http://www.who.int/features/qa/one-health/en/.

Xiao, K., Zhai, J., Feng, Y., Zhou, N., Zhang, X., Zou, J. J., ... Shen, Y. (2020). Isolation of SARS-CoV-2-related coronavirus from Malayan pangolins. *Nature, 583*(7815), 286–289.

Xiao, X., Newman, C., Buesching, C. D., Macdonald, D. W., & Zhou, Z. M. (2021). Animal sales from Wuhan wet markets immediately prior to the COVID-19 pandemic. *Scientific Reports, 11*, 11898. https://doi.org/10.1038/s41598-021-91470-2

Yu, V. L., & Madoff, L. C. (2004). ProMED-mail: An early warning system for emerging diseases. *Clinical Infectious Diseases, 39*(2), 227–232.

Zaki, A. M., Van Boheemen, S., Bestebroer, T. M., Osterhaus, A. D., & Fouchier, R. A. (2012). Isolation of a novel coronavirus from a man with pneumonia in Saudi Arabia. *New England Journal of Medicine, 367*(19), 1814–1820.

Zhang, Y., Bambrick, H., Mengersen, K., Tong, S., & Hu, W. (2018). Using Google Trends and ambient temperature to predict seasonal influenza outbreaks. *Environment International, 117*, 284–291.

Zhang, T., Wu, Q., & Zhang, Z. (2020). Probable pangolin origin of SARS-CoV-2 associated with the COVID-19 outbreak. *Current Biology, 30*(7), 1346–1351.

Zhou, P., Yang, X. L., Wang, X. G., Hu, B., Zhang, L., Zhang, W., ... Shi, Z. L. (2020). Discovery of a novel coronavirus associated with the recent pneumonia outbreak in humans and its potential bat origin. *BioRxiv*.

Zinsstag, J., Schelling, E., Crump, L., Whittaker, M., Tanner, M., & Stephen, C. (Eds.). (2020). *One health: The theory and practice of integrated health approaches*. CABI.

Chapter 3

Epidemiology of disease through the interactions between humans, domestic animals, and wildlife

Mariana Marrana
Preparedness and Resilience Department, World Organisation for Animal Health, Paris, France

1. Introduction

Disease emergence at the animal-human interface is an area of study that has gathered increasing interest over the last decades. Zoonotic diseases, or zoonoses, are diseases which are shared repeatedly between animals – including livestock, wildlife, and pets – and humans. Zoonoses can pose serious risks to both animal and human health and may have broad impact on economies and livelihoods. Zoonotic diseases are commonly spread at the animal-human interface – where people and animals interact with each other through a shared environment. Zoonotic diseases can be transmitted via direct contact with animals, or be foodborne, waterborne, vector-borne, and even transmitted indirectly by environmental contamination.

Research studies show us that 60% of existing human infectious diseases are zoonotic and that at least 75% of emerging infectious diseases are zoonotic (Taylor, Latham, & Woolhouse, 2001). These numbers suggest that controlling the emergence of disease at the animal source should be more effective than trying to mitigate the consequences of spillover events (when an animal disease is first transmitted to humans) and the spread of disease among the human population. It is important to note that many of the pathogens causing current human infectious diseases are not zoonotic, even though some of them may have descended from ancestors that spilled over from animals to our human relatives sometime in the past. Saying that all diseases had zoonotic origin would be going much too far. The point being made is the evident interconnectedness between humans and other animals in hosting similar pathogens.

One Health. https://doi.org/10.1016/B978-0-12-822794-7.00001-0

While, historically, dramatic epidemics killing large numbers of humans without attribution of cause have happened, nowadays we are better equipped to detect emerging infectious diseases earlier and closer to their source, and to use modern drugs and vaccines to help us slow down and in some cases halt the spread of disease. This means that, even though the opportunities for emergence of infectious diseases on the animal-human interface and their spread among the human population are more numerous than ever, we are also more likely to be able to mitigate the consequences of these events. However, at the present, humanity has created more occasions than ever for new diseases to emerge and to quickly spread among densely packed urban areas connected by airplanes, trains, and cars. For the last century, two new viruses per year have spilled from their natural hosts into humans (Woolhouse, 2012).

To a certain extent, it is known where emerging infectious diseases are likely to surface from. Practices and vulnerabilities in the food production and trade systems that increase the risk of infectious disease emergence have been identified. It is also known that the consequences of these events are devastating, not only from a public health perspective, but also for economies, productivity, and livelihoods. The cost of working towards prevention of disease emergence at the animal-human interface is minor in comparison to the protracted costs and effects caused by large scale epidemics. Therefore, building resilient systems capable of detecting animal and human disease events early should be emphasized, rather than putting more weight on bearing the costs of recovery and mitigation, which are addressed at a later stage.

2. Expansion of the interaction at the human-animal interface over the centuries

All animal species for which dates of domestication were revealed by archeological evidence have been domesticated between about 8000 and 2500 BC, apart from the dog, which appears to have been domesticated at least 2000 years beforehand (Diamond, 2019, pp. 231–256). The period 8000 to 2500 BC also represents the first few thousand years of the sedentary farming-herding societies that appeared after the end of the last Glacial Period, about 11,700 years ago. Many animal species were tamed by humankind before and after these dates. Yet, since then, there have been no significant changes to the list of main species to have been domesticated.

Taming is one of the first steps in the domestication of animals. However, in order to establish a population of domesticated animals that can serve work and food purposes, taming is not enough – a few mutually inclusive conditions apply, such as: the disposition to breed while in captivity, without which it is not possible to have a self-sustainable supply of animals; an adequate feed conversion ratio in the case of livestock, as there is no point in trying to herd, breed and slaughter animals that would be fed on expensive high-protein diets; adequate temperament, which means low aggressiveness and low propension to panic; and having a clear social structure when herded together, in order to avoid fights among the animals and attacks to the farmers (Diamond, 2019, pp. 231–256). There are fourteen species of large terrestrial herbivores that became the most important domesticated mammals. Out of these fourteen main species, five were widespread and domesticated around the world until the present days: cow, sheep, goat, pig, and horse.

The spillover of diseases from domestic, farmed, and wild animals to humans is not a new event. The inception of infectious diseases in human populations started with the rise of agriculture and herding practices and accelerated with the expansion of cities, several thousand years ago. During the period since the estimated appearance of *Homo sapiens,* 350,000 to 200,000 years ago, up to 10,000 BC it is thought that the total world population was way under one million individuals (Kremer, 1993; Mounier, 2019). Agriculture and farming allowed the global population to grow from 4 million in 10,000 BC to 190 million in year 0. Even though reliable resources other than archeological findings are not abundant, genetic, and molecular tests have convinced researchers that some diseases such as rabies, tuberculosis and bubonic plague occurred during antiquity (Nwanta et al., 2010; Vuorinen, 1997). The first reliable dates for infectious disease occurrence to have been determined are 1600 BC for smallpox, 400 BC for mumps, 200 BC for leprosy, 1840 for epidemic poliomyelitis and 1959 for AIDS (Matthews, 2014). The main reason why the rise of agriculture and animal herding practices are correlated to the emergence of infectious diseases in human and animal populations, is that agriculture sustains much higher human population densities than the hunting-gathering lifestyle did (Diamond, 2019, pp. 231–256).

Another reason for this rise in disease emergence is the closeness, proximity between humans and domesticated animals that results from herding and farming, which did not happen when humans were hunting wild animals on a need-basis. The rise of agriculture compelled human populations to settle around their lands in order to avoid losing their crops to competing humans or animals. This was the beginning of the first settled societies, which would later eventually evolve into cities with much more densely packed populations and usually worse sanitary conditions.

Another early push towards what we now call globalization was the development of world trade routes, which by the beginning of the current era (year 0) connected the populations of Europe, Asia and North Africa into one network. That was the time when smallpox reached Rome, known as the Plague of Antoninus, between A.D. 165 and 180 (Sabbatani, 2009). Shortly after, bubonic plague first appeared in Europe as the Plague of Justinian in A.D. 541–542 (Ligon, 2006). However, this disease did not hit Europe with all its might until the epidemics of the XIV century, when it traveled with Silk Road merchants from Asia all the way to Europe and North Africa (Ligon, 2006).

Historically, land and maritime trade and expeditions also favored the spread of smallpox between countries in Asia and north of Africa to Europe (Diamond, 2019, pp. 231–256). In the XV century, European expeditioners brought smallpox, measles, influenza, and typhus to the Americas. These diseases, followed closely by diphtheria, malaria, mumps, pertussis, plague, tuberculosis, and yellow fever, were one of the reasons why the native American population declined an estimated 95% in the 200 years following Columbus's arrival. Native Americans had never been exposed to the causative agents of these diseases and, therefore, they did not have acquired or genetic resistance. Smallpox ended up being eradicated many years later, in 1977, through strong international coordination in terms of implementing targeted vaccination campaigns. The first vaccine ever to be developed, in the XVIII century, was to treat smallpox and in the first experiments this was done by inoculating healthy people with pus from cowpox pustules presented by dairy maids.

Infectious diseases also require dense and numerous populations to spread among animals. Therefore, these affect mostly social animals living in herds. Evidence from genetic and molecular studies of infectious agents has allowed to trace back the ancestors of today's human pathogens and these often were causative agents of epizootics. For example, measles virus is most closely related to rinderpest virus – historically, the biggest killer of cattle and other artiodactyls, which has been declared as eradicated in 2011. The close similarity between measles and rinderpest viruses suggests that the latter was transmitted to humans at some point and then evolved into the measles virus by changing its properties while adapting to the new host. The spillover to human hosts is made possible by the fact that farmers frequently shared living quarters with cattle and their bodily fluids and excretions.

The most relevant epidemic diseases occurring during recent history are infectious diseases that, at some point, evolved from diseases of animals – smallpox, malaria, plague, measles and cholera – even though the infectious agents responsible for these diseases nowadays affect almost exclusively humans (this mechanism is described in Section 4). The more numerous the human population becomes - giving rise to larger and denser cities, with increased demand for animal products, combined with frequent contact with domestic and wild animals populations and encroaching into their ecosystems – the more opportunities will be created for emergence of disease at the animal-human interface.

Epidemiological data becomes more reliable towards the end of the XIX and beginning of the XX century, when large outbreaks of zoonotic influenza were recorded. In this period, influenza pandemics happened in 1889, 1918, 1957, 1968, and 2009 (Taubenberger & Morens, 2009). Out of those occurrences, the 1918 pandemic, usually called "Spanish Flu," has been found to have had origin in an avian reassorted virus (Worobey, Han, & Rambaut, 2014), while the 2009 H1N1 would have originated from a swine reassorted virus (CDC, 2009). The continuing spread of H5N1 highly pathogenic avian influenza (HPAI) viruses into poultry populations on several continents, associated with a growing number of spillover infections of humans, has amplified interest in pandemic prediction.

All throughout the XX century, as human activity increased its impact in natural ecosystems and the animals inhabiting them, a growing number of diseases which spilled over from animal populations were detected. Human Immunodeficiency virus (HIV), which spilled over from chimpanzees in West Africa in the beginning of the century (Sharp & Hahn, 2011), had a massive height in its spread in the 1980s. Nipah virus, which jumped from bats populations into pigs in South East Asia, began spreading in the 1990s, as did Hendra virus, which was transmitted from bats to horses and then to human populations in Australia in the same period.

In the XXI century we have seen even more frequent outbreaks of human disease emerging at the human-animal interface. Severe Acute Respiratory Syndrome (SARS) (2002), Middle East Respiratory Syndrome (MERS) (2015) and Coronavirus disease (COVID-19) (2019) are all coronaviruses causing respiratory disease which have emerged in human populations in the first decades of the XXI century. These viruses are also known to be present in bat populations, which act as reservoirs and, at least for SARS and MERS, other animals such as civets and camels acted as intermediary hosts which transmit the virus onto humans. The main reason for pandemic events to have become so much more frequent in the XX and XXI centuries when compared to the past is the dramatic change in population dynamics.

While at year 0 world population amounted to 190 million, by the year 1900, it was about to hit the 2 billion mark. At the time of this writing, world population has surpassed 7 billion and keeps growing, albeit its growth has been happening at a decreasing rate since the year 1968. United Nations experts estimate that the world population will stabilize somewhere around 11.2 billion people by the year 2100 (UN, 2017). The world is on its way to a new balance. The big global demographic transition that the world entered more than two centuries ago, triggered by agriculture and herding practices is coming to an end. The new set point is different from the one in the past, when it was the very high mortality rates that kept population growth in check. The new scenario it will be characterized by low fertility, from family planning choices, that keeps population from overgrowing (Rönnlund, Rosling, & Rosling, 2018).

The demand for protein, mainly in the form of animal protein, has been raising hand-in-hand with the world population count. Farmers and fishers have continuously increased production, to try to keep pace with population and income growth. At first, farming and fishing activities were expanded onto more and more land and water. Around the middle of the XX century, food production and agricultural land use became increasingly independent: between 1960 and today, world population more than doubled, global food production more than tripled, and agricultural land use increased by no more than 15%, indicating a more efficient use of land and better yield from crops (OECD, 2020). To date, agriculture, farming, and fishing outputs keep growing and will likely continue to grow once the world population stabilizes, unless a global shift in consumer preferences towards plant-based protein sources occurs. The forecast for continued growth of animal protein demand despite the projected stabilization of population numbers is due to the improvement of the economic situation in countries that were previously considered to be of low income. Fig. 3.1 illustrates how the increase in global meat production has followed the increase in global population. Knowing that Asia is the most populous continent, and where the most people have recently attained income levels that allow for the purchase of more expensive protein sources, it makes sense that meat production in Asia is more than double of any other continent.

An expanding population calls for expanding cities and for urban design that optimizes space, allows for fast commuting, and for people to access services with ease. In large cities, it is common to see higher-income residents making the choice between clustering together in apartment complexes near the city center or choosing the larger living spaces and nature of the suburbs, but having in exchange a longer often overcrowded commute. Lower-income residents and the population of low- and low to middle income countries often face explosive population growth and poverty which override urban planning efforts, and which are made worse by some governments. UN-Habitat estimates that, by 2030, 3 billion people, about 40% of the world's population, will need access to adequate housing. This translates into a demand for 96.000 new affordable and accessible housing units every day. Low- and low to

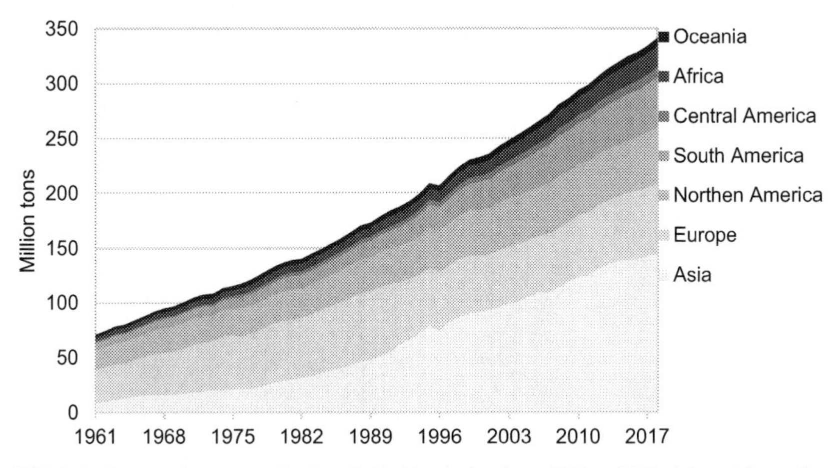

FIG. 3.1 Increase in meat production divided by region from 1961 to 2018. *Adapted from: Our World in Data ©.*

middle income countries cannot keep up with this level of demand for housing solutions fast enough, resulting in a significant part of the population being housed in slums. While poor-quality housing is not exclusive to low- and middle-income countries, the scale at which it happens in those countries in addition to concurrent challenges in access to sanitation and healthcare magnifies the problem. The United Nations defines slum as a type of housing that has the following aspects (UN-Habitat, 2012): 1) inadequate access to safe water; 2) inadequate access to sanitation and infrastructure; 3) poor structural quality of housing; 4) overcrowding; and 5) insecure residential status. Slum dwellers are increasingly vulnerable to environmental health problems, violence, and criminality. Likewise, people in slum areas suffer from water-borne diseases such as typhoid and cholera, as well as the opportunistic diseases that accompany HIV (UN-Habitat, 2003).

There is a parallel to be made with the animal side, between slums and factory farms. Due to the increasing demand for animal protein, animal farming activities have grown to levels previous unimaginable. Nowadays, 94% of mammal biomass (excluding humans) is livestock. This means livestock outweigh wild mammals by a factor of 15-to-1 (Bar-On, Phillips, & Milo, 2018). While preventive veterinary medicine and animal welfare science have contributed for the improvement of health status of farmed animal populations around the world, these remain far less than ideal in many places. Herding animals in numbers and densities unseen in nature while keeping them within confined housing has brought up a series of health issues that we have since learned to prevent and to treat. The contribution of appropriate nutrition and prophylactic treatments are of utmost importance in these settings. However, the latter can trigger problems of their own, as is the case of the misuse of antimicrobials and the use of these agents as growth promoters — a practice that has been banned in many countries. Finally, in order to feed a growing livestock contingent, occupying lands which are not appropriate for agricultural use is no longer enough. Agricultural land has been assigned to grow crops to feed livestock and for livestock to graze. As a result, there is a highly unequal distribution of land use between livestock and crops for human consumption. Half of the world's habitable land is used for agriculture (Ritchie et al., 2020); and if we combine pastures used for grazing with land used to grow crops for animal feed, livestock accounts for 77% of global farming land. While livestock takes up most of the world's agricultural land it only produces 18% of the world's calories and 37% of total protein (Poore et al., 2018).

We will see further along in the chapter that changes in land use, expansion of urban settlement, and the explosive growth in human and animal population numbers and densities are some of the most important drivers of disease emergence at the animal-human interface.

3. Drivers for disease emergence at the animal–human interface

The growth of human population and the increase in the demand for animal protein and agricultural land (mostly to feed livestock) correlate with an increase in the occurrence of outbreaks of diseases with zoonotic origin. Correlation does not imply causation – therefore, to explain how one leads to the other, hereafter, we will focus on the drivers that lead to disease emergence at the animal-human interface.

The large scale emerging infectious disease outbreaks happening in close succession during the last decades are connected. They are not just happening to us, humans, but they represent the consequences of our behavior as a species (Quammen, 2013). Human-caused ecological pressures and disruption of natural ecosystems are leading human populations to encounter animal pathogens that are new to them. Human technology and behavior help these pathogens to spread widely and quickly. Humans are mammals and therefore most zoonoses spillover to humans from other mammals.

3.1 Ecosystem disruption

The organisms that inhabit the same natural ecosystem are connected through interdependent relationships, which include mutualism, predation, parasitism, among others. These ecological relationships naturally limit the abundance and range of infectious agents. However, disruption of natural ecosystems causes infectious agents inhabiting that environment to meet new hosts and to seek alternative solutions to guarantee survival. With regards to zoonotic diseases, many transmission pathways involve human practices that bring people into contact with reservoirs or vectors often in places that were previously unexplored by humans or free from influence caused by human activities (Richardson et al., 2016). Some of these pathogens may adapt relatively fast to humans – the newly available and naïve host. If infectious agents are successful in being transmitted from animals to humans, and especially between humans, they will have available a population of hosts which is abundant, global, and interconnected, allowing the agent to travel around the world in less than two days.

Humankind's activities are causing the disintegration of natural ecosystems at an unprecedented rate. Activities such as logging, slash-and-burn agriculture, construction of buildings, roads and urban environments, hunting and eating wild animals, clearing forest for cattle pasture, mineral extraction, chemical pollution, plastics and microplastics pollution, exports of goods which production require any of the above, and other incursions related to urbanization are tearing natural ecosystems apart. When the total human population consisted of few cities and villages, the practices listed above would not be too harmful. However, with a population of 7 billion going on to

11.2 billion in 2100 (UN, 2017), the cumulative impacts are devastating. The more humans expand their ecological impact, encroach into natural ecosystems, alter these ecosystems to extract resources, intensify food production, and displace animals, people, commodities and their pathogens, the greater the potential for infectious diseases to emerge or re-emerge, and for pathogens and pests to spread (Bebber, Holmes, & Gurr, 2014; Jones et al., 2008). So far, human action has significantly impacted more than three-quarters of the Earth's land surface, destroyed more than 85% of wetlands and redirected more than a third of all land and almost 75% of available freshwater to crops and livestock production (Settele, Díaz, & Brondizio, 2020). Since 1900, the average abundance of native species in most major land-based habitats has fallen by at least 20% (IPBES, 2019). More than 40% of amphibian species, almost 33% of reef forming corals and more than a third of all marine mammals are threatened (IPBES, 2019). The picture is less clear for insect species, but available evidence supports a tentative estimate of 10% being threatened (IPBES, 2019). At least 680 vertebrate species had been driven to extinction since the XVI century and more than 9% of all domesticated breeds of mammals used for food and agriculture had become extinct by 2016, with at least 1000 more breeds still threatened (IPBES, 2019). In an attempt to attract political leverage and influence economic growth paradigms, the United Nations General Assembly has proclaimed 2021–2030 as the Decade on Ecosystem Restoration (UN, 2019).

Food production (Ritchie et al., 2020) is one of the human activities through which humankind is considered to have the largest environmental impact – it uses twice the amount of water compared to all other human activities combined. The risk of emergence of new pathogens and the spread of existing ones has also increased because of deep and global changes in the way that food is produced and consumed. A study published by the Intergovernmental Science-Policy Platform on Biodiversity and Ecosystem Services (IPBES) in May 2020 describes a global picture of a highly concerning human-driven landscape monotony spreading across the planet, as a small variety of "cash crops" and high-value livestock are replacing native forests and other biodiverse ecosystems (IPBES, 2019). In addition to favoring soil erosion, which causes a loss of fertility, these monocultures are more vulnerable to disease, drought, and other impacts of climate change due to their genetic uniformity.

The risk of disease spillover is greater for humans when ecosystem disruption facilitates encounters with species that are phylogenetically close: mammals. Focusing specifically on viruses, it has been found that the proportion of zoonotic viruses per species increases with host phylogenetic proximity to humans (Olival, 2017). For example, evolutionarily related species sharing host cell receptors and viral binding affinities and specific viral mutations that may expand host range in related mammal species may facilitate spillovers. However, mechanisms by which phylogeny affects spillover risk still need to be further explained. It is certain that there are many zoonotic diseases yet to be found, and many zoonotic disease agents that have not yet spilled over to humans. These pathogens are most likely found within reservoir hosts in undisturbed ecosystems, where biological diversity is high. Once the ecosystem's equilibrium is harmed or a few species are removed, ecological disturbance will cause the hiding pathogen to emerge.

Human encroachment into biodiverse ecosystems increases the risk of spillover of new infectious diseases and re-emergence of known ones by enabling new contacts between humans and wildlife. It has been determined that species in the primate and chiroptera orders were significantly more likely to host zoonotic viruses compared to all other orders (Johnson et al., 2020). Restoring damaged ecosystems and re-connecting isolated habitat patches can help bring back stability to natural environments. Ecosystem integrity and functioning, on land, freshwaters, and oceans, underpins human health, wellbeing, and resilience. At the same time, a greater diversity of species means it is more difficult for a single pathogen to dominate. Where native species diversity is high, infection rates for zoonotic diseases is likely to be lower (IPBES, 2019).

3.2 Wildlife hunting, farming, and trading

Domesticated animals like cattle, sheep, dogs, and goats may share the highest number of viruses with humans, with eight times more animal-borne viruses than wild mammal species. However, humans have built resistance to many of these viruses over time and have developed vaccines for many others. At the present time, the greatest risk for zoonotic spillover is that from threatened and endangered wild animals whose populations had declined largely due to hunting, wildlife trade and loss of habitat (Johnson et al., 2020). These human actions have simultaneously threatened conservation of endangered species and increased the risk of zoonotic pathogen spillover.

It is important to note that several teams of researchers have been undertaking work to map both what we know and what we do not know in terms of hotspots for spillover from wildlife populations, especially with regards to viruses that have not yet been identified ("missing viruses") and their zoonotic potential (Johnson et al., 2020; Karesh, Cook, Bennett, & Newcomb, 2005; Olival et al., 2017). Collectively, these works of research have led to interesting findings, such as that chiroptera (bats) carry a significantly higher proportion of zoonotic viruses than all other mammalian orders. The geographic regions with the largest estimated number of "missing viruses" and "missing zoonoses," and therefore of highest relevance for future surveillance have also been identified – in a study published on Nature in 2017 by Olival et al., geographic hotspots of "missing zoonoses" were described. These hotspots vary by host taxonomic order, with foci for carnivores and even-toed ungulates in eastern and southern Africa, bats in South and Central America and parts of Asia, primates in specific tropical regions in Central America, Africa, and southeast Asia, and rodents in pockets of North and South America and Central Africa. This group of researchers also identified geographic regions with large numbers of mammal species currently lacking any information regarding their viral diversity, and therefore of interest for further study. The maps developed by the authors of this paper can be used for cost-effective allocation of resources for viral discovery programs, ecosystem conservation efforts, and community-based surveillance.

The growing pool of knowledge concerning the outcomes of human encroachment into natural ecosystems has set the foundation to calls from researchers, governments, and non-governmental organizations not only for the conservation of these ecosystems but also for the regulation of hunting, farming, and trading of wild animals. At the time of this writing, trade and consumption of wild animals is a much discussed subject, due to the alleged origin of SARS-CoV-2 virus in a spillover or superspreading event which would have occurred in a "wet market." Wet markets and wildlife markets are not synonymous, though they are often used interchangeably. This semantic variance is driving a lot of the disagreement in the debate about whether to ban all of them (Maron, 2020; ACIAR, 2020). There are some open-air markets that sell only previously slaughtered animals and produce, others that sell commonly eaten live animals like chickens, and some that sell wild animals like bats, snakes, porcupines, crocodiles, etc. (Maron, 2020). All these types of markets are usually and confusingly referred to as wet markets. However, the gradation mentioned above also means a gradation in level of risk for zoonotic diseases.

In order to avoid confusion and misunderstandings, we will use the wording "wildlife wet markets." Those are open street markets where live animals, often wildlife, are sold or slaughtered on demand of the client. The real problem with these markets it is not that they are "wet," i.e. that they sell fresh meat and fish. It's rather that: 1) they are a melting pot of animal species that do not normally encounter each other in their natural habitats and are now in too closeness, proximity with each other - animals of different species are held in stacked cages, where the animals are likely to urinate, defecate, and potentially bleed or salivate on the animals below them; 2) animals are extremely stressed by the chaotic environment and the confinement situation, resulting in accelerated excretion of agents; 3) the lack of hygiene due to "slaughter in place" practices that cause surfaces to be contaminated with blood and excretions; and 4) animals may be consumed at the market, with questionable food safety precautions, or transported home alive or butchered. All these factors create the opportunity for interactions that would never happen in nature. For example, horseshoe bats, who live in caves when in the wild, would not come close to civets, who live on trees. At wildlife wet markets, not only do they share living quarters, but they are also in close contact with humans.

As long as the demand for wild animals and their products, such as skin, scales, nails, exists, these animals will continue to be farmed or captured. Whether they are legally sold at wildlife wet markets, or are illegally smuggled, will depend on the regulation in vigor at its enforcement. Having found that wildlife wet markets were the inception place of SARS due to the spillover of a new coronavirus from civets to humans, Chinese authorities banned the sale of civets it May 2003, alongside with other 53 wild animals usually sold at wildlife wet markets. However, due to the economic losses caused to animal farmers and their protests, the ban was lifted in late July of the same year. After the SARS epidemic, markets selling wild caught animals and farmed wildlife continue to exist, albeit they have become more discreet. Wildlife wet markets are not unique to China. They are common in many parts of the world, including Asian, African, and Latin American countries.

However, an all-out ban on wildlife wet markets and wild animal hunting may not be the wisest solution, as these types of places are engrained in local culture and numerous low-to middle-income families depend on them for their livelihoods (Reuters, 2020). An indiscriminate ban would bring about serious socioeconomic consequences onto those groups. Wild animals have been regarded as alternative sources of protein and sources of revenue in poor regions. "Bushmeat" may be the only meat that some impoverished populations in sub-Saharan Africa have access to. Smoked, dried, or cooked, the meat from animals hunted in the forest provides a valuable source of protein for people in rural communities where farming domesticated animals is too expensive or impractical. Hunting and selling bushmeat can also serve as an important source of income in this region (Actman, 2020). In Asia, consumption of wildlife as a source of protein became even more pertinent in the recent scenario caused by the spread of African swine fever, which caused the death of roughly half of the world's pigs effective and led to global concerns about shortages in pork supply. It should also be considered that some consumers ascribe traditional medicinal benefits to the wild animals – claims which are, at best, disputable. Also, to some extent, trade of rare, endangered species is driven by interests of subjects profiting from demand for such upper tier wildlife.

Production and sale of domesticated livestock on any country's farms is, compared to the sale of wildlife, subjected to far more regulation and inspection. Part of the problem with wildlife trade is the lack of regulation and its enforcement, despite the greater zoonotic risk, especially in wildlife wet markets. Outbreaks still occur in commercial farms, but they are identified and controlled faster. Wildlife trade works as a system of interconnected networks dependent on major trade hubs (Karesh et al., 2005). Rather than attempting to eradicate pathogens or the wild species that may carry them, a sensible approach would include decreasing the contact rate among species, including humans, at the interface created by the wildlife trade. Therefore, major hubs for trade in wildlife represent opportunities to maximize the implementation of regulatory efforts (Karesh et al., 2005). Growing calls for regulation of this sector coming from researchers and non-governmental organizations have amounted over the last few years (Harder, 2020). The 2020 pandemic caused by SARS-CoV-2 is causing policy-makers to take concerted positions, and standard-setting agencies like the World Health Organization (WHO) and the World Organization for Animal Health (OIE) to take steps to address the zoonotic risk brought about by wildlife hunting, farming and trade.

3.3 Globalization

As mentioned in the second section of this chapter, the creation of trade routes in Eurasia caused the spread of the first infectious diseases in the continent. The increase in trade exchanges between countries and regions via land, water, and air has brought populations closer over time and acted as a driver of advancements in science and technology (Diamond, 2019, pp. 231−256). Because of differences in geography, climate and population density, some parts of the world are better suited to produce food, agriculture, and fisheries commodities than others (OECD, 2020). While regions like North and South America have a lot of agricultural land and have emerged as major agricultural exporters, other regions such as the Middle East and North Africa have relatively little agricultural land and water. In addition, many agricultural products only grow in specific climates or soils. Developing countries, like China and Vietnam, have displaced advanced economies as the major source of fish products globally (OECD, 2020). Consumers enjoy fresh fruit and vegetables out of season and have access to meat and fish from foreign fields and oceans because these are grown and transported from elsewhere in the globe. At times, even the ingredients of the most traditional dishes of some cuisines, such as salt-cured codfish in Portugal, are not native of the country that has adopted them into their traditions. In the case of codfish, it is captured in the North Atlantic Sea, between Iceland, Norway and Scotland, and in some cases is cured with salt before being frozen at sea in order to simulate the old preservation process, Therefore, trade in livestock, animal products, and food is not just increasing, it has become global.

Among the changes seen in agricultural and food markets, there has been a significant increase in trade among emerging and developing countries, which are increasing in importance, both as suppliers and markets for agricultural and food products. Increasing trade has also been accompanied by deeper integration of the world's food system. A growing share of agricultural and food trade is taking place in global value chains (OECD, 2020) − processing value chains that are spread over several countries − linking sectors to other sectors of the economy from across the world. Countries in North and South America have emerged as major agricultural exporters. At the same time, agricultural imports have increased in Sub-Saharan Africa, in the Middle East and North Africa, and in South and East Asia, most notably in China. As a result, trade in food, agriculture, and fisheries products has never been as important as it is today. The food and clothing that consumers find in their local stores are increasingly made from a wider range of products, produced in a wider range of locations across the globe. These trends are expected to continue over the coming years, bringing domestic and international markets even closer together, and underscoring the growing importance of international trade for global food security.

While the principles underlying the evolution and ecological principles of disease emergence have not changed, changes in human activities have shifted the playing field on which these natural laws act (Karesh et al., 2012) The interconnectedness of our economies increases food security and contributes to global development through knowledge transfer, cultural, and scientific exchanges, and rapid disaster relief. However, in it also increases the risk of disease emergence. Large-scale human population displacements due to economic, political, and humanitarian crises represent another set of drivers for emergence and re-emergence of infections outside their zones of endemicity, drivers which are exacerbated by the changes in how livestock and food and are produced, transported, and consumed. With globalization and international travel, disease movement is now rapid, and the natural barriers that once prevented or delayed the spread of disease beyond its point of origin have become immaterial (Banks, Paini, Bayliss, & Hodda, 2015).

The capacity of present-day aircrafts to fly extremyly long distances have made even the longest intercontinental travel time (at this time, around 18 h and 30 min long from New York City to Singapore) briefer than the duration of the incubation period of most human infectious diseases. In 1918, H1N1 Influenza, the "Spanish Flu," was spread from the United States to Europe in Asia in roughly 3 months. This was at a time when international travel was made by land or boat. In 2003 SARS went halfway across the globe and back, between Hong Kong and Canada, in less than two weeks. It could have been quicker if, for example, one of the first infected patients had been an airline worker, and especially if they were a super spreader. Infectious disease outbreaks like SARS can be made even more concerning by the occurrence of superspreading events. In these occurrences, an individual, for some reason, directly infects far more people than the typical infected individual does. This is different from being an amplifier host (which gets infected but does not show clinical signs and allows the virus to replicate exponentially), that refers to a characteristic of a whole species. When superspreading events occur, the course of an infectious disease outbreak can change rapidly.

We are living in a super-connected reality, where livestock and food are produced and transported around the world, where travel for business and leisure are increasingly present in our normal routines, and where countries and regions depend on one another for increasing common resilience. The downside of this reality is that we open the way for infectious disease agents to travel the world rapidly alongside with us and our products. In this scenario, the only way to avoid the emergence and re-emergence of infectious diseases is to invest on preparedness and prevention measures, including early detection and early response systems.

3.4 Climate change

Undeniably interlinked with ecosystem disruption and globalization, climate change is the greatest challenge of this era. The 2019 Global assessment report on biodiversity and ecosystem services of the Intergovernmental Science-Policy Platform on Biodiversity and Ecosystem Services (IPBES) has ranked climate change in third place on the list of direct drivers of change in nature with the largest relative global impacts so far, right after changes in land and sea use, and direct exploitation of organisms. Climate change is increasing pressure on food production systems and will affect how and where food is produced and where known, emerging, and re-emerging diseases will appear next (OECD, 2020). Due to changes in climate, new habitats for living organisms are being created while others are destroyed. Biodiversity, including the distribution of crops, invasive species, as well as of pests and pathogens, is under great pressure from human activities associated with a changing climate.

Biodiversity provides numerous ecosystem services that are crucial to global equilibrium and human well-being (WHO, 2015). Climate is an integral part of ecosystem functioning, and human and animal health are impacted directly and indirectly by results of climatic conditions upon terrestrial and marine ecosystems. However, largely as a result of human actions, the biomass of wild mammals has fallen by 82%, natural ecosystems have lost about half their area, and a million species are at risk of extinction (IPBES, 2019). Marine biodiversity is affected by ocean acidification related to levels of carbon in the atmosphere and by rising temperatures and sea levels. Terrestrial biodiversity is influenced by climate variability, such as extreme and erratic weather events (i.e. drought, flooding) that directly influence ecosystem health and the productivity and availability of ecosystem goods and services for human use. Longer term changes in climate patterns and disruption of natural ecosystems are driving plants, pathogens, animals, and even humans to change their distribution and to find new ecological niches. This brings humans and animals closer, increasing opportunities for predation, parasitism, and disease emergence. Land-use change and agricultural industry changes are the two most commonly associated drivers of infectious disease emergence (Loh et al., 2015).

Another consequence of climate change is the modification of the distribution of vectors associated with the transmission of infectious diseases (Parham et al., 2015) – ticks, mosquitoes, and flies, shifting the incidence of vector-borne diseases. Water-borne disease incidence is also affected due to heavy rains, floods, and the consequent population movements in zones of precarious sanitary conditions. Changes in climate are likely to lengthen the transmission seasons of important vector-borne diseases and to alter their geographic range (WHO, 2018). For example, climate change is projected to significantly widen the area of China where the snail-borne disease schistosomiasis occurs. Malaria distribution is strongly influenced by climate. Transmitted by *Anopheles* mosquitoes, malaria kills over 400.000 people every year. The *Aedes* mosquito vector of Dengue and Zika viruses, is also highly sensitive to climate conditions, and studies suggest that climate change is likely to drive increased incidence of these diseases in new regions, previously unaffected.

Finally, pollution of ecosystems such as air and water pollution, is reducing human, animal and plant resilience to new diseases and other threats to their health and well-being. For example, and not surprisingly, a study at the time of the 2003 SARS outbreak in China found that people exposed to the highest level of air pollution were twice as likely to die from the disease as those who were not (Cui et al., 2003). Changes in climate can potentially alter the pattern of occurrence of high-impact zoonotic diseases, such as avian influenza, due to changes in breeding and migration patterns of waterfowl – the reservoir hosts of the virus (Gilbert, Slingenbergh, & Xiao, 2008).

3.5 Antimicrobial resistance

Preventing the emergence of infectious diseases at the animal source at a farm setting is ideally achieved through good husbandry practices. These include adapting animal density to avoid overcrowding, investing in good facilities and adequate nutrition, and implementation of strict biosecurity measures on the farm. Good husbandry practices promote animal health and welfare, increasing the return on investment for producers. However, when good husbandry practices are not upheld, or when animals become sick, livestock producers may make use of antimicrobials to control diseases. Antimicrobials are medicinal products that kill or stop the growth of living microorganisms. Excessive or inappropriate use can lead to the emergence of resistant bacteria which do not respond to antibiotic treatment. Antimicrobials include, among others, antibacterials (also called antibiotics), antimycobacterials, antivirals, antifungals and antiparasitics.

Antimicrobial resistance is not a disease but a characteristic of microorganisms that is acquired when they adapt to a new environment (ECDC, 2020) – any use of an antimicrobial forces microorganisms to either adapt or die. New resistance mechanisms are emerging and spreading globally, threatening the ability to treat common infectious diseases, resulting in prolonged illness, disability, and death (WHO, 2015). Antimicrobial resistance poses a threat to disease control throughout the world and is a critical concern for human and animal health (IACG, 2019).

The two major drivers for antimicrobial resistance are: 1) the use of antimicrobials in humans, animals, and crops, which exerts an ecological pressure on microorganisms and contributes to emergence and selection of antimicrobial-resistant microorganisms in populations, and 2) the spread and cross-transmission of antimicrobial-resistant microorganisms between humans, between animals, and on the interface between humans, animals and the environment (ECDC, 2020). Therefore, the key areas for management, control and prevention of antimicrobial resistance are the prudent use of antimicrobials, the implementation of good hygienic precautions for the control of cross-transmission of antimicrobial-resistant microorganisms at human healthcare settings, and the uphold of good husbandry practices at livestock farm level.

The misuse and overuse of antimicrobials is accelerating appearance of resistances. In many places, antibiotics are misused and overused in people and animals, and often given without professional oversight. Examples of misuse include when they are taken by people with viral infections like colds and flu, and when they are given as growth promoters in animals or used prophylactically. When animals are given antibiotics for growth promotion or to increase feed efficiency, bacteria are exposed to low doses of these drugs over a long period of time (Wegener, 1999). This is inappropriate antibiotic use and can lead to the development of resistant bacteria. In 2015, a study found that worldwide antimicrobial consumption is expected to rise 67% percent between 2010 and 2030 (Van Boeckel et al., 2015). According to the authors, five countries—Brazil, Russia, India, China, and South Africa – would experience a growth of 99% in antibiotic consumption, compared with an expected 13% growth in their human populations over the same period. In numerous countries, including high-income countries, antimicrobial agents are widely available to the public with hardly any restrictions on appropriate conditions for the importation, production, distribution and use of veterinary products (OIE, 2019). These antimicrobial agents potentially circulate freely in the market and like ordinary goods and there is the risk that they may be falsified or substandard. (OIE, 2019).

Many countries have prohibited the use of medically important antibiotics for growth promotion or feed efficiency, such as the member countries of the European Union. Additionally, some countries are leading the way in showing that it is possible to farm livestock with minimal doses or even without antibiotics. Denmark, which is among the world's top pork exporters, has proved that a country can build a thriving industry while aggressively cutting back on antibiotic use in pigs (Dupont, Fertner, Kristensen, Toft, & Stege, 2016). In 2018, there were about 200.000 pigs being produced without antibiotics, and the government hopes to reach the 1.5 million pigs mark by 2021 through a government-supported 3-year project (Jacobs, 2019). The changes in Denmark were achieved through progressively implemented regulations and by removing financial incentives to antibiotic prescription (DVFA, 2016).

The OIE, in collaboration with its Tripartite partners – FAO (Food and Agriculture Organization of the United Nations) and WHO (World Health

Organization) – published a list of antimicrobial agents of veterinary importance in order to promote their prudent and responsible use in animal health settings (OIE, 2019). Although the European Union, through the European Medicines Agency, regularly implements surveys to monitor the consumption of antimicrobials for veterinary use (EMA, 2017), there is no harmonized surveillance system to monitor the use and circulation of antimicrobial agents in animals worldwide. The OIE has been conducting an annual survey among its 182 member countries to monitor the quantities of veterinary antimicrobials for used in animals since 2016. In the fourth questionnaire, relative to the year 2018, 77% of the 153 responding countries did not use any antimicrobial agents for growth promotion in animals in their countries, either with or without legislation or regulations (OIE, 2019). The remaining 23% countries reported use of antimicrobials for growth promotion; of these, 57% countries had a regulatory framework that either provided a list of antimicrobials that can be used as growth promoters or provided a list of those that should not be used as growth promoters. When differentiated by OIE Region, the Americas and Asia, Far East and Oceania have the highest proportions of countries using antimicrobial growth promoters (Fig. 3.2).

The matter of antimicrobial resistance has been gathering increasing levels of attention. A political declaration endorsed at the UN General Assembly in September 2016 signaled the commitment to taking a broad, coordinated approach to address the root causes of antimicrobial resistance across multiple sectors. The UN Secretary-General has established an Interagency Coordination Group on Antimicrobial Resistance (IACG) which is co-chaired by the

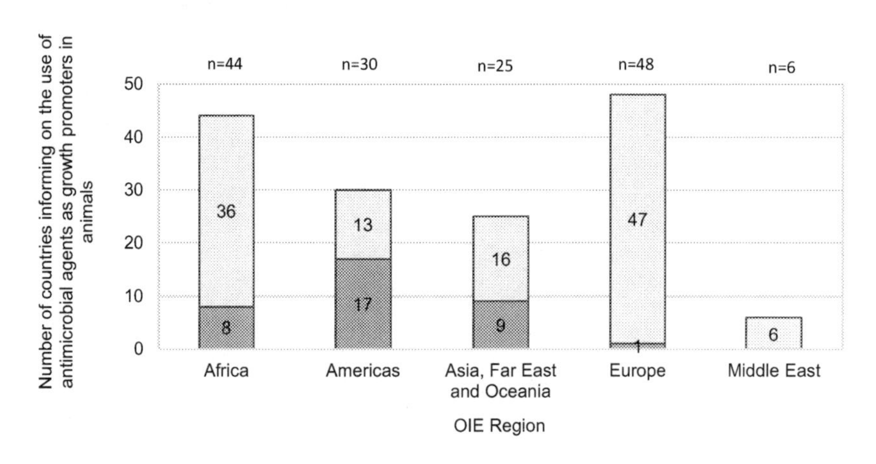

FIG. 3.2 Number of countries using antimicrobial agents for growth promotion in animals in 2018, out of 153 responding countries, divided by World Organization for Animal Health (OIE) region.

UN Deputy Secretary-General and the Director General of WHO and includes high level representatives of relevant UN agencies, other international organizations, and individual experts across different sectors. FAO, OIE and WHO are working closely under a One Health approach to promote best practices to avoid the emergence and spread of antibiotic resistance. This high-level One Health partnership needs to be mirrored at regional and national levels to ensure its effective implementation.

4. Mechanisms of disease emergence at the animal–human interface

Nearly all zoonotic diseases result from infection or infestation with one of six kinds of pathogens: viruses, bacteria, fungi, protists, prions, and worms. Natural ecosystems are full of infectious agents that coexist with animals often without causing any harm. These agents have evolved alongside their natural hosts for hundreds of thousands of years and both agents and hosts have suffered the effects of natural selection and largely arrived at a state of equilibrium. The opportunity for spillover is created when this equilibrium is disturbed. A disturbance can, for example, take the form of a species being introduced into a new ecosystem – as seen in the cases of bringing large numbers of goats into the Netherlands and the consequent rise of the bacteria causing Q Fever, the first cases of Hendra in horses that were brought to Australia in the previous century – or humans encroaching into an ecosystem where they do not belong – as it was the case when HIV first spilled over to humans in West Africa in the early XX century. Occasionally, the spillover occurs without major consequences. Other times, a large-scale outbreak occurs. Outbreaks usually end either if sufficient immunity in the population is achieved, if a vaccine or a treatment are found, or if the source is found and contained, among other less self-evident options. At times, rather worryingly, the pathogen adapts well to the newfound host population, not killing it for a while, and manages to establish itself and become endemic – this was the case of HIV.

Viruses are the most numerous biological entities and can be found in every ecosystem on earth. Even though other kinds of pathogens can be equally hard to detect or eliminate, viruses are especially insidious. The genetic material of viruses can be DNA or RNA. This translates into one of the most important differences among viruses: their mutation rate. RNA viruses mutate much more frequently that DNA viruses since there is not a proofreading system in place when RNA strands are replicating. This is the reason why RNA viruses are known to change over time. While the majority of the mutations cause disfunctions to the virus bringing it to an evolutionary end of road, some allow it to gain new functions or to infect new hosts. RNA viruses are able to evolve and adapt faster than DNA viruses and other infectious agents. Examples of RNA viruses are filoviruses (Ebola, Marburg), henipaviruses (Nipah, Hendra), coronaviruses (SARS, MERS, SARS-CoV-2), influenza virus, and rabies virus. That being said, viruses can be considered more problematic than the other types of pathogens mentioned above. They evolve quickly, they are not affected by most antimicrobials, they can be elusive and versatile, and can cause high fatality rates while being deceitfully simple in their constitution.

In order to describe how and why some diseases are transmitted by animals to humans and can establish themselves on the human population, an explanation based on an article published in Nature by Wolfe et al. in 2017 will be given. This group of researchers proposes five stages that describe how an animal pathogen can evolve into a specialized pathogen of humans. In this explanation there is no inevitable progression of microorganisms from Stage 1 to Stage 5: at each stage, any agents can remain stationary. Skipping stages is also possible. The agents causing nearly half of the 25 important diseases that were selected for this analysis have not reached Stage 5.

1. The first stage comprises microorganisms that are present in animals but that have never been detected in humans under natural conditions (i.e. excluding modern technologies that can inadvertently transfer microorganisms, such as blood transfusion, organ transplants, or hypodermic needles).

   Examples of agents grouped in this stage are most malarial plasmodia, which tend to be specific to one host species or to a closely related group of host species, and the viruses causing African swine fever, African horse sickness, Porcine reproductive and respiratory syndrome, among many others. These agents may be present in more than one species of animals but are not known to have affected humans to date.
2. The second stage refers to pathogens of animals that, under natural conditions, have been transmitted from animals to humans ("primary infection") but have not been transmitted between humans ("secondary infection"). Examples of pathogens of animals capable of causing primary infection in humans are anthrax and tularemia bacteria, Nipah, Hendra, rabies, Rift Valley and West Nile viruses.

   Anthrax, which is known for its potential to be used as biological weapon, does not typically spread from animal to animal nor from human to human. Herbivores usually become infected by ingesting spores in the soil and in feed, while carnivores can be infected by eating meat from an infected animal.

   Tularemia can cause severe disease in humans after infection through direct contact with urine and feces from infected lagomorphs or contaminated surfaces, through the bite of arthropods (primarily ticks and mosquitoes), or even inhalation or consumption infected water or food. The causative agent of tularemia is not directly transmitted between humans. (OIE, 2018).

Nipah and Hendra viruses emerged in the last decade of the XX century as the causes of outbreaks of respiratory and neurological disease that affected a number of animal species and humans. Fruit bats, also known as flying foxes, of the genus Pteropus are natural reservoir hosts of both Nipah and Hendra viruses (OIE, 2009). These viruses can cause severe disease and death in humans, although they cannot be transmitted directly between humans. For these viruses to spill over from bats to humans, in both cases there was an intermediary host involved – pigs, in the case of Nipah, and horses, in the case of Hendra.

Rabies is one of the deadliest zoonoses and can affect all mammals. Each year, rabies kills nearly 59,000 people worldwide, mostly children in developing countries. Over 95% of human cases of rabies are due to dog bites and unlike many other diseases, dog-mediated human rabies could theoretically be eradicated through the vaccination of dogs (OIE, 2020).

Rift Valley fever is an arthropod-borne viral disease affecting domestic animals (e.g. cattle, camels, goats) in which human infection results from direct or indirect contact with the blood or organs of infected animals, such as during slaughter or veterinary care. Transmission from human to human has not been documented. (OIE, 2020).

Finally, West Nile virus causes disease in humans, horses, and several species of birds. The reservoir hosts of West Nile virus are birds. Mosquitoes become infected when they bite an infected bird (OIE, 2020). The mosquitoes act as vectors, spreading the virus from an infected bird to other birds and to other animals. Infection of other animals (e.g. horses, and humans) is incidental to the cycle in birds since most mammals do not develop enough virus in the bloodstream to spread the disease.

3. The third stage includes animal pathogens that can undergo only a few cycles of secondary transmission between humans, so that occasional human outbreaks triggered by a primary infection soon die out. Examples are Ebola, Marburg, and monkeypox viruses.

   Ebola virus disease is a severe, often fatal illness affecting humans and other primates. The virus is transmitted to humans from wild animals (such as fruit bats, porcupines and non-human primates) and then spreads in the human population through direct contact with the blood, secretions, organs or other bodily fluids of infected people, and with surfaces and materials contaminated with these fluids. The first EVD outbreaks occurred in remote villages in Central Africa, near tropical rainforests (WHO, 2020). It is thought that fruit bats of the Pteropodidae family are natural Ebola virus hosts.

   Marburg virus disease is a highly virulent disease that causes hemorrhagic fever, with a case fatality ratio of up to 88%. It is in the same family as the virus that causes Ebola virus disease. The first recorded outbreak was associated with laboratory work using African green monkeys imported from Uganda. Human infection with Marburg virus disease normally results from prolonged exposure to mines or caves inhabited by Rousettus bat colonies. Once an individual is infected with the virus, Marburg can spread through human to human transmission via direct contact with the blood, secretions, organs or other bodily fluids of infected people, and with surfaces and materials contaminated with these fluids (WHO, 2020).

Monkeypox virus causes a disease with symptoms similar, but less severe, to smallpox. While smallpox was eradicated in 1980, monkeypox continues to occur in countries of Central and West Africa close to tropical rainforests where there are animals that carry the virus. Human to human transmission occurs but is limited. It can be transmitted through contact with bodily fluids, lesions on the skin or on internal mucosal surfaces, such as in the mouth or throat, respiratory droplets, and contaminated objects (WHO, 2020).

4. The fourth stage, as proposed by N. D. Wolfe et al., has separate sub-stages. The fourth stage includes diseases that exist in animals and that have a natural (sylvatic) cycle of infecting humans through primary transmission from the animal hosts, but that also can undergo long sequences of secondary transmission between humans without the involvement of animal hosts.

a. In this sub-stage, the sylvatic cycle is much more important than direct human to human transmission. Examples are Chagas disease and yellow fever.

Chagas disease is caused by *Trypanosoma cruzi*, a protozoan parasite that infects a broad range of triatomines and mammalian species, including humans (WHO, 2020). The main route of transmission is vector-borne through the bite and subsequent contamination of the wound caused by insects from the triatomine sub-family.

In the case of Yellow fever virus, the causative agent is transmitted by infected mosquitoes (most frequently Aedes species) which acquire and transmit the virus by feeding on primates (human and non-human), having a three transmission cycles: (i) jungle (sylvatic), where mosquitos acquire the virus from non-human primates and transmit to humans visiting the jungle; (ii) intermediate (savannah), where vectors acquire and transmit the virus from and to human or non-human primates, affecting humans living or working in jungle border areas, (iii) urban, where the virus is brought from a viremic human into an urban setting and continuously transmitted between humans and urban mosquitos (e.g. *Aedes aegypti*) (CDC, 2019). Eradication of yellow fever is not feasible since it is not possible to control the virus in the natural animal hosts (non-human primates). However, a human vaccine which is safe and confers lifelong immunity is available.

b. In sub-stage 4b, both the sylvatic cycle and direct transmission from human to human are important. The main example is Dengue fever in forested areas of West Africa and Southeast Asia.

Dengue is a potentially fatal mosquito-borne viral infection, common in warm, tropical climates, which is spread through the bites of infected *Aedes* species mosquitoes (*Ae. aegypti* or *Ae. Albopictus*). Dengue epidemics tend to have seasonal patterns, with transmission often peaking during and after rainy seasons. There are several factors contributing to this increase and they include high mosquito population levels, susceptibility to circulating serotypes, favorable air temperatures, precipitation and humidity, all of which affect the reproduction and feeding patterns of mosquito populations, as well as the dengue virus incubation period (WHO, 2020). Changes in climate patterns around the world are impacting the distribution of *Aedes* mosquitoes, bringing Dengue, Zika and other diseases to areas that were previously free from infection. Dengue can also be transmitted congenitally, from infected pregnant woman to her fetus during pregnancy or around the time of birth, and through infected blood, laboratory, or healthcare setting exposures (CDC, 2019). Monkeys act as a reservoir host in west Africa and south-east Asia (WHO, 2020).

c. In the last sub-stage, 4c, the most relevant mode of transmission is between humans. Examples of diseases where animal transmission is possible but human-to-human transmission is far more relevant are cholera, influenza A, typhoid fever and West African sleeping sickness.

Cholera, caused by the bacterium *Vibrio cholerae*, is a potentially deadly acute diarrheal illness transmitted by ingestion of contaminated water or food, and is one of the first human diseases for which there is accurate epidemiological information. (WHO, 2020). *V. cholerae* started by infecting the bodies of copepods, a marine crustacean (Huq et al., 1983), with human transmission dependent on the contact with this reservoir. With time, benefitting from a mutation which originated a flagellum that allowed for the bacteria to adhere to the wall of the human gut and colonize it, *V. cholerae* became a zoonotic agent, causing now human disease without intervention of animal reservoirs, vectors, or intermediary hosts.

From the four types of influenza viruses (A, B, C, D), only influenza A viruses are capable of causing flu pandemics when a strain emerges that is capable of infecting humans and spreading efficiently, while influenza C causes mild illness without epidemics, and influenza D primarily affects cattle (CDC, 2019). A pandemic can occur when a new and very different influenza A virus emerges that both infects people and can spread efficiently between individuals. Influenza A viruses are divided into subtypes based on two proteins on the surface of the virus: hemagglutinin (H) and neuraminidase (N), which can combine in up to

198 subtype combinations. Sporadic pandemic outbreaks involve influenza A strains of zoonotic origin, such as the 2009 H1N1 virus originating from pigs that caused a global pandemic and now propagates in the human population without the intervention of reservoir hosts. The currently circulating influenza A (H1N1) viruses are related to the pandemic 2009 H1N1 virus that emerged in the spring of 2009 from an influenza A virus reassortant (a new virus resulting from switching of viral RNA segments in cells infected with two different influenza virus) that was previously circulating in pigs and caused a global pandemic (Krammer et al., 2018). Although the main reservoir of diverse strains and subtypes of influenza A virus is wild birds, especially migratory ducks and geese, the circulation of influenza viruses such as H1N1 is sustained in the human population without the intervention of reservoir hosts.

Typhoid fever is a life-threatening systemic infection caused by the bacterium *Salmonella enterica* serovar Typhi (commonly known as *Salmonella typhi*), usually spread through the ingestion of contaminated food or water. Once *Salmonella typhi* bacteria are eaten or drunk, they multiply and spread into the bloodstream (WHO, 2018). Urbanization and climate change have the potential to increase the global burden of typhoid. In addition, increasing resistance to antibiotic treatment is making it easier for typhoid to spread through overcrowded populations in cities and inadequate and/or flooded water and sanitation systems. Humans are the only reservoir for *Salmonella typhi* (which is the most serious), whereas *Salmonella paratyphi* also has animal reservoirs. Some humans can carry the bacteria in the gut for very long periods (chronic carriers) and transmit the bacteria to other people either directly or via food or water contamination (ECDC, 2020).

5. In the fifth and final stage, pathogens are exclusive to humans. Examples are the agents causing falciparum malaria, measles, mumps, rubella, smallpox, and syphilis. It was proposed by N. D. Wolfe et al. that pathogens belonging to Stage 5 could have become confined to humans in either of two ways:
   - either an ancestral pathogen already present in the common ancestor of chimpanzees and humans could have infected both species long ago (cospeciation), when the chimpanzee and human lineages diverged around five million years ago;
   - or else an animal pathogen could have colonized humans more recently and evolved into a specialized human pathogen.

   The cospeciation theory accounts well for the distribution of simian foamy viruses of non-human primates, which are lacking and presumably lost in humans: each virus is restricted to one primate species, but related viruses occur in related primate species. While both interpretations are still debated for falciparum malaria, the latter interpretation of recent origins is widely preferred for most other human Stage 5 diseases.

For an epidemic of a Stage 5 disease to persist, a few conditions must be observed. As the disease infects only humans and lacks an animal or environmental reservoir, each infected human introduced into a large population of susceptible individuals must on average give rise during their contagious lifespan to an infection in at least one other individual (basic reproductive number, Ro, is of 1). Ro can be a mathematical expression of the difference between a zoonotic pathogen and one that has spilled over, evolved, and became a human pathogen. For a pathogen to establish a chain of sequential infections (epidemic or pandemic depending on the extent of the spread), its basic reproductive number must be greater than 1. Persistence of a Stage 5 disease depends on factors such as the duration of a host's infectivity; the rate of infection of new hosts; rate of development of host protective immunity; and host population density, size and structure permitting the pathogen's regional persistence despite temporary local extinctions. Still less understood are two of the critical transitions between stages: one is the transition from Stage 1 to Stage 2, when a pathogen initially confined to animals first infects humans, and the other is the transition from Stage 2 to Stages 3 and 4, when a pathogen of animal origin that is nevertheless transmissible to humans evolves the ability to sustain many cycles of human to human transmission, rather than just a few cycles before the outbreak dies out, as seen in recent Ebola outbreaks.

It is worth noting that transmission of diseases from humans to animals and then between animals is also possible. This occurrence is called a reverse zoonosis or anthroponosis. Great apes (gorillas, orangutans, chimpanzees, and bonobos) are most at risk of this type of occurrence due to their genetic similarity to us, humans. Gorillas share 98.3% of their DNA with humans, second only to chimpanzees and bonobos, which each share 98.7% of their DNA with humans (Scally et al., 2012). Mountain gorillas who live in selected ecosystems in central Africa are a prime example of a species that is endangered by reverse zoonoses. These animals are greatly appreciated by tourists who bring anthroponotic infections, such as measles along with them. Other infectious diseases that threaten great apes are tuberculosis, chickenpox, and Ebola virus disease.

5. Intersectoral cooperation

Emergence and re-emergence of infectious diseases are local phenomena with global repercussion and consequences. While there will always be locations remaining unaffected or only mildly affected by a new or re-emerging disease, there is no place that exists in isolation from the rest of the world. As it has been discussed in previous sections, countries and regions are connected by trade, business, tourism, and ecosystems where pathogens do not know borders.

In order to prevent new diseases from emerging and to be prepared for the eventuality of that event, there should be a global awareness regarding drivers of infectious disease emergence and a joint effort to systematically analyze emergence risk. Drivers for disease emergence at the animal-human interface, having been identified and well characterized, allow us to predict, to a certain extent, where future spillover events may take place based on patterns of ecosystem change. Most of the drivers for disease emergence are not solely specific to either human, animal, or environmental health. Therefore, analysis of change to ecosystems should be conducted from a One Health perspective, considering the inherent connections between people, animals, and the ecosystems they exist in, including its infectious agents.

Identification of new infectious agents is an important component of preparedness for disease emergence at the animal-human interface. To quantify and reduce the risk of spillover of pathogens requires viral discovery in wildlife and testing of human and livestock populations in regions of high disease emergence risk (Dobson, 2020). The characterization of the pathways for transmission of these agents from animals to humans complements the identification of new infectious agents by finding which human practices cause pathogens to move from their natural settings into other systems. Monitoring such pathways can help in determining not only the transmission and emergence patterns, but also the agents' evolutionary trajectory, as well as providing targets of intervention for prevention, control, and eradication. An example of this is the work of EcoHealth Alliance, a New York based nonprofit organization that is dedicated to protecting wildlife and public health from the emergence of disease, working to prevent pandemics in global hotspot regions across the world.

National- and regional-level efforts in agent identification and surveillance of known infectious agents and their transmission pathways is the cornerstone for global preparedness and prevention. The US Agency for International Development (USAID) PREDICT project analyzed the spillover of viruses in people with high wildlife contact in 31 countries over 10 years, and included community education programs, identification of high-risk behaviors, and serology surveys to examine seasonal patterns of risk. Similarly, the Global Virome Project is a 10-year collaborative scientific initiative launched in 2016 to discover unknown zoonotic viral threats and stop future pandemics.

In order for early detection systems to translate into better preparedness and improved control of disease, good integration of these systems on a global level is necessary. An example of a well-integrated global early warning system is the Program for Monitoring Emerging Diseases (ProMED) from the International Society for Infectious Diseases. ProMED was launched in 1994 as a service to identify unusual health events related to emerging and re-emerging infectious diseases and toxins affecting humans, animals, and plants. At the present time, ProMED is the largest publicly available system conducting global reporting of infectious diseases outbreaks. It is used daily by international public health leaders, governments, physicians, veterinarians and other healthcare workers, researchers, private companies, journalists, and the general public. Other systems of the same kind are available such as: the Global Public Health Intelligence Network (GPHIN), which was formed in the late 1990s by the Government of Canada in collaboration with the World Health Organization and is headquartered at the Public Health Agency of Canada. GPHIN is only accessible to members of the network; Bluedot, a private global early warning system that provides its clients with tailored infectious disease information using cutting-edge technology; among many other public and private systems. On the animal health side, the World Animal Health Information System (WAHIS), which belongs to the OIE, processes data on animal disease notifications in real-time and informs the international community of animal disease events, as well as doing analysis of trends in animal health. WAHIS is only as efficient as OIE Member Countries are prompt and transparent in notifying animal disease events within their borders.

All these systems work towards a common objective: early detection of infectious diseases in order to prevent and mitigate their spread. After spill-over, a there is a critical window of opportunity for the prevention of large outbreaks. Early cases of HIV/AIDS, hantavirus pulmonary syndrome, Nipah, SARS, and COVID-19 went undetected for weeks, months, or years (in the case of HIV) before pathogen identification (Dobson, 2020). Pathogens can travel with humans, animals, their food and feed, and therefore early warning systems work to prevent local disease events from becoming of epidemic proportions. The economic cost of disease outbreaks far outweighs the cost of preventing them. Surveillance alone would realize substantial cost savings, even in the context of pandemic outbreaks much less severe than COVID-19 (Dobson, 2020). Epidemics do long-term damage to economies, as well as to people's health and productivity, by diverting resources from non-emergency health care as well as from other public goods and services. A recent article published in Foreign Affairs (Machalaba & Karesh, 2020), illustrates this point perfectly by reporting that during the Ebola virus disease (EVD) epidemic in West Africa, GDP growth in Liberia collapsed from 8.7% to 0.7% between 2013 and 2014. Sierra Leone experienced a similar decline in growth, as did neighboring Guinea. The official death toll from the 2013–16 EVD outbreak in West Africa was 11.300, but more than 10.600 additional deaths occurred because of untreated HIV/AIDS, malaria, and tuberculosis. EVD cases led the region's already strained health-care sector to collapse, causing an 8% reduction in Liberia's health-care workforce and leaving human resources gaps that persist to this day. Months-long school closures disrupted education and left children vulnerable to exploitation, causing a spike in sexual violence and unintended teen pregnancies. The costs of these and other secondary effects of EVD will be felt for years if not decades to come.

Researchers, governments, intergovernmental organizations and non-profit organizations can and should tighten up their links in order to implement evidence-based policies that work towards preparedness, prevention and early detection of new and re-emerging infectious disease outbreaks. It is not possible to pinpoint the exact timing and location of these events. However, the abovementioned strategies would reduce their likelihood and lessen the damage caused by them.

6. Applying the One Health concept to COVID-19

The One Health concept has gathered increasing attention over the last 10 years. Although at the time of its inception it was very much a veterinary-pushed effort, the public health and environment sectors have been progressively brought together and numerous joint task forces have been created at national and international levels (OHC, 2015). One Health principles have been applied to the control and elimination of diseases as diverse as dog-mediated human rabies, Ebola virus disease, avian influenza, Lyme disease, MERS, etc. Considering the global public health situation at the time of writing this chapter, this section will focus on One Health collaboration in the scope of the 2019 SARS-CoV-2 outbreak.

In December 2019, human cases of pneumonia of unknown origin were reported in Wuhan City, Hubei Province of China. A new coronavirus was identified as the causative agent by Chinese Authorities. Since then, human cases have been reported by almost all countries around the world and the COVID-19 event has been declared by WHO as a pandemic in January 2020. Current evidence suggests that SARS-CoV-2 emerged from an animal source (Denis, 2020). Genetic sequence data reveals that SARS-CoV-2 is a close relative of other CoV found circulating in *Rhinolophus* bat (Horseshoe Bat) populations (Denis, 2020). However, to date, there is not enough scientific evidence to identify the source of SARS-CoV-2 or to explain the original route of transmission to humans (which may have involved an intermediate host). This would be consistent with past investigations that have demonstrated that SARS-CoV was transmitted from civets to humans, and MERS-CoV from dromedary camels to humans.

Researchers and international organizations agree that the emergence of SARS-CoV-2 has been brought about by human behavior and practices that cause encroachment and destruction of natural ecosystems combined with unregulated breeding, hunting, and selling wildlife species (Dobson, 2020). The timeline between virus spillover to humans and the declaration of a pandemic by WHO is likely to have been much shorter than in previous epidemic events due to the normalization of global travel and the long time that most countries took to implement effective measures to contain the global spread of the disease.

Acknowledging the zoonotic nature of SARS-CoV-2, investigations concerning animal hosts are of great importance to the study of the event in order to identify susceptible animal species and possible zoonotic and anthroponotic transmission patterns. At this point, investigations are needed to find the animal source, to determine how the virus entered the human population, and to determine the role of animals in this event. While it is clear that the pandemic is being driven by human-to-human transmission, the role of animals should still be clarified. In January 2020, a short time after the disease was reported to WHO by the Chinese Authorities and the first SARS-CoV-2 sequences were published by Chinese researchers to the public domain, environmental investigations were conducted on the Wuhan Seafood Market, which at the time was thought to be the epicenter of the epidemic. Although a few surfaces presented contamination with SARS-CoV-2, the sampling was not conclusive. The fact that a number of early cases without epidemiological links to the market was identified has suggested that the virus had already been circulating before the December 2019 (Denis, 2020).

Although each country is responsible for the measures implemented within its territory, WHO is the leader and coordinator of the international response to COVID-19 and issues recommendations that should be applied by all its Member States. In February 2020, WHO has activated a research and development (R&D) Blueprint to accelerate the development of diagnostics, vaccines, and therapeutics for COVID-19. The R&D Blueprint aims to improve coordination between scientists, health professionals and governments, accelerate the research and development process, and develop new norms and standards to improve the global response (WHO, 2020). One of the workstreams of the R&D Blueprint is related to animal and environmental research on the virus origin, and to management measures at the human-animal interface. Although the investigation regarding the animal source has not yield conclusions, animal infection studies and reports of natural infection have shown that fruit bats are susceptible to SARS-CoV-2 infection and that felids (domestic and big cats), mustelids (ferrets and mink) and some rodents are not only susceptible to infections with SARS-CoV-2, but can transmit the virus among themselves and potentially to humans, in the case of mink. SARS-CoV-2 replicates poorly in dogs, cattle, pigs, chickens, and ducks (Schlottau, 2020; Denis, 2020; Griffin, 2020; Qiang, 2020).

At the moment of writing, SARS-CoV-2 has affected all countries around the world either due to COVID-19 cases or to the economic impacts resulting from travel and trade restrictions. While the investigations regarding susceptible animal species continue, animal health laboratories have supported the public health response in more ways than just experimental infection studies. In numerous countries, these laboratories have supported their public health partners to meet the surge in demand by testing human samples (OIE, 2020).

Veterinary laboratories are well set up to do this since they are used to scaling up testing capacity for animal disease outbreaks. Veterinary services supported the public health response in other ways too, by providing much needed equipment such as ventilators and personal protective equipment (PPE) to mitigate shortages and contributed with epidemiology expertize for epidemiological studies and contact tracing. Animal health expertize was also crucial in performing risk assessments to human health or animal health associated with international trade in animals or animal products.

In the longer term, veterinary services must play a central role in reducing the risks of future pandemics by assessing the risks of disease emergence from animals, including wildlife, and managing those risks through better surveillance and regulation of high-risk practices alongside its enforcement. By jointly analyzing laboratory and epidemiological information on human and animal events collected by public health and veterinary services, so-called 4-way linking, the understanding of the epidemiology of emerging infectious diseases and potential transmission between humans and animals will be greatly enhanced.

7. Conclusion

Emergence of diseases at the interface between animals and humans has been occurring since agricultural and farming practices allowed humans to live in sedentary communities, close to their animals and crops. However, this occurrence has become increasingly relevant since the beginning of the XX century, a time since when we have seen an increase in outbreaks of zoonotic diseases. Although animal-to-human diseases already cause an estimated 700.000 deaths each year, the potential for future pandemics is vast (Settele et al., 2020). As many as 1.7 million unidentified viruses of the type known to be zoonotic are believed to exist in mammals and water birds. Growing numbers of spillover events are connected to the way food is procured or produced, in order to feed a growing human population; to the way cities, agricultural, and mining lands expand to cater such population, encroaching into natural ecosystems; to the increasing globalization of trade, business, and tourism; as well as to climate change and antimicrobial resistance, that generate more opportunities for disease emergence and amplify its consequences.

The drivers and mechanisms that cause diseases to emerge between animals and humans are complex. Therefore, a whole of systems approach is needed to prevent and prepare for future disease outbreaks and to create a food production system that is resilient to crisis. Scientists are calling for more robust and effective prevention strategies translated into strengthening and harmonization of national and international laws and standards which govern land use, to ensure that high-risk agricultural and food production practices are made safer (Machalaba & Karesh, 2020). At the same time, there are calls for modernization of food production systems in order to reduce the number of live animals moving through import/export platforms and food markets, which in turn will reduce the risk of future infectious disease outbreaks. These are important top-down approaches. However, the choices made by individuals and communities as consumers, and which are reflected on market demand, are pivotal in shaping food systems towards a more sustainable and resilient paradigm. To a certain extent, emergence of new or known infectious diseases can be slowed down or prevented, but to do so the global community must rethink its approach to economic growth, lifestyle choices, and the industries that cater to them (Spinney, 2020).

In the wake of SARS-CoV-2 pandemic, in 2020, the United Nations Environment Program (UNEP) has put forward a list of six priority actions to sustainably manage global ecosystems and thereby reduce the risk of new and known infectious diseases from emerging:

1. **Protection of wildlife habitats**: the destruction and conversion of natural ecosystems can increase the transmission of zoonotic diseases through the displacement of species which then come into closer contact with humans. Protecting remaining natural habitats, such as primary forests, can significantly reduce the chances of zoonotic disease transmission. Biodiversity in natural ecosystems and habitats is not only crucial from the viewpoint of preventing infectious diseases from emerging, but also from the public health and welfare perspective. Sudden loss of animal communities can negatively affect food security regionally. Abrupt losses of biodiversity from climate change represent a significant threat to local communities in a first instance, and later to the global population. In many regions, a large percentage of people rely on their immediate natural environment for livelihoods. Sudden disruption of local ecosystems would negatively affect their ability to earn an income and feed themselves, pushing them into poverty.
2. **Restoring ecosystem integrity**: research has shown that fragmented habitats may stimulate more rapid evolutionary processes and diversification of diseases. At the present, we see primary forests being cut down to give space to monocultures of cash crops. Restoring damaged ecosystems and re-connecting isolated habitat patches can help bring back stability to natural environments. Ecosystem integrity and functioning, on land, freshwaters, and in our oceans, underpins human health, wellbeing, and community resilience. While deforestation appears to be on the decline in some countries, it remains disturbingly high in others—including Brazil and Indonesia—and a serious threat to the world's most valuable forests remains. The main cause of deforestation is agriculture (poorly planned infrastructure is emerging as an important threat) and the main cause of forest degradation is illegal logging. Agricultural products, such as soy and palm oil, are used in an ever-increasing list of products, from animal feed

to lipstick and biofuels. Rising demand has created incentives to convert forests to monocultural farmland and cattle pasture. The world is losing forests at a rate equivalent to 27 soccer fields every minute. For example, in the Amazon around 17% of the forest has been lost in the last 50 years, mostly due to forest conversion for cattle pasture (WWF, 2020).

3. **Safeguarding natural species diversity:** in areas where native species diversity is high, infection rates with zoonotic pathogens are usually lower. A greater diversity of species means it is more difficult for a single pathogen to dominate. A good example is the occurrence of Lyme disease: the risk of contracting Lyme disease seems to go up as the roster of native species goes down in a given area. Numerous local conservation measures (such as the establishment of protected areas, indigenous areas, and community conserved areas) can help protect species diversity and, by extension, the overall ecosystem health.
4. **Ensuring sustainable, safe, and legal wildlife trade.** Banning wildlife trade is not a solution that fits all scenarios. There are many communities that rely on bushmeat for their livelihoods and whose food security would be harmed by an all-out ban. At the same time, for those who consume wild animals for reasons of social status and for its alleged health benefits, monetary power would allow them to still procure wild animals and their products despite a ban on trade. Thereby, wise stewardship of wildlife resources and adherence to appropriate regulation ensures that wildlife trade is sustainable, legal, and safe are essential to conservation of wild animal species and their natural ecosystems. Moreover, illegal and unsustainable wildlife trade brings humans into contact with the hosts of pathogens with zoonotic potential. Concomitantly, surveillance activities that include animals and humans in close contact situations will advance preparedness for outbreaks of emerging infectious diseases during "peace time," assist in prioritizing studies needed to understand epidemiological patterns in pathogen transmission, and optimize disease prevention actions (Johnson et al., 2020).
5. **Greening supply chains**: sustainable trade of commodities can reduce the displacement of wildlife that results from damaging interventions such as deforestation, unsustainable land use practices and pollution. Having the natural capital at the heart of economic and business decision making will open the way for cleaner and greener supply chains. There is a growing urgency for governments to roll back policy measures that can harm the environment by encouraging unsustainable production (OECD, 2020). Doing so would release financial resources that could be devoted to high-return investments in innovation systems, in physical and digital infrastructure, in targeted climate and environmental measures, and in risk management systems. Changing policies for food, agriculture, and fisheries is rarely easy, as the potential impacts on rural livelihoods, income distribution, and competitiveness will create significant uncertainties. Policy

reform processes must be designed to enable those affected to have a voice, to anticipate negative impacts, and to have in place transition support to facilitate structural adjustments. Governments can work with public agencies, non-governmental organizations, businesses, and civil society groups to strengthen domestic cooperation, both to facilitate policy reform and to address local problems as regards the environment. A key element of more sustainable future policies is the evolution of global financial and economic systems towards a global sustainable economy, steering away from the current paradigm of economic growth (Settele et al., 2020).

6. **Agreeing on ambitious international environmental policies**: multilateral environmental agreements with measurable targets, strengthened accountability, and associated means for implementation can help safeguard the ecosystems on which the whole world depends. This type of policies should be based on international cooperation and transparency, both to create better information on shared risks (climate and other environmental changes and on research and development gaps) and to identify potential solutions of widespread interest (climate-resilient infrastructure, land and water management, weather-related risk mitigation, and new digital applications) that will improve international cooperation, and regional preparedness, and resilience (OECD, 2020).

There is an ever-increasing amount of scientific evidence that illustrates how human actions have led to the disturbance and destruction of natural ecosystems, and how these events have driven the emergence of new infectious diseases at the interface between animals and humans. Academia, international, and non-governmental organizations have repeatedly done calls to action that relate to halting damaging action that result in climate change. It is time for national governments, regional organizations, businesses, and citizens to each do their part in shaping and implementing policies towards conservation of natural ecosystems, regulating trade of wild animals, promoting responsible consumption and travel, using antimicrobials with care, and striving for the enforcement of these measures and for the accountability of those who should implement them.

References

Actman, J. (2020). *"What is bushmeat?" National Geographic*. https://www.nationalgeographic.com/animals/reference/bushmeat-explained/.

Australian Centre for International Agricultural Research (ACIAR). (June 7, 2020). *Wet markets, not so cut and dry*. https://reachout.aciar.gov.au/wet-markets-not-so-cut-and-dry.

Banks, N. C., Paini, D. R., Bayliss, K. L., & Hodda, M. (2015). The role of global trade and transport network topology in the human-mediated dispersal of alien species. *Ecology Letters, 18*(2), 188–199. https://doi.org/10.1111/ele.12397

Bar-On, Y., Phillips, R., & Milo, R. (2018). The biomass distribution on Earth. *Proceedings of the National Academy of Sciences, 115*(25), 6506–6511. https://doi.org/10.1073/pnas.1711842115

Bebber, D. P., Holmes, T., & Gurr, S. J. (2014). The global spread of crop pests and pathogens. *Global Ecology and Biogeography, 23*, 1398–1407. https://doi.org/10.1111/geb.12214

Van Boeckel, T. P., Brower, C., Gilbert, M., Grenfell, B. T., Levin, S. A., ... Laxminarayan, R. (2015). Global trends in antimicrobial use in food animals. *Proceedings of the National Academy of Sciences of the United States of America, 112*(18), 5649–5654. https://doi.org/10.1073/pnas.1503141112

Centers for Disease Control (CDC). (January 15, 2019). *Transmission of yellow fever virus*. https://www.cdc.gov/yellowfever/transmission/index.html (Consulted on 31 May 2020).

Centers for Disease Control (CDC). (18 November). *Types of Influenza Viruses*. https://www.cdc.gov/flu/about/viruses/types.htm. (Consulted on 31 May 2020).

Centers for Disease Control (CDC). (November 25, 2009). *Origin of 2009 H1N1 flu (swine flu)*. Questions and Answers https://www.cdc.gov/h1n1flu/information_h1n1_virus_qa.htm (Consulted on 4 April 2020).

Centers for Disease Control (CDC). (September 26, 2019). *Dengue*. https://www.cdc.gov/yellowfever/transmission/index.html (Consulted on 31 May 2020).

Chandran, R. (2020). *Traditional markets blamed for virus outbreak are lifeline for Asia's poor*. Thomson Reuters Foundation. https://www.reuters.com/article/southeast-asia-health-markets/traditional-markets-blamed-for-virus-outbreak-are-lifeline-for-asias-poor-idUSL8N2A5201.

Cui, Y., Zhang, Z., Froines, J., Zhao, J., Wang, H., Yu, S. Z., & Detels, R. (2003). Air pollution and case fatality of SARS in the People's Republic of China: An ecologic study. *Environmental Health, 2*, 15. https://doi.org/10.1186/1476-069X-2-15

Danish Veterinary and Food Administration (DVFA). (2016). *Low use of antibiotic in Denmark*. https://www.foedevarestyrelsen.dk/english/Animal/MRSA/Pages/Low_use_of_antibiotic_in_Denmark.aspx.

Denis, M., Vanderweerd, V., Verbeeke, R., Laudisoit, A., Reid, T., Hobbs, E., ... Van der Vliet, D. (2020). *COVIPENDIUM: Information available to support the development of medical countermeasures and interventions against COVID-19 (version 2020-08–04). Transdisciplinary insights*.

Diamond, J. (2019). *Guns, germs, and steel* (Vintage Classics).

Dobson, A. P., Pimm, S. L., Hannah, L., Kaufman, L., Ahumada, J. A., Ando, A. W., ... Vale, M. M. (2020). Ecology and economics for pandemic prevention. *Science (New York, N.Y.), 369*(6502), 379–381. https://doi.org/10.1126/science.abc3189

Dupont, N., Fertner, M., Kristensen, C. S., Toft, N., & Stege, H. (2016). Reporting the national antimicrobial consumption in Danish pigs: Influence of assigned daily dosage values and population measurement. *Acta Veterinaria Scandinavica, 58*(1), 27. https://doi.org/10.1186/s13028-016-0208-5

European Centre for Disease Prevention and Control (ECDC). *Factsheet for experts - Antimicrobial resistance*. https://www.ecdc.europa.eu/en/antimicrobial-resistance/facts/factsheets/experts. (Consulted on 31 May 2020).

European Centre for Disease Prevention and Control (ECDC). *Facts about typhoid and paratyphoid fever*. https://www.ecdc.europa.eu/en/typhoid-and-paratyphoid-fever/facts. (Consulted on 31 May 2020).

European Medicines Agency (EMA). (2017). *Sales of veterinary antimicrobial agents in 30 European countries in 2015. Trends from 2010 to 2015.* Seventh ESVAC report. EMA/184855/2017 https://www.ema.europa.eu/en/documents/report/seventh-esvac-report-sales-veterinary-antimicrobial-agents-30-european-countries-2015_en.pdf.

Gilbert, M., Slingenbergh, J., & Xiao, X. (2008). Climate change and avian influenza. *Revue scientifique et technique (International Office of Epizootics), 27*(2), 459–466.

Griffin, B. D., Chan, M., Tailor, N., Mendoza, E. J., Leung, A., Warner, B. M., Duggan, A. T., Moffat, E., He, S., Garnett, L., Tran, K. N., Banadyga, L., Albietz, A., Tierney, K., Audet, J., Bello, A., Vendramelli, R., Boese, A. S., Fernando, L., Lindsay, R., Jardine, C. M., Wood, H., Poliquin, G., Strong, J. E., Drebot, M., Safronetz, D., Embury-Hyatt, C., & Kobasa, D. (2020). *bioRxiv.* https://doi.org/10.1101/2020.07.25.221291, 2020.07.25.221291.

Harder, A. (2020). *"Citing coronavirus, lawmakers call for a ban on wildlife markets".* Axios. https://www.axios.com/coronavirus-crisis-lawmakers-call-to-ban-wildlife-markets-244e9447-d44c-4a66-add6-8d5311de856f.html?utm_source=newsletter&utm_medium=email&utm_campaign=newsletter_axiosfutureofwork&stream=future.

Huq, A., Small, E. B., West, P. A., Huq, M. I., Rahman, R., & Colwell, R. R. (1983). Ecological relationships between *Vibrio cholerae* and planktonic crustacean copepods. *Applied and Environmental Microbiology, 45*(1), 275–283.

Interagency Coordination Group on Antimicrobial Resistance (IACG). (April 2019). *No time to wait: Securing the future from drug-resistant infections report to the Secretary-general of the United Nations.* https://www.who.int/antimicrobial-resistance/interagency-coordination-group/IACG_final_report_EN.pdf?ua=1.

IPBES. (2019). In E. S. Brondizio, J. Settele, S. Díaz, & H. T. Ngo (Eds.), *Global assessment report on biodiversity and ecosystem services of the intergovernmental science-policy platform on biodiversity and ecosystem services.* Germany: IPBES secretariat, Bonn.

Intergovernmental science-policy platform on biodiversity and ecosystem services (IPBES) media release: "Nature's Dangerous Decline 'Unprecedented' Species Extinction Rates 'Accelerating'.(2020). https://ipbes.net/news/Media-Release-Global-Assessment (Consulted on 10 May 2020).

Jacobs, A. (2019). *"Denmark raises antibiotic-free pigs. Why can't the US?".* The New York Times. https://www.nytimes.com/2019/12/06/health/pigs-antibiotics-denmark.html.

Johnson, C. K., Hitchens, P. L., Pandit, P. S., Rushmore, J., Evans, T. S., Young, C., & Doyle, M. M. (2020). Global shifts in mammalian population trends reveal key predictors of virus spillover risk. *Proceedings Biological sciences, 287*(1924), 20192736. https://doi.org/10.1098/rspb.2019.2736

Jones, K. E., Patel, N. G., Levy, M. A., Storeygard, A., Balk, D., Gittleman, J. L., & Daszak, P. (2008). Global trends in emerging infectious diseases. *Nature, 451*(7181), 990–993. https://doi.org/10.1038/nature06536

Karesh, W. B., Cook, R., Bennett, E. L., & Newcomb, J. (2005). Wildlife trade and global disease emergence. *Emerging Infectious Diseases, 11*(7), 1000–1002. https://doi.org/10.3201/eid1107.050194

Karesh, W. B., Dobson, A., Lloyd-Smith, J. O., Lubroth, J., Dixon, M. A., Bennett, M., … Heymann, D. L. (2012). Ecology of zoonoses: Natural and unnatural histories. *Lancet (London, England), 380*(9857), 1936–1945. https://doi.org/10.1016/S0140-6736(12)61678-X

Krammer, F., Smith, G., Fouchier, R., Peiris, M., Kedzierska, K., Doherty, P., … García-Sastre, A. (2018). Influenza. *Nature Reviews Disease Primers, 4*, 3. https://doi.org/10.1038/s41572-018-0002-y

Kremer, M. (1993). Population growth and technological change: One million B.C. To 1990. *Quarterly Journal of Economics, 108*(3), 681–716. https://doi.org/10.2307/2118405. August 1993.

Ligon, B. L. (2006). Plague: A review of its history and potential as a biological weapon. *Seminars in Pediatric Infectious Diseases, 17*(3), 161–170. https://doi.org/10.1053/j.spid.2006.07.002

Loh, E. H., Zambrana-Torrelio, C., Olival, K. J., Bogich, T. L., Johnson, C. K., Mazet, J. A., ... Daszak, P. (2015). Targeting transmission pathways for emerging zoonotic disease surveillance and control. *Vector Borne and Zoonotic Diseases (Larchmont, N.Y.), 15*(7), 432–437. https://doi.org/10.1089/vbz.2013.1563

Machalaba, C., & Karesh, W. (2020). Fight pandemics like wildfires. *Foreign Affairs*. https://www.foreignaffairs.com/articles/2020-03-06/fight-pandemics-wildfires.

Maron, D. F. (2020). *Wet markets' likely launched the coronavirus. Here's what you need to know.* National Geographic. https://www.nationalgeographic.com/animals/2020/04/coronavirus-linked-to-chinese-wet-markets/.

Mounier, A., & Mirazón Lahr, M. (2019). Deciphering African late middle Pleistocene hominin diversity and the origin of our species. *Nature Communications, 10*(1), 3406. https://doi.org/10.1038/s41467-019-11213-w

Nwanta, J., Anaelom, Ikechukwu, Onunkwo, Ezema, Wilfred, ... Umeononigwe. (2010). Zoonotic tuberculosis: A review of epidemiology, clinical presentation, prevention and control. *Journal of Public Health and Epidemiology, 2*(6), 118–124.

Organisation for Economic Co-operation and Development (OECD). *Global value chains and agriculture*. http://www.oecd.org/agriculture/topics/global-value-chains-and-agriculture/. (Consulted on 03 May 2020).

Olival, K. J., Hosseini, P. R., Zambrana-Torrelio, C., Ross, N., Bogich, T. L., & Daszak, P. (2017). Host and viral traits predict zoonotic spillover from mammals. *Nature, 546*(7660), 646–650. https://doi.org/10.1038/nature22975

One Health Commission (OHC). (September 2015). *History of the OHC*. https://www.onehealthcommission.org/en/why_one_health/history/ (Consulted on 03 August 2020).

Organisation for Economic Co-operation and Development (OECD). *How we feed the world today*. http://www.oecd.org/agriculture/understanding-the-global-food-system/how-we-feed-the-world-today/. (Consulted on 03 May 2020).

Organisation for Economic Co-operation and Development (OECD). *What is the future of food and farming*. http://www.oecd.org/agriculture/understanding-the-global-food-system/what-is-the-future-of-food-and-farming/. (Consulted on 03 May 2020).

Organisation for Economic Co-operation and Development (OECD). *Climate change and the policy implications for agriculture and fisheries*. http://www.oecd.org/agriculture/topics/climate-change-and-food-systems/. (Consulted on 07 June 2020).

Parham, P. E., Waldock, J., Christophides, G. K., Hemming, D., Agusto, F., Evans, K. J., ... Michael, E. (2015). Climate, environmental and socio-economic change: Weighing up the balance in vector-borne disease transmission. Philosophical transactions of the Royal society of London. Series B. *Biological Sciences, 370*(1665), 20130551. https://doi.org/10.1098/rstb.2013.0551

Poore, J., & Nemecek, T. (2018). Reducing food's environmental impacts through producers and consumers. *Science, 360*(6392), 987–992. https://doi.org/10.1126/science.aaq0216

Qiang, Z., Zhang, H., Huang, K., Yang, Y., Hui, X., Gao, J., ... Jin, M. (2020). SARS-CoV-2 neutralizing serum antibodies in cats: A serological investigation. *bioRxiv*. https://doi.org/10.1101/2020.04.01.021196. preprint.

Quammen, D. (2013). *Spillover: Animal infections and the next human pandemic*. W. W. Norton Company.

Richardson, J., Lockhart, C., Pongolini, S., Karesh, W. B., Baylis, M., Goldberg, T., ... Poppy, G. (2016). Special issue: Drivers for emerging issues in animal and plant health. *EFSA Journal, 14*(S1), 11. https://doi.org/10.2903/j.efsa.2016.s0512. s0512.

Ritchie, H., & Roser, M. (2020). *Environmental impacts of food production*. Published online at OurWorldInData.org https://ourworldindata.org/environmental-impacts-of-food (Consulted on 31 May 2020).

Rönnlund, A. R., Rosling, H., & Rosling, O. (2018). Factfullness. *Sceptre*.

Sabbatani, S., & Fiorino, S. (2009). La peste antonina e il declino dell'Impero Romano. Ruolo della guerra partica e della guerra marcomannica tra il 164 e il 182 d.c. nella diffusione del contagio [The Antonine Plague and the decline of the Roman Empire]. *Infezioni in Medicina, Le, 17*(4), 261–275.

Scally, A., Dutheil, J., Hillier, L., Jordan, L., Goodhead, G., Herrero, J., ... Durbin, R. (2012). Insights into hominid evolution from the gorilla genome sequence. *Nature, 483*, 169–175. https://doi.org/10.1038/nature10842

Schlottau, K., Rissmann, M., Graaf, A., Schön, J., Sehl, J., Wylezich, C., Höper, D., Mettenleiter, T. C., Balkema-Buschmann, A., Harder, T., Grund, C., Hoffmann, D., Breithaupt, A., & Beer, M. (July 7, 2020). SARS-CoV-2 in fruit bats, ferrets, pigs, and chickens: an experimental transmission study. *The Lancet Microbe*. https://doi.org/10.1016/S2666-5247(20), 30089–6.

Settele, J., Díaz, S., Brondizio, E., & Daszak, P. (2020). *IPBES expert guest article COVID-19 stimulus measures must save lives, Protect Livelihoods, and Safeguard Nature to Reduce the Risk of Future Pandemics*. https://ipbes.net/covid19stimulus (Consulted on 12 May 2020).

Sharp, P. M., & Hahn, B. H. (2011). Origins of HIV and the AIDS pandemic. *Cold Spring Harbor Perspectives in Medicine, 1*(1), a006841. https://doi.org/10.1101/cshperspect.a006841

Spinney, L. (2020). We need to rethink our food system to prevent the next pandemic. *Time*. https://time.com/5819801/rethink-industrialized-farming-next-pandemic/.

Taubenberger, J. K., & Morens, D. M. (2009). Pandemic influenza—including a risk assessment of H5N1. *Revue scientifique et technique (International Office of Epizootics), 28*(1), 187–202. https://doi.org/10.20506/rst.28.1.1879

Taylor, L. H., Latham, S. M., & Woolhouse, M. E. (2001). Risk factors for human disease emergence. Philosophical transactions of the Royal Society of London. Series B. *Biological Sciences, 356*(1411), 983–989. https://doi.org/10.1098/rstb.2001.0888

United Nations Department of Economic and Social Affairs. (2017). *World population projected to reach 9.8 billion in 2050, and 11.2 billion in 2100*. https://www.un.org/development/desa/en/news/population/world-population-prospects-2017.html (Consulted on 01 May 2020).

United Nations General Assembly. (March 2019). *Seventy-third session, 73/284. United Nations Decade on Ecosystem Restoration (2021–2030)*. https://undocs.org/A/RES/73/284.

United Nations Human Settlements Programme (UN-Habitat). (2003). *The challenge of slums: Global Report on Human Settlements*. https://www.un.org/ruleoflaw/files/Challenge%20of%20Slums.pdf (Consulted on 01 May 2020).

United Nations Human Settlements Programme (UN-Habitat). (2012). *Housing & slum upgrading*. https://oldweb.unhabitat.org/urban-themes/housing-slum-upgrading/ (Consulted on 01 May 2020).

Vuorinen, H. S. (1997). Taudeista antiikin aikana (Diseases in the ancient world). *Hippokrates*, 74–97.

Wegener, H. C., Aarestrup, F. M., Jensen, L. B., Hammerum, A. M., & Bager, F. (1999). Use of antimicrobial growth promoters in food animals and Enterococcus faecium resistance to therapeutic antimicrobial drugs in Europe. *Emerging Infectious Diseases, 5*(3), 329–335. https://doi.org/10.3201/eid0503.990303

World Health Organization (WHO). *Health Topics, Marburg virus disease*. https://www.who.int/health-topics/marburg-virus-disease/#tab=tab_1. (Consulted on 31 May 2020).

Woolhouse, M., Scott, F., Hudson, Z., Howey, R., & Chase-Topping, M. (2012). Human viruses: Discovery and emergence. *Philosophical Transactions of the Royal Society of London - Series B: Biological Sciences, 367*(1604), 2864–2871. https://doi.org/10.1098/rstb.2011.0354

World Health Organization (WHO). *Health Topics, Ebola virus disease*. https://www.who.int/health-topics/ebola#tab=tab_1. (Consulted on 31 May 2020).

World Health Organization (WHO). *Health Topics, Monkeypox*. https://www.who.int/health-topics/monkeypox/#tab=tab_1. (Consulted on 31 May 2020).

World Health Organization (WHO). *Health Topics, Chagas disease (also known as American trypanosomiasis)*. https://www.who.int/news-room/fact-sheets/detail/chagas-disease-(american-trypanosomiasis). (Consulted on 31 May 2020).

World Health Organization (WHO). *Health Topics, Dengue and severe dengue*. https://www.who.int/health-topics/dengue-and-severe-dengue#tab=tab_1. (Consulted on 31 May 2020).

World Health Organization (WHO). *Health Topics, Cholera*. https://www.who.int/health-topics/cholera#tab=tab_1. (Consulted on 31 May 2020).

World Health Organization (WHO). (2018). *Climate change and health*. https://www.who.int/news-room/fact-sheets/detail/climate-change-and-health (Consulted on 15 May 2020).

World Health Organization (WHO). (2020). *A coordinated global research roadmap: 2019 novel coronavirus*. https://www.who.int/publications/m/item/a-coordinated-global-research-road-map (Consulted on 13 August 2020).

World Health Organization (WHO). (2018). *Antimicrobial resistance*. https://www.who.int/news-room/fact-sheets/detail/antimicrobial-resistance (Consulted on 10 May 2020).

World Health Organization (WHO). (2015). *Global action plan on antimicrobial resistance*. https://apps.who.int/iris/bitstream/handle/10665/193736/9789241509763_eng.pdf?sequence=1.

World Health Organization (WHO). (2015). *Biodiversity and health*. https://www.who.int/news-room/fact-sheets/detail/biodiversity-and-health (Consulted on 12 May 2020).

World Health Organization (WHO). (2018). *Health topics, typhoid*. https://www.who.int/news-room/fact-sheets/detail/typhoid (Consulted on 31 May 2020).

World Organisation for Animal Health (OIE). *Rabies portal*. https://www.oie.int/en/animal-health-in-the-world/rabies-portal/. (Consulted on 31 May 2020).

World Organisation for Animal Health (OIE). *Rift Valley fever*. https://www.oie.int/en/animal-health-in-the-world/animal-diseases/Rift-Valley-fever/. (Consulted on 31 May 2020).

World Organisation for Animal Health (OIE). *West Nile fever*. https://www.oie.int/en/animal-health-in-the-world/animal-diseases/West-Nile-fever/. (Consulted on 31 May 2020).

World Organisation for Animal Health (OIE). (2009). *Technical disease card for Nipah*. https://www.oie.int/fileadmin/Home/eng/Animal_Health_in_the_World/docs/pdf/Disease_cards/NIPAH.pdf.

World Organisation for Animal Health (OIE). (2018). *Terrestrial manual health code*.

World Organisation for Animal Health (OIE). (2019). *4th OIE annual report on the use of anti-microbial agents intended for use in animals*. https://www.oie.int/fileadmin/Home/eng/Our_scientific_expertise/docs/pdf/A_Fourth_Annual_Report_AMU.pdf.

World Organisation for Animal Health (OIE). (July 2019). *List of antimicrobial agents of veterinary importance*. https://www.oie.int/fileadmin/Home/eng/Our_scientific_expertise/docs/pdf/AMR/A_OIE_List_antimicrobials_July2019.pdf.

World Organisation for Animal Health (OIE). (May 2020). *OIE news: Special edition on COVID-19*. https://mailchi.mp/oie.int/the-oies-role-in-global-efforts-to-combat-covid-19.

World Wildlife Fund (WWF). *Deforestation and Forest Degradation*. https://www.worldwildlife.org/threats/deforestation-and-forest-degradation. (Consulted on 07 June 2020).

Worobey, M., Han, G. Z., & Rambaut, A. (2014). Genesis and pathogenesis of the 1918 pandemic H1N1 influenza A virus. *Proceedings of the National Academy of Sciences of the United States of America, 111*(22), 8107–8112. https://doi.org/10.1073/pnas.1324197111

Chapter 4

Risks and benefits of the interaction with companion animals

Katia C. Pinello[a,b,c], Chiara Palmieri[d], Joelma Ruiz[e], Maria Lúcia Zaidan Dagli[f] and João Niza-Ribeiro[b,c,g]

[a]*Departamento de Estudo de Populações, ICBAS, Instituto de Ciências Biomédicas Abel Salazar, Universidade do Porto, Porto, Portugal;* [b]*Laboratório para a Investigação Integrativa e Translacional em Saúde Populacional (ITR), Porto, Portugal;* [c]*EPIUnit - Instituto de Saúde Pública, Universidade do Porto, Porto, Portugal;* [d]*The University of Queensland, School of Veterinary Science, QLD, Australia;* [e]*Joelma Ruiz Psychology Center, São Paulo, Brazil;* [f]*Laboratory of Experimental and Comparative Oncology, School of Veterinary Medicine and Animal Science, University of São Paulo, São Paulo, Brazil;* [g]*Departamento de Estudos de População, Vet-OncoNet, ICBAS, Instituto de Ciências Biomédicas Abel Salazar, Universidade do Porto, Porto, Portugal*

1. Domestic animals, companion animals and wild animals

Companion animals (pet animals) within the group of domestic animals, are defined as "domesticated, domestic-bred or wild-caught animals, permanently living in a community and kept by people for company, amusement, work (e.g. support for blind or deaf people, police or military dogs) or psychological support – including dogs, cats, horses, rabbits, ferrets, guinea pigs, reptiles, amphibians, birds and ornamental fish" (Day, 2016a).

However, not all animal species or even breeds are fit for domestic purposes. Wild animals are definitely ancestors of domestic animals, but domestic breeds have taken a different evolutionary pathway from their wild ancestors, through the domestication process which has an estimated beginning date 15,000 years B.P. for the dog (Driscoll, Macdonald, & O'Brien, 2009). Domestic animals are crucial to modern human society and they have adapted and progressively changed to be kept for human purposes (National Geographic Society, 2020), with the first domesticated animal being the dog (Axelsson et al., 2013). For many centuries, breeds have been specifically selected to serve as work animals, food source or companion animals. Domestication may be easily confused with taming. Domestication implies a permanent genetic

One Health. https://doi.org/10.1016/B978-0-12-822794-7.00012-5

modification of a breed lineage. Taming is the conditioned behavioral modification of a wild-born animal that is becoming relatively tolerant to the human presence (Driscoll et al., 2009).

More specifically, humans and dogs have established an ongoing partnership for thousands of years now. The domestication of wolves, and the subsequent genetic transformations that dogs underwent until they reached the breed diversity as of today, is the result of a process that evolved for approximately 11000 years and would have occurred simultaneously with the beginning and evolution of agriculture (Axelsson et al., 2013).

Such a domestication process has most likely started with the humans offering food to the animals and receiving, in return, protection against predators. In this relationship, the most docile species was "adopted," which made dogs increasingly domesticable through a natural selection process to the point of behaving like today's dogs (Axelsson et al., 2013).

The "domestication syndrome" is also most likely associated with the progressive acquisition of several behavioral and phenotypic traits differentiating modern dogs from wolves, such as reduced aggressiveness, altered social cognition capabilities, reduced skull, teeth and brain sizes (Axelsson et al., 2013). In a whole genome resequencing study conducted on dogs and wolves, 36 genomic regions were identified as possible target for selection during the domestication process. Nineteen of these regions contained genes important to the cerebral function, in particular pathways of the nervous system development, that may explain the behavioral changes associated with dog domestication. Ten genes with an important role in starch digestion and fat metabolism also showed mutations providing support for an increased starch digestion occurring in dogs compared to wolves, a process being considered a crucial step in the early domestication of dogs (Axelsson et al., 2013).

Dogs have been bred for different purposes to help human communities in improving daily tasks like hunting, shepherding livestock, guarding houses or properties, companionship (Axelsson et al., 2013). Hundreds of canine breeds are registered around the world. Nowadays, dogs are primarily kept offering companionship to their human counterparts, living in residential buildings and in urban areas. Dogs have also been involved in other services, such as animal assisted interventions (AAI) or sentinel tool for disease surveillance, which will be discussed later in this chapter.

Domestic cats are the other popular companion animals, but the process of their domestication is not well known (Hu et al., 2014). The process of feline domestication of eventually started when cats were needed as rodent and small animal predators in human settlements over 10,000 years ago (Hu et al., 2014).

Capture, breed and release of wild animals intended to be kept as pets is also an issue of great concern. Companion animals should have specific physical, emotional and behavioral characteristics that allow them to adapt easily and quickly to the living conditions of the human urban life. Examples of pets that should not be considered as companion animals include snakes, primates, rodents and reptiles. Exotic pets, defined as all animals kept as companion animals excluding dogs, cats and horses, makes up between 34% and 64% of the pet population (Schuppli, Fraser, & Bacon, 2014). Keeping and trading exotic companion animals is a controversial issues due to animal welfare, public health and conservation concerns (Pasmans et al., 2017). Once kept as pets, it is extremely difficult to provide the adequate welfare conditions to these animals.

Capturing wild species also contributes to the interference and sometimes the disruption of the original ecosystem (Europe Union, 2017). Known as "alien species," these animals have been deliberately introduced through human into a new natural environment from other parts of the world. The Pallas and gray squirrel, raccoon dog, Siberian chipmunk are examples of the introduction of invasive species in the ecosystem of different continents or regions with great damage to the autochthone fauna and flora (Europe Union, 2017).

Due to the high diversity of companion animals nowadays and because dogs and cats are still considered the main companion animals, this chapter reflects a revision of recent studies on the beneficial aspects of dogs and cats, at the same time highlighting the emerging risks of this growing relationship.

2. The contemporary role of companion animals

In the last decades, companion animals – specifically dogs – are increasingly designated as "man's best friends." In an era where the concept of family is quite broad and diverse, these four-legged "man's best friends" are considered more than friends, becoming instead true family members. This level up is reflected by the impressive number of the population of cats and dogs worldwide, as well as the continuous growth of pet services and products. According to a report by the European Federation of the Pet Food Industry (FEDIAF, 2019), in 2019 the pet services and products sector was valued at 19.7 billion euros, with a billing of 21 billion euros in pet food, 8.7 billion spent on accessories and 11 billion on services with an average annual growth of 2.6% and annual sales of 8.5 million tons of feed. Thirty-eight percent of European households have at least one pet for a total of 66 million cats, 61 million dogs, 39 million ornamental birds, 6 million horses and 9 million aquaria. The food sector alone has been able to employ 100,000 people directly and 900,000 indirectly in Europe (Day, 2016a; FEDIAF, 2019). The change in attitude and perception of the companion animals makes the concept of One Health increasingly important for human health. Prof. Michael J. Day (*in memoriam*), founder of the World Small Animal Veterinary Association's (WSAVA) One Health Committee and great passionate of the One Health approach, wrote that: "In a viable One Health partnership, physicians and veterinarians would be known to each other and communicate professionally for the benefit of the families they serve" (Day, 2016b). In a very recent revision addressing the question *Do companion animals bring us more benefits or risks?* the authors concluded that companion animals indeed and above all have a positive effect on human health (Overgaauw, Vinke, Hagen, & Lipman, 2020). However, this change of attitude toward companion animals being family members, and thus living together and closer to humans can at the same time increase existing risks and/or create new ones.

3. Human–animal bond–*Zooeyia*

Throughout the years, the relationship between humans and animals, mainly dogs and cats, has increasingly become stronger, thus providing enormous benefits for both species and promoting a relationship of greater respect and complicity. The coexistence of humans and domestic animals is quite significant and occupies a distinct economic and cultural place in societies of all times. More than 90% of dog's and cat's owners expressed that their pets make them happy (Overgaauw et al., 2020).

In the last 25 years, scientific studies on Human–Companion Animal Bond (HCAB) have progressively increased in number, making this a multidisciplinary field of great interest (Takashima & Day, 2014). HCAB has been defined as "the dynamic relationship between people and animals such that each influences the psychological and physiological state of the other" (College of Veterinary Medicine). Being aware that more people own a companion animal, in particular in large urban centers, the power of the human-animal bond is extremely important, and this emotional connection has numerous benefits for the mental and physical health (Friedman & Krause-Parello, 2018; Takashima & Day, 2014). Therefore, a new concept was created: *Zooeyia*. It is the opposite of zoonosis, since Zooeyia is focused on the positive benefits of animal-human bond (Hodgson & Darling, 2011).

4. Benefits for the human health

Health may be defined as "a state of complete physical, mental and social well-being and not merely the absence of disease or infirmity" (International Health, 2002). Pets may contribute to the maintenance of a good health status in humans, hence promoting health. They may also contribute to educate children, prevent mental diseases or assist in therapeutic treatments of different medical conditions.

4.1 Psychological effects

Despite the low mortality rate associated with mental disorders – depression, bipolar disorder, schizophrenia and other psychoses, dementia, and developmental disorders including autism – these diseases have significant deleterious consequences on the quality of life, substantially affecting the Disability Adjusted Life Years (DALY) and, especially, the Years Lived with Disability (YLD), two recent objective indicators to measure the burden of disease (WHO, 2019). There are effective strategies for preventing mental disorders and pet animals play an important role in this field (WHO, 2019).

A pet is defined as such, regardless of the species it belongs to, as providing "fun and companionship" to humans (Hodgson and Darling, 2011). The significant increase in the levels of oxytocin (the "attachment hormone") of tutors when watched by their pets provides scientific support to this feeling of bonding, which contributes to a state of psychological well-being (Hodgson & Darling, 2011).

Several studies (Table 4.1) have shown, in general, that companion animals are associated with reduced anxiety levels in humans, as suggested by cardiovascular, behavioral and psychological indicators. Several researchers have explored these beneficial effects, concluding that they can be achieved by physical contact but also simply by eye contact with pets (Friedman & Krause-Parello, 2018; Nagasawa, Kikusui, Onaka, & Ohta, 2009).

The bond between man and companion animals, in addition to the feeling of the connections experienced, contributes also to the reduction of loneliness and isolation or depression, which seem highly important in certain stages of life, such as old age (Hui Gan, Hill, Yeung, Keesing, & Netto, 2019; Wood, Giles-Corti, & Bulsara, 2005). Studies carried out in this area of human health have shown less depressive signs in individuals, as well as increased self-esteem and perceived well-being (Friedman & Krause-Parello, 2018; Purewal et al., 2017).

Companion animals play also a fundamental role in humans social life, not only because of the connections they develop with their tutors, but also by facilitating socialization with other human beings, considering that less than 5% of the 48 million dogs and 7% of the 27 million cats kept as pets have tutors who live alone (Krause-Parello, 2012). The increase in socialization is reflected in an improvement of the community relations, and in an increase of individual health, due to the direct impact on the social well-being and the secondary repercussions on mental and/or physical health mainly in older adults, people with autism spectrum disorders, cancer, or in wheelchairs (Hodgson & Darling, 2011). This positive relationship transcends individual differences like gender, race, religion (Jacobson & Chang, 2018).

TABLE 4.1 Recent studies on the effects of companion animals on human health.

Health aspects	Psychological effects	Reference
Sense of belonging	Oxytocin ("attachment hormone") levels in pet owners increases significantly when their dogs gaze at them.	(Nagasawa et al., 2009)
Loneliness and socialization	"Pets emerged as an inferred antidote to loneliness — not just as direct companions but by virtue of the social contact and interactions."	(Wood et al., 2005)
	Companion animal ownership is associated with less loneliness in adult women.	(Krause-Parello, 2012)
	Pet ownership may benefit community-dwelling older adults by providing companionship, giving a sense of purpose and meaning, reducing loneliness and increasing socialization. These benefits may also increase resilience in older adults against mental health disorders, which may positively influence their mental health outcomes.	(Hui Gan et al., 2019)
Depression	Among older adults in England, those with more depressive symptoms are more likely to own a dog.	(Sharpley et al., 2020)
	Depressive signs decrease in individuals with pets.	(Friedman & Krause-Parello, 2018)
Child/ adolescent development	Owning pets may provide children with opportunities to control their emotions, and lead to a lower prevalence of poor emotional expression. Pet ownership in toddlerhood may contribute to the development of self-expression.	(Sato, Fujiwara, Kino, Nawa, & Kawachi, 2019)
	Increase social competence, network, interaction and play behavior. Less absenteeism from school. Self-esteem and perceived well-being.	(Purewal et al., 2017)

TABLE 4.1 Recent studies on the effects of companion animals on human health.—cont'd

Health aspects	Psychological effects	Reference
Adolescents with autism spectrum disorder (ASD)	Taking care of a pet may facilitate better social-emotional adjustments among adolescents with ASD.	(Ward, Arola, Bohnert, & Lieb, 2017)
Lack of animals mourning	Post-traumatic stress disorders due to loss of companion animals. Symptoms such as deep pain, sadness, emptinaess, longing and bitterness are frequent.	(Tzivian, Friger, & Kushnir, 2015; Adrian, Deliramich, & Frueh, 2009; Reisbig, Hafen, Siqueira Drake, Girard, & Breunig, 2017)
Health effects	**Physical activity and health outcomes**	
Cardiovascular (CV)	Using a large national database, pet ownership is associated with a decreased prevalence of systemic hypertension.	(Krittanawong et al., 2020)
	Pet ownership protects against the development of hypertension in children.	(Yeh et al., 2019)
	Pet ownership is associated with a lower adjusted CV mortality in the general population and a lower CV disease risk in patients with established cardiovascular diseases (CVD).	(Yeh et al., 2019)
	Pet ownership is associated with an improved cardiovascular disease survival rate in a treated elderly hypertensive population.	(Chowdhury et al., 2017)
Physical activity	Dog owners could be more likely to exercise by walking their dogs.	(Mein & Grant, 2018)
	Pet owners show better sleep patterns, therefore feeling better and healthier. The most plausible explanation, proposed by the authors, is the greater practice of exercise	(Headey et al., 2008)

Continued

TABLE 4.1 Recent studies on the effects of companion animals on human health.—cont'd

Health effects	**Physical activity and health outcomes**	
Obesity	No direct association between pet ownership and obesity.	(Miyake et al., 2020)
Smoking	Concerns about second-hand smoke affecting pets has been found to be a strong motivator to stop smoking, to stop smoking at home, and to encourage other household members to quit smoking.	(Milberger et al., 2009)
Development of the immune system	Having an animal during the first year of life may protect against asthma and allergy	(Purewal et al., 2017)
Microbial diversity	Pet ownership is associated with differences in the human gut microbiota.	(Kates et al., 2020)
	Close interactions with pets appear protective against the recurrence of community-acquired *Clostridium difficile* infection (CDI)	(Redding et al., 2020)
Health effects	**Animal-assisted interventions**	
Cancer treatment	Called "pet effect" pet AATs are considered a valuable addition to the care for cancer patients therapy, given their effect on emotional well-being.	(Chan & Tapia Rico, 2019)
Nursing home	Visiting dogs in nursing home residents helps to improve mood and decreases distress, depression, dementia and loneliness	(Vegue Parra, Hernandez Garre, & Echevarria Perez, 2021)
Emergency department	Reduction of patient anxiety in the emergency department	(Kline, Fisher, Pettit, Linville, & Beck, 2019)
Surgery	Reduction of children's pain perception after surgery	(Sobo, Eng, & Kassity-Krich, 2006)
Mental disorders	Improvement of the state of children and adults with mental disorders, such as attention deficit hyperactivity disorder (ADHD) and autism spectrum disorder (ASD).	(Hoagwood, Acri, Morrissey, & Peth-Pierce, 2017)

TABLE 4.1 Recent studies on the effects of companion animals on human health.—cont'd

Health effects	Animal-assisted interventions	
Children's reading disorders	Dogs from AAE support children with poor reading skills and who are reluctant to read aloud.	(Lane, 2013)
Post-traumatic stress disorder (PTSD)	AAI is a complementary treatment option for traumas.	(O'Haire et al., 2015)
	Treatment of PTSD in military veterans	(van Houtert, Endenburg, Wijnker, Rodenburg, & Vermetten, 2018)

4.1.1 The importance of companion animals mourning

As an integral part of the family, pets are therefore involved in a caring relationship, which encourages attachment behavior, defined as "any form of behavior that results in a person reaching and maintaining proximity to some other individual, considered better able to deal with the world" (Bowlby, 2002). In this context, the relationship between humans and animals is part of the Theory of Attachment. Although it was originally used in the context of human relationships, this concept can be applied to this interspecies relationship since the main function of the attachment behavior is exactly protection.

"Source of unlimited love, affection and companionship" – these are the most commonly highlighted aspects of the relationship between humans and pets in the literature. The reason is that our relationship with other human beings may be deep and rewarding, but it is still subject to whims, moods, different obligations and pressures of everyday life. Companion animals, instead, are always there, loving and willing to give and receive affection. However, when companion animals die, a gap in the owner's life opens, with suffering, pain and loss emerging, risks that are as inevitable as the attachment to the animal (Oliveira & Franco, 2015).

Such suffering is natural and healthy in the face of breaking an emotional bond, thus initiating the grieving process. However, these feelings are often not recognized either by the society or even by the main caregiver and their family members. With the lack of welcome and space to express this gap, owners experience their pain without social support, considering a protective factor. As a consequence, psychosomatic illnesses can develop (Casselato, 2015; Adrian, Deliramich, & Frueh, 2009; Reisbig et al., 2017; Tzivian, Friger, & Kushnir, 2015).

This does not only apply to the owners, but the health team involved in animal care deals with the social pressure of not being allowed to mourn their patients. As a result, they develop behaviors such as lack of empathy and absence from work, which distance them from communicating with owners and aggravate the risk of developing physical and/or psychological disorders in the medium and long term, creating a vicious circle that needs to be broken (Adrian et al., 2009; Reisbig et al., 2017; Tzivian et al., 2015).

Such unpreparedness of veterinarians in the face of the owners grief underlines the need for a better psychological and emotional training of the team. When communicating with family members, the veterinary medical team needs to be prepared to deal with the emotions of others and, at the same time, take their own emotions and feelings under control and this double weight may be challenging. Therefore, it is extremely important to focus the attention on the emotional reception and information about the grieving process, actions that can be approached by the health team as a preventive measure (Casselato, 2015; Reisbig et al., 2017).

4.2 Physical activity and health outcomes

There is ample evidence that human–animal interaction (HAI) is associated with health improvement in a number of ways, as outlined in Table 4.1.

Sharing life with a pet leads to a decreased risk of coronary artery disease (Krittanawong et al., 2020), decreased stress levels and increased physical activity (mainly through walking with dogs) (Headey, 1999; Mein & Grant, 2018). For example, considering the specific case of older adults affected by cardiovascular diseases and undergoing treatments for hypertension, having a companion animal has been associated with longer survival and lower blood pressure during stressful activities (Yeh, Lei, Liu, & Chien, 2019).

Among other markers, β-endorphin, oxytocin and dopamine increase in both humans and dogs during affectionate interactions (Nagasawa et al., 2009). At the same time, circulating levels of cholesterol and triglycerides decrease. Currently, the most supported theory to explain this phenomenon is based on the decrease in the autonomic nervous response, together with a better vascular function, resulting from improved mental well-being, increased physical exercise, and most likely decrease in body weight (Arhant-Sudhir, Arhant-Sudhir, & Sudhir, 2011). Therefore, it is widely accepted that the time that human and their pets spend together is physiologically beneficial for both species.

Pets play also a positive role on smokers who may be concerned for the health of their animals and therefore are more inclined to quit smoking (Hodgson & Darling, 2011; Milberger, Davis, & Holm, 2009). Primary and second-hand exposure to tobacco smoke is a recognized risk factor for many diseases, including cancer, and cardiovascular and respiratory diseases in human and animals (Zierenberg-Ripoll et al., 2018). In a smoking cessation study, concerns about second-hand smoke affecting pets was found to be a strong motivator to stop smoking at home and to encourage other household members to quit smoking (Milberger et al., 2009). This pilot study provided a new support for health care professionals to amplify smoking cessation messages, of particular significance for smokers who live alone with a pet. Veterinarians could be an important influence on smoking cessation (Milberger et al., 2009).

Another important effect concerns the development of the immune system. Exposure to dogs and cats during childhood is correlated with lower prevalence of allergic and asthmatic diseases (Ownby, Johnson, & Peterson, 2002). However, this is a controversial field of research. In previous studies, exposure to dogs and cats during infancy has been thought to increase the risk of subsequent allergy to these animals (Lindfors, van Hage-Hamsten, Rietz, Wickman, & Nordvall, 1999; Ownby et al., 2002). However, other studies have suggested that this type of exposure reduces risk of allergic disease (Hesselmar, Aberg, Aberg, Eriksson, & Bjorksten, 1999; Ownby et al., 2002) and, also, children growing up on farms with animals were less likely to be allergic than children growing up in urban environments (Riedler, Eder, Oberfeld, & Schreuer, 2000). Some studies suggest that the protective effect of dogs and cats is related to increased exposure to bacterial endotoxin associated with household pets preventing the allergic sensitization that is normally induced by the allergens (Gereda et al., 2000; Ownby et al., 2002).

Other hypothesis relies on the diversity of the gastrointestinal microbiome which is becoming an emerging field of study due to its influence on health and diseases. The population structure of microorganisms and their associated genes may be influenced by environmental factors, such as pet ownership (Kates et al., 2020). In the early 1980s, one study demonstrated that pets and their owners share common intestinal bacteria (Caugant, Levin, & Selander, 1984). In more recent studies, children who live with pets experience increased richness and diversity of the gut microbiome and decreased risk of developing atopic diseases, allergies and metabolic disorders (Ownby et al., 2002; Tun et al., 2017).

As an indirect effect of the increase of activity level, owners seem to have lower number of annual medical appointments and a lower probability of medications required, especially later in life. This reflects into decreased costs of the health system (Headey, 1999; Hodgson & Darling, 2011; Siegel, 1990) as demonstrated in a survey study in Australia with approximately $1 billion of savings for the financial year 1994–1995 (Headey, 1999) and in UK with

£2.45 billion of savings per year (Hall, 2017). Another study concluded that dog owners in China miss fewer days of work due to illness compared to people without a pet (Headey, Na, & Zheng, 2008).

Thus, companion animals may play an important role in reducing the incidence and therefore mortality of different conditions that are strictly related to modifiable risk factors, such as arterial hypertension, hypercholesterolemia, hypertriglyceridemia, diabetes, physical inactivity, overweight, smoking and alcoholism. These effects can then influence the DALYs.

These beneficial effects of companion animals may increase in popularity, becoming a hot topic for current and future research studies, although a comprehensive research approach may be required before attributing a "clinical role" to companion animals (Silva, 2020).

4.3 Animal assisted interventions

Animal Assisted Intervention is delivered in different ways, across a variety of settings and a range of domesticated species. Animal Assisted Interventions include human-animal teams in formally recognized human services such as Animal Assisted Therapy (AAT), Animal Assisted Education (AAE) and Animal Assisted Activity (AAA) (Society of Companion Animals Studies, 2019).

The AAT with the planned inclusion of an animal in a patient's treatment plan is increasingly adopted in hospitals as a co-adjuvant therapy, which involves a trained animal (mainly dogs) to improve patients' health. It is a recent approach mainly applied in the field of cancer, pediatric oncology, trauma and other behavioral and mental disorders listed on Table 4.1 (Chan & Tapia Rico, 2019; Fine, 2019; O'Haire, Guerin, & Kirkham, 2015).

Since 1980s, AAE is applied in many countries with teachers introducing animals into their classes for promoting cognitive, physiological and social skills through interaction with animals. Having an animal in the classroom positively influences children's social behavior, socio-emotional competence and empathy and improves the classroom environment, motivation and discipline (Nakajima, 2017). Systematic guidelines for introducing animals into the classroom have been developed (The International Association of Human-Animal Interaction Organizations (IAHAIO), 2014) and AAE has been used in different programs. The canine-assisted reading program support children with poor reading skills with many global initiatives such as the "Reading Education Assistance Dogs (READ)," "All Ears Reading and Share program," "Literacy Education Assistance Pups (LEAP)" and "Paws to Read" (Lane, 2013; Nakajima, 2017).

AAA is the involvement of domestic animals in recreational and visitation programs to help people with special needs in a wide range of interventions, such as visiting hospitals and long-term health care facilities, "cell dog" programs in prisons, "wounded warrior" therapy for veterans with ADHD (Mossello et al., 2011; Romero, Cepeda, Quinn, & Underwood, 2018).

5. Comparative medicine

Companion animals have become "new best friends of translational scientists" (Kol et al., 2015) since naturally occurring diseases in companion animals might better reflect the complex genetic, environmental, and physiological variations of human diseases compared to laboratory animals (Macy & Horvath, 2017). Thus, this field is gaining particular attention from the perspective of using companion animals in clinical studies and trials to improve human therapies (LeBlanc & Mazcko, 2020) to the concept of disease surveillance using animals as sentinels for environmental risks (Ali et al., 2013; Aslan, Saini Stockman & Ryan, 2020; Backer, Grindem, Corbett, Cullins, & Hunter, 2001; Pastorinho & A.C.A, 2020; Pastorinho & Sousa, 2020; Poma, G, and, Covaci, 2020; Reif, 2011; Rial-Berriel, H.-H. L. A., Luzardo 2020; Schmidt, 2009; Shan & Tan, 2017). Besides these advantages, animals have shorter life span and shorter disease latencies, particularly important with chronic diseases, thus enabling faster observations and accelerated outcomes (Pinello, Niza-Ribeiro, Fonseca, & de Matos, 2019).

In the oncology field, cancers in companion animal may provide useful insights into genetics, carcinogenesis, progression, immunology and therapy of human cancers (Di Cerbo, Palmieri, De Vico, & Iannitti, 2014; Garden, Volk, Mason, & Perry, 2018; Pinho, Carvalho, Cabral, Reis, & Gartner, 2012). It is estimated that more than 40 million of new canine cancer cases are diagnosed every year elevating the dog as a powerful model (Kol et al., 2015). In 2007 the US National Cancer Institute launched the Comparative Oncology Trials Consortium (COTC), which is an active network of 20 academic comparative oncology centers working in designing and executing multi-institutional clinical trials in dogs to assess novel therapies for human cancer patients (Gordon, Paoloni, Mazcko, & Khanna, 2009) (LeBlanc and Mazcko, 2020).

Osteoarthritis (Black et al., 2007), spinal cord injury (Levine, Levine, Porter, Topp, & Noble-Haeusslein, 2011; Safra et al., 2013), keratoconjunctivitis sicca (Williams, 2018), chronic inflammatory disorders (Cerquetella et al., 2010), cardiomyopathies (Fox, Basso, Thiene, & Maron, 2014), diabetes, neurodegenerative diseases (Levine et al., 2011), polycystic kidney disease (Brizi, Benedetti, Lavecchia, & Xinaris, 2019) are additional examples of studies conducted in companion animals with the aim of improving human health (Kol et al., 2015; Stroud et al., 2016).

5.1 Animals as sentinels of environmental contamination

The concept that animals can be sentinels of environmental hazards to public health is not new (Backer, Grindem, Corbett, Cullins, & Hunter, 2001; Reif, 2011). The image of the canary at the entrance of coal mines to detect levels of carbon monoxide described in the last century (1914) (Burrell & Seibert, 1914) still remains a strong representation of how important animals are in this field (Reif, 2011). Animals in different habitats can be used to monitor environmental hazards (Reif, 2011; Shan Neo & Tan, 2017). In 2006, the phenomenon "birds falling from the sky" was the alert for a case of environmental poisoning in Australia (Gulson et al., 2009). The neurological disorder known as "dancing cat fever" represented a wake-up call when people began to experience similar symptoms followed by the discovery of mercury contamination of the water from local factories in Minamata, Japan (Takeuchi, D'Itri, Fischer, Annett, & Okabe, 1977).

Through sharing the same environment, companion animals are potentially exposed to several environmental agents that are considered harmful to people (Pastorinho & Sousa, 2020; Pinello et al., 2019; Reif, 2011). Considering the similarities between human and animal diseases, including etiology and pathogenesis, it is easily inferred that animals can be seen as sensitive indicators of environmental hazards, thus providing an early warning system for public health interventions that may be more effective and closer to reality, compared to laboratory studies of animal experimentation (Reif, 2011). The advantages of using domestic animals as sentinels or comparative models of human diseases are to some extent due to the relative absence of competing exposures, bias or confounders as those commonly present in humans (Reif, 2011). For example, the influence of tobacco, alcohol or occupational exposures can mask or mimic the harmful effect of exposure to other environmental contaminants (Reif, 2011). Since the accuracy of the exposure assessment is one of the greatest challenges of environmental epidemiology, it can be argued that the restricted mobility, the low frequency of migrations and travel and the short life expectancy of animals are additional assets for the efficiency and credibility of the model (Reif, 2011).

The effects of exposure to xenobiotics (chemical compounds that are foreign to and absorbed by an organism or biological system) (Kumar, Abbas, Aster, & Perkins, 2018), environmental pollution, heavy metals, organochlorine compounds, mercury, pesticides and other contaminants have been widely documented in these animals (Pastorinho & Sousa, 2020; Reif, 2011). Companion animals can also help to identify early food contamination, infectious disease transmission, environmental contamination, bio or chemical terrorism events (Schmidt, 2009) (Table 4.2).

TABLE 4.2 Summary of recent findings on the role of companion animals in comparative medicine and as sentinels of diseases and environmental hazards.

Comparative medicine		References
Cancer	Improved knowledge of cancer biology and accelerated development of novel anti-cancer therapies for non-Hodgkin's lymphoma, osteosarcoma, mammary carcinoma, melanoma, gliomas, urothelial carcinoma and soft tissue sarcoma.	(Bentley, Ahmed, Yanke, Cohen-Gadol, & Dey, 2017; Garden et al., 2018; Knapp et al., 2019)
Musculoskeletal diseases	Studies in dogs contribute to improving treatment of chronic osteoarthritis.	(Black et al., 2007)
Spinal cord injuries	Canine models of intervertebral disk herniation and neural tube defects	(Levine et al., 2011; Safra et al., 2013)
Inflammatory bowel disease	The parallel study of IBD in animals could also lead to important information in man	(Cerquetella et al., 2010)
Cardiomyopathies	Primary myocardial diseases occurring spontaneously in domestic cats are remarkably similar to restrictive nondilated and nonhypertrophic cardiomyopathy in man.	(Fox et al., 2014)
Domestic animals as sentinels		
Indoor pollutants	Metals, persistent organic pollutants, flame retardants, and polycyclic aromatic hydrocarbons	(Poma et al., 2020)
Polycyclic aromatic hydrocarbons (PAHs)	Benzo(a)pyrene (BaP)	(Rial-Berriel et al., 2020)
Neurotoxic metals	Mercury, lead, and cadmium in pets Fur	(Pastorinho & A.C.A., 2020)
Pesticides	Herbicides, rodenticides, pesticides mainly in water and food	(Aslan et al., 2020)
Organic contaminants associated to common household dust.	Indoor products, textiles, building materials, flame retardants, plasticizers, and surfactants.	(Weiss & Jones, 2020)

Continued

TABLE 4.2 Summary of recent findings on the role of companion animals in comparative medicine and as sentinels of diseases and environmental hazards.—cont'd

	Comparative medicine	References
Human infectious disease	Dogs could be excellent sentinels for certain infectious pathogens as Chikungunya virus, West Nile virus, Lyme borreliosis, *Rickettsia* spp., *Ehrlichia* spp., and *Dirofilaria immitis*	(Bowser & Anderson, 2018)
Scent detection		
Fecal contamination of foods	Consumption of raw produce commodities has been associated with foodborne outbreaks and dogs can improve the detection of contaminations.	(Partyka et al., 2014)
Breast cancer	The superior olfactory ability of dogs can discriminate the odor originating from metabolic waste produced by cancer cells compared to benign cells	(Seo et al., 2018)
Colorectal cancer	In this study, the sensitivity and specificity of canine scent detection of stool samples was 0.97 and 0.99, respectively. The accuracy of canine scent detection was high even in case of early cancer.	(Sonoda et al., 2011)
Prostate cancer	Detection of prostate cancer specific volatile organic compounds in urine samples with high sensitivity and specificity	(Sonoda et al., 2011)
Skin cancer (melanoma)	Dogs can be trained to select odors associated with melanomas which produce different odors from those associated with basal cell carcinoma (BCC), benign naevi and healthy skin.	(Elliker & Williams, 2016)
Hypoglycemia	Diabetes alerting dogs (DADs) have shown capacity to perceive blood glucose variations.	(Lippi & Plebani, 2019)

5.2 Scent detection

Another emerging role of dogs is their ability to be effective "scent detectors" for a quicker identification of human diseases. Some studies have shown the unique canine's olfactory acuity to distinguish persons with or without diseases like different types of cancer (Elliker & Williams, 2016; Seo et al., 2018; Sonoda et al., 2011), malaria (Guest et al., 2019), diabetes (Lippi & Plebani, 2019), fecal contamination of foods (Partyka et al., 2014) bacterial (Taylor et al., 2018), and viral infections (Angle et al., 2015), as well as more recently COVID-19 infection (Jendrny et al., 2020) (Table 4.2).

The human body emits hundreds of volatile organic compounds (VOCs) and these components usually reflect the metabolic condition of an individual (Shirasu & Touhara, 2011). Therefore, contracting an infectious or metabolic disease often results in the change of the body odor (Shirasu & Touhara, 2011). Specific VOCs are produced during particular diseases and can be used as diagnostic olfactory biomarkers as mentioned above, detected by trained dogs.

Biomedical scent detection has many advantages, such as relative cost-effectiveness, safety and the sensitivity and specificity of the dog's nose. However, this method has also various challenges, in particular its clinical effectiveness and the standardization of the training techniques, both intra- and inter-dog reproducibility (Hackner & Pleil, 2017; Reeve, 2018).

6. Risks for human health

Living with a companion animal is not free of risks. Companion animals and human share much of their evolutionary history and therefore, they can act as hosts for pathogenic organisms, which may be readily transmitted from one species to another. Therefore, the infections transmitted by pets are a potential cause of serious diseases in owners (Sterneberg-van der Maaten, Turner, Van Tilburg, & Vaarten, 2016).

6.1 Zoonotic diseases

The changing attitude to companion animals and their environment can increase the risk of transmission of zoonotic infections. This is mainly associated with increased proximity and specific habits such as sleeping with pets, allowing pets to lick the face or wounds and bite accidents (P. A. M. Overgaauw et al., 2020).

Zoonosis is referred to "any disease or infection that is naturally transmitted from vertebrate animals to man," as defined by the World Health Organization (WHO). It is estimated that zoonoses are responsible for one billion cases of illness and millions of deaths worldwide and 60% of emerging infectious diseases that are reported globally are actually zoonotic (WHO).

From a One Health perspective, companion animals can serve as sources of zoonotic diseases, through direct contact, transmission by vectors or as intermediate hosts between wildlife reservoirs and humans, thus being sentinels for emerging disease surveillance (Chomel, 2014; Day, 2016b; Day et al., 2012; Halliday et al., 2007). The main zoonoses of dogs and cats are summarized in Table 4.3, with a total of 37 diseases listed and more than half caused by bacteria.

Reverse zoonoses (zooanthroponoses) occur when diseases are transmitted from the human reservoir to the animal (Table 4.3) (Day et al., 2012; Messenger, Barnes, & Gray, 2014; Overgaauw et al., 2020) and they are considered a worldwide disease threat since described in every continent except Antarctica (Table 4.3).

6.2 Antimicrobial resistance

There is increasing concern about the rapid emergence of antimicrobial resistance (AMR) which is the resistance of microorganisms to an antimicrobial agent to which they were at first sensitive. The issue of AMR in household companion animals has increased substantially in the last few years due to better veterinary care, increased close cohabitation and strengthened by the genetic similarities between MDR isolates from human infections and household animals (Damborg et al., 2016; Leite-Martins et al., 2014, 2015; Walther, Tedin, & Lubke-Becker, 2017).

Methicillin-resistant *Staphilococcus aureus* (MRSA) is the perfect prototype, since humans can contract this infection by contact with pets, that represent a reservoir of this pathogen for humans (Bouchiat et al., 2017). In contrast, there are evidences that companion animals can acquire MRSA from individuals since the MRSA types isolated from dogs and cats often correspond to the clones observed in the human population (Vincze et al., 2014). In addition to the risks of zoonotic transmission, MDR infections in companion animals have a negative emotional and social impact on the owners and their families (Bengtsson & Greko, 2014; Damborg et al., 2016).

Thus, a more rational antimicrobial use is an essential measure to prevent further spread of MDR bacteria in companion animals. Veterinarians play an important role in educating the owners to follow best hygiene practices, to choose the right antimicrobials, to avoid the critical ones and to perform antimicrobial susceptibility testing whenever possible.

TABLE 4.3 Zoonotic diseases of dogs and cats.

Route of human exposure	Disease	Agent	Type	References
Direct contact				
Bites, scratches, or contact with exudates	Cat-scratch disease[a]	*Bartonella henselae*	Bacterium	(Baranowski & Huang, 2020)
	Wound infections	*Capnocytophaga canimorsus*	Bacterium	(Zajkowska, Krol, Falkowski, Syed, & Kamienska, 2016)
	Cat bites	*Pasteurella multocida*	Bacterium	(Overgaauw et al., 2009)
	Tularemia	*Francisella tularensis*	Bacterium	(Petersson & Athlin, 2017)
	Dermatophytosis (ringworm)	Dermatophytes	Fungi	(Day, 2016b)
	Rabies[a,b,c]	*Lyssavirus*	Virus	(Fooks et al., 2017)
	Sporotrichosis	*Sporothrix schenkii*	Fungus	(Gremiao, Miranda, Reis, Rodrigues, & Pereira, 2017)
	Scabies	*Sarcoptes scabiei*		(Day, 2016b)
Indirect contact				
Contact with infected feces	Campylobacteriosis[a,c]	*Campylobacter jejuni* and *C. coli*	Bacteria	(Acke, 2018)
	Escherichiosis	*Escherichia coli*	Bacterium	(Damborg et al., 2016)
	Salmonellosis[a,c]	*Salmonella* spp.	Bacterium	(Day, 2016b)
	Helicobacteriosis	*Helicobacter* spp.	Bacterium	(Priestnall et al., 2004)

Continued

TABLE 4.3 Zoonotic diseases of dogs and cats.—cont'd

Route of human exposure	Disease	Agent	Type	References
	Yersiniosis[c]	*Yersinia enterocolitica*	Bacterium	(Neubauer, Sprague, Scholz, & Hensel, 2001)
	Amebiasis	*Entamoeba histolytica*	Ameba	(Dado et al., 2012)
	Hookworm	*Ancylostoma braziliense*	Hookworm	(Bowman, Montgomery, Zajac, Eberhard, & Kazacos, 2010)
	Toxocarids	*Toxocara canis* (dogs) and *T. cati* (cats)	Roundworm	(Lee, Schantz, Kazacos, Montgomery, & Bowman, 2010)
	Strongyloidiasis	*Strongyloides stercoralis*	Threadworm	(Jaleta et al., 2017)
	Alveolar Echinococcosis[a,b,c]	*Echinococcus multilocularis*	Cestode	(Conraths et al., 2017)
	Cyst hydatid disease or dog tapeworm[a,b,c]	*Echinococcus granulosus*	Cestode	(Craig, Mastin, van Kesteren, & Boufana, 2015)
	Cryptosporidiosis	*Cryptosporidium* spp.	Coccidian	(Moreira et al., 2018)
	Toxoplasmosis[a,c]	*Toxoplasma gondii*	Coccidian	(Mareze et al., 2019)
	Giardiasis[a,c]	*Giardia* spp.	Flagellate	(Bouzid, Halai, Jeffreys, & Hunter, 2015)

Contact with infected respiratory or ocular secretions	Bordetellosis (epizootic pneumonia)	*Bordetella bronchiseptica*	Bacterium	(Clements, McGrath, & McAllister, 2018; Egberink et al., 2009)
	Chlamydiosis	*Chlamydophila felis*	Bacterium	(Gruffydd-Jones et al., 2009)
	Tularemia[b,c]	*Francisella tularensis*	Bacterium	(Foley & Nieto, 2010)
	Plague[c]	*Yersinia pestis*	Bacterium	(Sousa, Alencar, Almeida, & Cavalcanti, 2017)
Contact with infected genital secretions	Canine Brucellosis	*Brucella canis*	Bacterium	(Kauffman & Petersen, 2019)
	Q fever[b,c]	*Coxiella burnetti*	Rickettsia	(Egberink et al., 2013)
Contact with infected urine	Leptospirosis[a,b]	*Leptospira* spp.	Bacteria	(Schuller et al., 2015)
Vector-borne				
Flea-borne	Bartonellosis	*Bartonella* spp.	Bacterium	(Alvarez-Fernandez, Breitschwerdt, & Solano-Gallego, 2018)
	Plague	*Yersinia pestis*	Bacterium	(Pennisi et al., 2013)
	Rickettsiosis	*Rickettsia felis*	Rickettsia	(AAngelakis, Mediannikov, Parola, & Raoult, 2016)
	Murine typhus	*Rickettsia typhi*	Rickettsia	(Nogueras et al., 2013)
Tick-borne	Lyme Borreliosis[c]	*Borrelia burgdorferi*	Bacterium	(Littman et al., 2018)
	Anaplasmosis	*Anaplasma phagocytophilium*	Rickettsia	(Pawelczyk, Asman, & Solarz, 2019)
	Ehrlichiosis	*Ehrlichia* spp.	Rickettsia	(Little, 2010)
	Rocky Mountain spotted fever (RMSF)	*Rickettsia rickettsii*	Rickettsia	(Alhassan et al., 2019)
	Tularemia[c]	*Francisella tularensis*	Bacterium	(Zellner & Huntley, 2019)

Continued

TABLE 4.3 Zoonotic diseases of dogs and cats.—cont'd

Route of human exposure	Disease	Agent	Type	References
Sandfly-borne	Leishmaniasis[a,c]	*Leishmania infantum* and *L. chagasi*	Protozoa	(Dantas-Torres et al., 2019)
Reverse zoonoses				
	H_1N_1[c]	Influenza A	Virus	(Borland, Gracieux, Jones, Mallet, & Yugueros-Marcos, 2020)
	Tuberculoses[b,c]	*Mycobacterium tuberculosis*	Bacterium	(Hackendahl, Mawby, Bemis, & Beazley, 2004)
Multidrug-resistant pathogens				
	MRSA	Methicillin-resistant *Staphylooccus aureus*	Bacterium	(Walther et al., 2012)
	MRSP	Methicillin-resistant *Staphylococcus pseudointermedius*	Bacterium	(Gronthal et al., 2014)
		Carbapenemase-producing *E. coli*	Bacterium	(Yousfi et al., 2016)
		Carbapenemase-producing *Pseudomonas aeruginosa*	Bacterium	(Fernandes et al., 2018)
		MDR *Klebsiella pneumoniae*	Bacterium	(Carvalho et al., 2020)

		MDR *Acinetobacter baumannii*	Bacterium	
		AMR *Bordetella bronchiseptica*	Bacterium	(Kadlec & Schwarz, 2018)
	ESBL[a]	Extended spectrum beta-lactamase-producing *Escherichia coli*	Bacterium	(Bortolami et al., 2019)
		MDR *Salmonella* serovars	Bacterium	(Leonard et al., 2012)
		Macrolide-resistant *Streptococcus pyogenes*	Bacterium	(Samir, Abdel-Moein, & Zaher, 2020)

[a]*Highlighted in the CALLISTO (Companion Animals multisectorial. interprofessional. and Interdisciplinary Strategic Think tank On zoonoses) project.*
[b]*Notifiable to the OIE-World Organization for Animal Health.*
[c]*ECDC (European Center for Disease Prevention and Control) communicable diseases.*
Font: Adapted from Day, M.J., Breitschwerdt, E., Cleaveland, S., Karkare, U., Khanna, C., & Kirpensteijn, J. (2012). Surveillance of zoonotic infectious disease transmitted by small companion animals. *Emerging Infectious Diseases*. https://doi.org/10.3201/eid1812.120664.

6.3 Risk surveillance

As outlined in Table 4.3, companion animals may lead to the transmission of diverse zoonotic infectious diseases. Thus, it becomes critical to reduce the risk of transmission associated with the closer integration of companion animals into people's life. This may require an integrated approach involving veterinary and human healthcare stakeholders, pet industry, governments and owners, in the establishment of a surveillance program that captures data on zoonoses occurring in humans and animals (Day, 2016a).

The surveillance of zoonotic diseases is not mandatory in every country: none of the international health agencies are mandated to coordinate surveillance of diseases in companion animals. Reporting systems exist only for a couple of diseases affecting companion animals, namely rabies and leishmaniasis (Day et al., 2012).

The OIE - World Organization for Animal Health, that is the intergovernmental organization responsible for improving animal health worldwide, hosts a database called WAHID (World Animal Health Information Database), which captures specific zoonotic human infections reported annually by member countries. The American (CDC) and European Center for Disease Prevention and Control (ECDC) are other institutions collecting data on zoonoses.

The project entitled CALLISTO (Companion Animal multisectoriaL interprofessionaL and Interdisciplinary Strategic Think tank On zoonoses) funded by the European Union (EU) has been running from 2012 to 2014 with the aim of investigating the risk of occurrence and transmission of diseases among companion animals, and from companion animals to man (zoonoses) or to livestock (Rijks, Cito, Cunningham, Rantsios, & Giovannini, 2016). The process led to the selection of 15 diseases of prime public health relevance in a Europe perspective. Of these, ten diseases are related to companion animals (dogs and cats) listed and marked (*) in Table 4.3: rabies, campylobacteriosis, leptospirosis, cat scratch disease, infection due to extended spectrum beta lactamase (ESBL)-producing bacteria, cystic echinococcosis, leishmaniosis, toxoplasmosis, alveolar echinococcosis and giardiasis. Most of them are currently notifiable to the OIE ("#" in Table 4.3) (OIE, 2020) and/or to the ECDC ("&" in Table 4.3) (ECDC, 2020). Cat Scratch Disease and toxocarosis, two important diseases, are not yet notifiable to ECDC or OIE (Day, 2016a).

Local and international small animal diseases surveillances systems initiatives are becoming a reality in the era of data science. Big data projects like SAVSNET (Small Animal Veterinary Surveillance Network) (Jones et al., 2014; Radford et al., 2011; Sanchez-Vizcaino et al., 2015) and VetCompass (McGreevy et al., 2017) are active, essential and important players for a bigger veterinary evidence base and zoonosis surveillance ("The power of data," 2020).

Surely, surveillance and risk assessment are key tools for the control of zoonoses. However, the education of pet owners, the promotion of the concept of responsible pet ownership and the continuing education of relevant professionals and general society are considered an important cufflink in this process as highlighted in CALLISTO report (Day, 2016a; Rijks et al., 2016).

7. Bias of the human-companion animal bond

7.1 Anthropomorphism and companion animals

As previously mentioned, companion animals are increasingly perceived as family members and sometimes treated like human beings. This attribution of human personification or humanization to the animal is called anthropomorphism (Overgaauw et al., 2020). This human behavior seems to originate from the affectionate bond with animals. Despite the human-animal bond has several beneficial effects on humans, an excessive bond can cause welfare problems to the animal and lack of provision of their essential needs as true animals (Overgaauw et al., 2020). Obesity, vegan feeding and chocolate intoxication in companion animals can be somewhat attributed to this behavior.

Moreover, an exacerbation of this behavior can increase the incidence of zoonotic diseases. A study carried out in the Netherlands showed that 50% of companion animals lick the owners' face, 45%–60% are allowed to stay on the bed, 18%–30% sleep with the owner and 45% of cats jump onto the kitchen sink. In addition, approximately 39% of the dog's owners never clean up the feces of their dog and only 15% of the dog's owners and 8% of the cat's owners consistently wash their hands after contact with the animal (Overgaauw et al., 2009).

This alarming behavior and close contact between the owners and their animals is becoming extremely common and poses an increased risk of transmission of zoonotic pathogens.

7.2 Abandon and stray animals

Approximately 6.5 million companion animals are admitted to animal shelters every year in the United States, 16,000 in the UK and between 100,000 and 200,000 pets are abandoned in France every year, with a peak of 60% in summer (ASPCA; Eurogroupforanimals, 2020). The reasons behind companion-animal relinquishment or abandonment are many and diverse and vary according to the country and the socioeconomic conditions (Coe et al., 2014), although most of them are absolutely preventable (ASPCA). When abandoned, pets are forced to fend for themselves and become stray or feral. A feral dog is defined as a dog that has not had any interaction with humans especially during the critical puppy development phases. Feral dogs usually live in groups and are adapted to find food wherever they can, often scavenging in the garbage or attacking wild animals. They are now considered a major threat to wildlife (BBC, 2019). These animals also represent a threat to human health due to physical treats with attacks and even property destruction. Stray animals also increase the potential exposure to zoonotic diseases, as those listed in Table 4.3. In most cases, feral animals are challenging to be handled and socialize enough to be re-introduced to a new owner (Seidman, 2001).

8. The importance of the veterinary medicine and its professionals in the one health space

The Veterinary Medicine with its large spectrum of fields and activities play a vital role in the security and economic and social well-being of humanity, and are therefore a key component of the One Health concept (Stratton et al., 2019). To improve the health of companion animals, control, prevent and study zoonoses and help owners to appropriately interact with these new family members (Fig. 4.1).

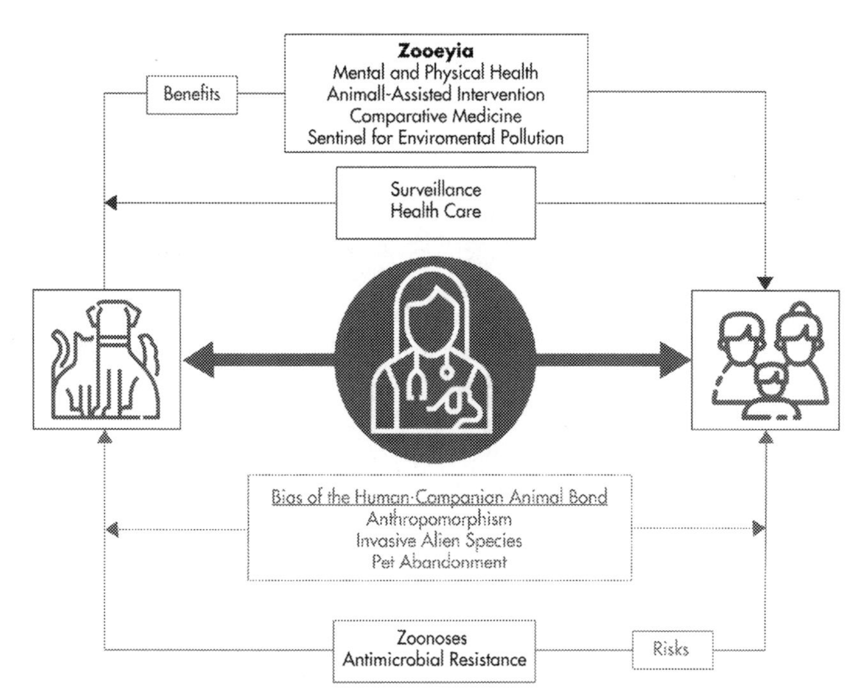

FIG. 4.1 Risks and benefits of companion animals to humans and vice-versa. The important role of veterinary medicine in the One Health space as an important mediator between animals and humans. *Blue line (dark gray in print version)* represent the benefits; *red lines (light gray in print version)* the risks and *green lines (gray in print version)* the biases in the human-companion animal relationship.

The COVID-19 outbreak has highlighted the potential of the animal-human interface to act as the primary source of emerging zoonotic diseases and how veterinary medicine is important. Veterinarians are the first and best line of defense against animal diseases that could threaten public health. The preventive measures aligned by veterinary authorities, if applied correctly and with awareness of their importance might avoid these pandemic situations, death of people and severe economic losses.

Veterinarians are not just restricted to treat sick or injured animals. These multifaceted professionals have the unique knowledge and experience to address preventive care and keep the animals and society healthy.

Acting on several fronts as clinical practitioners, epidemiologists, pathologists, toxicologists, food safety and security or ecological experts, veterinarians are essential to advancing One Health and protecting the health and safety of its three pillars— animals, people, and the environment.

Physicians and veterinarians are thus expected to increasingly collaborate across common challenges (Wilkes, Conrad, & Winer, 2019). This notion represents a natural sequel to Virchow's maxim stating: "*between animal and human medicine there is no dividing line – nor should there be.*" One Health serves as an ideal framework for developing problem-focused that promote inter-professional teamwork. All types of One Health issues - zoonotic diseases, water pollution, food safety and security - require leadership across various health-related fields to share knowledge and take the correct action (Wilkes et al., 2019).

An increasingly worrying problem in the One Health approach is surely the antimicrobial resistance (AMR) considered nowadays a global threat to public health (Ferri, Ranucci, Romagnoli, & Giaccone, 2017). The misuse and abuse of antimicrobials in veterinary and human medicine have accelerated this growing phenomenon (Ferri et al., 2017). The antimicrobial usage in companion animal practice together with the closer interactions of these animals with humans, poses a concerning alarming risk (Leite-Martins et al., 2014, 2015; Smith et al., 2018).

This represents a perfect example of a One Health approach. Responding to the AMR challenges includes a) studies that allow to evaluate and quantify the use of antibiotics in clinical practice (Alcantara, Pinello, Severo, & Niza-Ribeiro, 2021) enabling to designing more effective preventive measures to reduce the use of antimicrobials; 2) strengthening of the AMR surveillance system in animal and human populations; and c) increased awareness of human and veterinarian stakeholders on the prudent use of antibiotics.

9. Conclusions

Companion animals play an important role in human well-being and health, and this has been well-known for decades. However, in an era of changing paradigms, the misunderstandings of animal well-being and the limitations of this relationship can bring old and new risks for humans and their environment. Veterinarians in its broadest spectrum of action are essentials to look for the right balance and to provide the correct education to the community.

References

Acke, E. (2018). Campylobacteriosis in dogs and cats: A review. *New Zealand Veterinary Journal, 66*(5), 221–228. https://doi.org/10.1080/00480169.2018.1475268

Adrian, J. A., Deliramich, A. N., & Frueh, B. C. (2009). Complicated grief and posttraumatic stress disorder in humans' response to the death of pets/animals. *Bulletin of the Menninger Clinic, 73*(3), 176–187. https://doi.org/10.1521/bumc.2009.73.3.176

Alhassan, A., Liu, H., McGill, J., Cerezo, A., Jakkula, L., Nair, A. D. S., ... Ganta, R. R. (2019). Rickettsia rickettsii whole-cell antigens offer protection against rocky mountain spotted fever in the canine host. *Infection and Immunity, 87*(2). https://doi.org/10.1128/IAI.00628-18

Alcantara, G. L. C., Pinello, K. C., Severo, M., & Niza-Ribeiro, J. (2021). Antimicrobial resistance in companion animals – veterinarians' attitudes and prescription drivers in Portugal. *Comparative Immunology, Microbiology and Infectious Diseases, 76*(June). https://doi.org/10.1016/j.cimid.2021.101640

Ali, N., Malik, R. N., Mehdi, T., Eqani, S. A., Javeed, A., Neels, H., & Covaci, A. (2013). Organohalogenated contaminants (OHCs) in the serum and hair of pet cats and dogs: Biosentinels of indoor pollution. *The Science of the Total Environment, 449*, 29–36. https://doi.org/10.1016/j.scitotenv.2013.01.037

Alvarez-Fernandez, A., Breitschwerdt, E. B., & Solano-Gallego, L. (2018). Bartonella infections in cats and dogs including zoonotic aspects. *Parasit Vectors, 11*(1), 624. https://doi.org/10.1186/s13071-018-3152-6

Angle, T. C., Passler, T., Waggoner, P. L., Fischer, T. D., Rogers, B., Galik, P. K., & Maxwell, H. S. (2015). Real-time detection of a virus using detection dogs. *Frontiers in Veterinary Science, 2*, 79. https://doi.org/10.3389/fvets.2015.00079

Angelakis, E., Mediannikov, O., Parola, P., & Raoult, D. (2016). Rickettsia felis: The complex journey of an emergent human pathogen. *Trends in Parasitology, 32*(7), 554–564. https://doi.org/10.1016/j.pt..2016.04.009

Arhant-Sudhir, K., Arhant-Sudhir, R., & Sudhir, K. (2011). Pet ownership and cardiovascular risk reduction: Supporting evidence, conflicting data and underlying mechanisms. *Clinical and Experimental Pharmacology and Physiology, 38*(11), 734–738. https://doi.org/10.1111/j.1440-1681.2011.05583.x

Aslan, B., Viola, L., Saini, S. S., Stockman, J., & Ryan, E. P. (2020). Pets as sentinels of human exposure to pesticides and Co-exposure concerns with other contaminants/toxicants. In S. A. Pastorinho M (Ed.), *Pets as sentinels, forecasters and promoters of human health*. Cham: Springer.

ASPCA. *American Society for the Prevention of Cruelty to Animals*. Retrieved from www.aspca.org.

Axelsson, E., Ratnakumar, A., Arendt, M. L., Maqbool, K., Webster, M. T., Perloski, M., ... Lindblad-Toh, K. (2013). The genomic signature of dog domestication reveals adaptation to a starch-rich diet. *Nature, 495*(7441), 360–364. https://doi.org/10.1038/nature11837

Backer, L. C., Grindem, C. B., Corbett, W. T., Cullins, L., & Hunter, J. L. (2001). Pet dogs as sentinels for environmental contamination. *The Science of the Total Environment, 274*(1–3), 161–169. https://doi.org/10.1016/s0048-9697(01)00740-9

Backer, L. C., Grindem, C. B., Corbett, W. T., Cullins, L., & Hunter, J. L. (2001). Pet dogs as sentinels for environmental contamination. *Science of the Total Environment, 274*(1−3), 161−169. https://doi.org/10.1016/s0048-9697(01)00740-9

Baranowski, K., & Huang, B. (2020). Cat scratch disease. In *StatPearls*. Treasure Island (FL).

BBC. (2019). *Dogs' becoming major threat' to wildlife.*

Bengtsson, B., & Greko, C. (2014). Antibiotic resistance–consequences for animal health, welfare, and food production. *Upsala Journal of Medical Sciences, 119*(2), 96−102. https://doi.org/10.3109/03009734.2014.901445

Bentley, R. T., Ahmed, A. U., Yanke, A. B., Cohen-Gadol, A. A., & Dey, M. (2017). Dogs are man's best friend: In sickness and in health. *Neuro-Oncology, 19*(3), 312−322. https://doi.org/10.1093/neuonc/now109

Black, L. L., Gaynor, J., Gahring, D., Adams, C., Aron, D., Harman, S., … Harman, R. (2007). Effect of adipose-derived mesenchymal stem and regenerative cells on lameness in dogs with chronic osteoarthritis of the coxofemoral joints: A randomized, double-blinded, multicenter, controlled trial. *Veterinary Therapeutics, 8*(4), 272−284. Retrieved from https://www.ncbi.nlm.nih.gov/pubmed/18183546.

Borland, S., Gracieux, P., Jones, M., Mallet, F., & Yugueros-Marcos, J. (2020). Influenza A virus infection in cats and dogs: A literature review in the light of the "one health" concept. *Frontiers in Public Health, 8*, 83. https://doi.org/10.3389/fpubh.2020.00083

Bortolami, A., Zendri, F., Maciuca, E. I., Wattret, A., Ellis, C., Schmidt, V., … Timofte, D. (2019). Diversity, virulence, and clinical significance of extended-spectrum beta-lactamase- and pAmpC-producing *Escherichia coli* from companion animals. *Frontiers in Microbiology, 10*, 1260. https://doi.org/10.3389/fmicb.2019.01260

Bouzid, M., Halai, K., Jeffreys, D., & Hunter, P. R. (2015). The prevalence of Giardia infection in dogs and cats, a systematic review and meta-analysis of prevalence studies from stool samples. *Veterinary Parasitology, 207*(3−4), 181−202. https://doi.org/10.1016/j.vetpar.2014.12.011

Bouchiat, C., Curtis, S., Spiliopoulou, I., Bes, M., Cocuzza, C., Codita, I., … Staphylococcal, I. (2017). MRSA infections among patients in the emergency department: A European multicentre study. *Journal of Antimicrobial Chemotherapy, 72*(2), 372−375. https://doi.org/10.1093/jac/dkw431

Bowlby, J. (2002). In E. S. Paulo (Ed.), *Apego e perda: a natureza do vínculo* (Vol. 3).

Bowman, D. D., Montgomery, S. P., Zajac, A. M., Eberhard, M. L., & Kazacos, K. R. (2010). Hookworms of dogs and cats as agents of cutaneous larva migrans. *Trends in Parasitology, 26*(4), 162−167. https://doi.org/10.1016/j.pt.2010.01.005

Bowser, N. H., & Anderson, N. E. (2018). Dogs (*Canis familiaris*) as sentinels for human infectious disease and application to Canadian populations: A systematic review. *Journal of Veterinary Science, 5*(4). https://doi.org/10.3390/vetsci5040083

Brizi, V., Benedetti, V., Lavecchia, A. M., & Xinaris, C. (2019). Engineering kidney tissues for polycystic kidney disease modeling and drug discovery. *Methods in Cell Biology, 153*, 113−132. https://doi.org/10.1016/bs.mcb.2019.04.011

Burrell, G., & Seibert, F. (1914). Experiments with small animals and carbon monoxide. *Journal of Industrial and Engineering Chemistry*, (6), 241−244.

Carvalho, I., Alonso, C. A., Silva, V., Pimenta, P., Cunha, R., Martins, C., … Poeta, P. (2020). Extended-Spectrum beta-lactamase-producing *Klebsiella pneumoniae* isolated from healthy and sick dogs in Portugal. *Microbial Drug Resistance, 26*(6), 709−715. https://doi.org/10.1089/mdr.2019.0205

Casselato, G. (2015). *O resgate da empatia: suporte psicológico ao luto não reconhecido*. São Paulo: Summus Editorial.

Caugant, D. A., Levin, B. R., & Selander, R. K. (1984). Distribution of multilocus genotypes of *Escherichia coli* within and between host families. *Journal of Hygiene, 92*(3), 377–384. https://doi.org/10.1017/s0022172400064597

Cerquetella, M., Spaterna, A., Laus, F., Tesei, B., Rossi, G., Antonelli, E., … Bassotti, G. (2010). Inflammatory bowel disease in the dog: Differences and similarities with humans. *World Journal of Gastroenterology, 16*(9), 1050–1056. https://doi.org/10.3748/wjg.v16.i9.1050

Chan, M. M., & Tapia Rico, G. (2019). The "pet effect" in cancer patients: Risks and benefits of human-pet interaction. *Critical Reviews in Oncology/Hematology, 143*, 56–61. https://doi.org/10.1016/j.critrevonc.2019.08.004

Chomel, B. B. (2014). Emerging and re-emerging zoonoses of dogs and cats. *Animals (Basel), 4*(3), 434–445. https://doi.org/10.3390/ani4030434

Chowdhury, E. K., Nelson, M. R., Jennings, G. L., Wing, L. M., Reid, C. M., & Committee, A. M. (2017). Pet ownership and survival in the elderly hypertensive population. *Journal of Hypertension, 35*(4), 769–775. https://doi.org/10.1097/HJH.0000000000001214

Clements, J., McGrath, C., & McAllister, C. (2018). Bordetella bronchiseptica pneumonia: Beware of the dog. *BMJ Case Reports*. https://doi.org/10.1136/bcr-2018-224588, 2018.

Coe, J. B., Young, I., Lambert, K., Dysart, L., Nogueira Borden, L., & Rajic, A. (2014). A scoping review of published research on the relinquishment of companion animals. *Journal of Applied Animal Welfare Science, 17*(3), 253–273. https://doi.org/10.1080/10888705.2014.899910

College of Veterinary Medicine, P. U. Center for the Human Animal Bond. Retrieved from https://www.purdue.edu/vet/chab/.

Conraths, F. J., Probst, C., Possenti, A., Boufana, B., Saulle, R., La Torre, G., … Casulli, A. (2017). Potential risk factors associated with human alveolar echinococcosis: Systematic review and meta-analysis. *PLoS Neglected Tropical Diseases, 11*(7), e0005801. https://doi.org/10.1371/journal.pntd.0005801

Craig, P., Mastin, A., van Kesteren, F., & Boufana, B. (2015). Echinococcus granulosus: Epidemiology and state-of-the-art of diagnostics in animals. *Veterinary Parasitology, 213*(3–4), 132–148. https://doi.org/10.1016/j.vetpar.2015.07.028

Dado, D., Izquierdo, F., Vera, O., Montoya, A., Mateo, M., Fenoy, S., … Miro, G. (2012). Detection of zoonotic intestinal parasites in public parks of Spain. Potential epidemiological role of microsporidia. *Zoonoses Public Health, 59*(1), 23–28. https://doi.org/10.1111/j.1863-2378.2011.01411.x

Dantas-Torres, F., Miro, G., Baneth, G., Bourdeau, P., Breitschwerdt, E., Capelli, G., … Otranto, D. (2019). Canine leishmaniasis control in the context of one health. *Emerging Infectious Diseases, 25*(12), 1–4. https://doi.org/10.3201/eid2512.190164

Day, M. J. (2016). Pet-related infections. *American Family Physician, 94*(10), 794–802. Retrieved from: https://www.ncbi.nlm.nih.gov/pubmed/27929279.

Damborg, P., Broens, E. M., Chomel, B. B., Guenther, S., Pasmans, F., Wagenaar, J. A., … Guardabassi, L. (2016). Bacterial zoonoses transmitted by household pets: State-of-the-Art and future perspectives for targeted research and policy actions. *Journal of Comparative Pathology, 155*(1 Suppl. 1), S27–S40. https://doi.org/10.1016/j.jcpa.2015.03.004

Day, M. J. (2016a). The CALLISTO project: A summary. *Journal of Comparative Pathology, 155*(1 Suppl. 1), S1–S7. https://doi.org/10.1016/j.jcpa.2015.01.005

Day, M. J. (2016b). Pet-related infections. *American Family Physician, 94*(10), 794–802. Retrieved from https://www.ncbi.nlm.nih.gov/pubmed/27929279.

Day, M. J., Breitschwerdt, E., Cleaveland, S., Karkare, U., Khanna, C., Kirpensteijn, J., et al. (2012). Surveillance of zoonotic infectious disease transmitted by small companion animals. *Emerging Infectious Diseases*. https://doi.org/10.3201/eid1812.120664

Di Cerbo, A., Palmieri, B., De Vico, G., & Iannitti, T. (2014). Onco-epidemiology of domestic animals and targeted therapeutic attempts: Perspectives on human oncology. *Journal of Cancer Research and Clinical Oncology, 140*(11), 1807–1814. https://doi.org/10.1007/s00432-014-1664-9

Driscoll, C. A., Macdonald, D. W., & O'Brien, S. J. (2009). From wild animals to domestic pets, an evolutionary view of domestication. *Proceedings of the National Academy of Sciences of the United States of America, 106*(Suppl. 1), 9971–9978. https://doi.org/10.1073/pnas.0901586106

ECDC. *European Centre for Disease Prevention and Control*. Retrieved from https://www.ecdc.europa.eu/en/zoonoses.

ECDC. (2020). *Diseases and special health issues under EU surveillance*. Retrieved from https://www.ecdc.europa.eu/en/all-topics-z/surveillance-and-disease-data/surveillance/diseases-and-special-health-issues-under-eu.

Egberink, H., Addie, D., Belak, S., Boucraut-Baralon, C., Frymus, T., Gruffydd-Jones, T., … Horzinek, M. C. (2009). Bordetella bronchiseptica infection in cats. ABCD guidelines on prevention and management. *Journal of Feline Medicine & Surgery, 11*(7), 610–614. https://doi.org/10.1016/j.jfms.2009.05.010

Egberink, H., Addie, D., Belák, S., Boucraut-Baralon, C., Frymus, T., Gruffydd-Jones, T., … Horzinek, M. C. (2013). Coxiellosis/Q fever in cats: ABCD guidelines on prevention and management. *J. Feline Med. Surg, 15*(7), 573–575. https://doi.org/10.1177/1098612X13489216. PMID: 23813818.

Elliker, K. R., & Williams, H. C. (2016). Detection of skin cancer odours using dogs: A step forward in melanoma detection training and research methodologies. *British Journal of Dermatology, 175*(5), 851–852. https://doi.org/10.1111/bjd.15030

Euro group for animals. (2020). *Why the French are 'European champions' at abandoning pets*. Retrieved from https://www.eurogroupforanimals.org/.

Europe Union. (2017). *Invasive alien species of union concern*. Luxembourg: Publications Office of the European Union. Retrieved from https://ec.europa.eu/environment/nature/pdf/IAS_brochure_species.pdf.

FEDIAF. (2019). *European Federation of the pet food industry*. Retrieved from http://www.fediaf.org/.

Ferri, M., Ranucci, E., Romagnoli, P., & Giaccone, V. (2017). Antimicrobial resistance: A global emerging threat to public health systems. *Critical Reviews in Food Science and Nutrition, 57*(13), 2857–2876. https://doi.org/10.1080/10408398.2015.1077192

Fernandes, M. R., Sellera, F. P., Moura, Q., Carvalho, M. P. N., Rosato, P. N., Cerdeira, L., & Lincopan, N. (2018). Zooanthroponotic transmission of drug-resistant *Pseudomonas aeruginosa*, Brazil. *Emerging Infectious Diseases, 24*(6), 1160–1162. https://doi.org/10.3201/eid2406.180335

Fine, A. H. (2019). *Handbook on animal-assisted therapy : Foundations and guidelines for animal-assisted interventions* (5). SanDiego: Elsevier.

Foley, J. E., & Nieto, N. C. (2010). Tularemia. *Veterinary Microbiology, 140*(3–4), 332–338. https://doi.org/10.1016/j.vetmic.2009.07.017

Fooks, A. R., Cliquet, F., Finke, S., Freuling, C., Hemachudha, T., Mani, R. S., … Banyard, A. C. (2017). Rabies. *Nature Reviews Disease Primers, 3*, 17091. https://doi.org/10.1038/nrdp.2017.91

Fox, P. R., Basso, C., Thiene, G., & Maron, B. J. (2014). Spontaneously occurring restrictive nonhypertrophied cardiomyopathy in domestic cats: A new animal model of human disease. *Cardiovascular Pathology, 23*(1), 28–34. https://doi.org/10.1016/j.carpath.2013.08.001

Friedman, E., & Krause-Parello, C. A. (2018). Companion animals and human health: Benefits, challenges, and the road ahead for human-animal interaction. *Revue scientifique et technique, 37*(1), 71–82. https://doi.org/10.20506/rst.37.1.2741

Garden, O. A., Volk, S. W., Mason, N. J., & Perry, J. A. (2018). Companion animals in comparative oncology: One Medicine in action. *Veterinary Journal, 240*, 6–13. https://doi.org/10.1016/j.tvjl.2018.08.008

Gereda, J. E., Leung, D. Y., Thatayatikom, A., Streib, J. E., Price, M. R., Klinnert, M. D., & Liu, A. H. (2000). Relation between house-dust endotoxin exposure, type 1 T-cell development, and allergen sensitisation in infants at high risk of asthma. *Lancet, 355*(9216), 1680–1683. https://doi.org/10.1016/s0140-6736(00)02239-x

Gordon, I., Paoloni, M., Mazcko, C., & Khanna, C. (2009). The comparative oncology trials Consortium: Using spontaneously occurring cancers in dogs to inform the cancer drug development pathway. *PLoS Med, 6*(10), e1000161. https://doi.org/10.1371/journal.pmed.1000161

Gremiao, I. D., Miranda, L. H., Reis, E. G., Rodrigues, A. M., & Pereira, S. A. (2017). Zoonotic epidemic of sporotrichosis: Cat to human transmission. *PLoS Pathogens, 13*(1), e1006077. https://doi.org/10.1371/journal.ppat.1006077

Gronthal, T., Moodley, A., Nykasenoja, S., Junnila, J., Guardabassi, L., Thomson, K., & Rantala, M. (2014). Large outbreak caused by methicillin resistant staphylococcus pseudintermedius ST71 in a finnish veterinary teaching hospital–from outbreak control to outbreak prevention. *PLoS One, 9*(10), e110084. https://doi.org/10.1371/journal.pone.0110084

Gruffydd-Jones, T., Addie, D., Belak, S., Boucraut-Baralon, C., Egberink, H., Frymus, T., … Horzinek, M. C. (2009). Chlamydophila felis infection. ABCD guidelines on prevention and management. *Journal of Feline Medicine and Surgery, 11*(7), 605–609. https://doi.org/10.1016/j.jfms.2009.05.009

Guest, C., Pinder, M., Doggett, M., Squires, C., Affara, M., Kandeh, B., … Lindsay, S. W. (2019). Trained dogs identify people with malaria parasites by their odour. *Lancet Infectious Diseases, 19*(6), 578–580. https://doi.org/10.1016/S1473-3099(19)30220-8

Gulson, B., Korsch, M., Matisons, M., Douglas, C., Gillam, L., & McLaughlin, V. (2009). Windblown lead carbonate as the main source of lead in blood of children from a seaside community: An example of local birds as "canaries in the mine. *Environmental Health Perspectives, 117*(1), 148–154. https://doi.org/10.1289/ehp.11577

Hackendahl, N. C., Mawby, D. I., Bemis, D. A., & Beazley, S. L. (2004). Putative transmission of *Mycobacterium tuberculosis* infection from a human to a dog. *Journal of the American Veterinary Medical Association, 225*(10), 1573–1577. https://doi.org/10.2460/javma.2004.225.1573, 1548.

Hackner, K., & Pleil, J. (2017). Canine olfaction as an alternative to analytical instruments for disease diagnosis: Understanding 'dog personality' to achieve reproducible results. *Journal of Breath, 11*(1), 012001. https://doi.org/10.1088/1752-7163/aa5524

Hall, S., Dolling, L., Bristow, K., Fuller, T., & Mills, D. (2017). Indirect costs: Extending the scope of economic value. In *Companion animal economics: The economic impact of companion animals in the UK.: Research report.*

Halliday, J. E., Meredith, A. L., Knobel, D. L., Shaw, D. J., Bronsvoort, B. M., & Cleaveland, S. (2007). A framework for evaluating animals as sentinels for infectious disease surveillance. *Journal of the Royal Society Interface, 4*(16), 973–984. https://doi.org/10.1098/rsif.2007.0237

Headey, B. (1999). Health benefits and health cost savings due to pets: Preliminary estimates from an Australian national survey. *Social Indicators Research, 47*, 233–243. https://doi.org/10.1023/A:1006892908532

Headey, B., Na, F., & Zheng, R. (2008). Pet dogs benefit owners' health: A 'natural experiment' in China. *Social Indicators Research, 87*, 481–493. https://doi.org/10.1007/s11205-007-9142-2

Hesselmar, B., Aberg, N., Aberg, B., Eriksson, B., & Bjorksten, B. (1999). Does early exposure to cat or dog protect against later allergy development? *Clinical & Experimental Allergy, 29*(5), 611–617. https://doi.org/10.1046/j.1365-2222.1999.00534.x

Hoagwood, K. E., Acri, M., Morrissey, M., & Peth-Pierce, R. (2017). Animal-assisted therapies for youth with or at risk for mental health problems: A systematic review. *Applied Developmental Science, 21*(1), 1–13. https://doi.org/10.1080/10888691.2015.1134267

Hodgson, K., & Darling, M. (2011). Zooeyia: An essential component of "one health. *Canadian Veterinary Journal, 52*(2), 189–191. Retrieved from https://www.ncbi.nlm.nih.gov/pubmed/21532829.

van Houtert, E. A. E., Endenburg, N., Wijnker, J. J., Rodenburg, B., & Vermetten, E. (2018). The study of service dogs for veterans with post-traumatic stress disorder: A scoping literature review. *European Journal of Psychotraumatology, 9*(Suppl. 3), 1503523. https://doi.org/10.1080/20008198.2018.1503523

Hu, Y., Hu, S., Wang, W., Wu, X., Marshall, F. B., Chen, X., … Wang, C. (2014). Earliest evidence for commensal processes of cat domestication. *Proceedings of the National Academy of Sciences of the United States of America, 111*(1), 116–120. https://doi.org/10.1073/pnas.1311439110

Hui Gan, G. Z., Hill, A. M., Yeung, P., Keesing, S., & Netto, J. A. (2019). Pet ownership and its influence on mental health in older adults. *Aging in Mental Health*, 1–8. https://doi.org/10.1080/13607863.2019.1633620

International Health, C. (2002). Constitution of the World Health Organization 1946. *Bulletin of the World Health Organization, 80*(12), 983–984. Retrieved from https://www.ncbi.nlm.nih.gov/pubmed/12571729.

Jacobson, K. C., & Chang, L. (2018). Associations between pet ownership and attitudes toward pets with youth socioemotional outcomes. *Frontiers in Psychology, 9*, 2304. https://doi.org/10.3389/fpsyg.2018.02304

Jaleta, T. G., Zhou, S., Bemm, F. M., Schar, F., Khieu, V., Muth, S., … Streit, A. (2017). Different but overlapping populations of Strongyloides stercoralis in dogs and humans-Dogs as a possible source for zoonotic strongyloidiasis. *PLoS Neglected Tropical Disease, 11*(8), e0005752. https://doi.org/10.1371/journal.pntd.0005752

Jendrny, P., Schulz, C., Twele, F., Meller, S., von Kockritz-Blickwede, M., Osterhaus, A., … Volk, H. A. (2020). Scent dog identification of samples from COVID-19 patients - a pilot study. *BMC Infectious Diseases, 20*(1), 536. https://doi.org/10.1186/s12879-020-05281-3

Jones, P. H., Dawson, S., Gaskell, R. M., Coyne, K. P., Tierney, A., Setzkorn, C., … Noble, P. J. (2014). Surveillance of diarrhoea in small animal practice through the Small Animal Veterinary Surveillance Network (SAVSNET). *Veterinary Journal, 201*(3), 412–418. https://doi.org/10.1016/j.tvjl.2014.05.044

Kadlec, K., & Schwarz, S. (2018). Antimicrobial resistance in bordetella bronchiseptica. *Microbiology Spectrum, 6*(4). https://doi.org/10.1128/microbiolspec.ARBA-0024-2017

Kates, A. E., Jarrett, O., Skarlupka, J. H., Sethi, A., Duster, M., Watson, L., … Safdar, N. (2020). Household pet ownership and the microbial diversity of the human gut microbiota. *Frontiers in Cellular and Infection Microbiology, 10*, 73. https://doi.org/10.3389/fcimb.2020.00073

Kauffman, L. K., & Petersen, C. A. (2019). Canine brucellosis: Old foe and reemerging scourge. *Veterinary Clinics of North America Small Animal Practice, 49*(4), 763–779. https://doi.org/10.1016/j.cvsm.2019.02.013

Kline, J. A., Fisher, M. A., Pettit, K. L., Linville, C. T., & Beck, A. M. (2019). Controlled clinical trial of canine therapy versus usual care to reduce patient anxiety in the emergency department. *PLoS One, 14*(1), e0209232. https://doi.org/10.1371/journal.pone.0209232

Knapp, D. W., Dhawan, D., Ramos-Vara, J. A., Ratliff, T. L., Cresswell, G. M., Utturkar, S., ... Hahn, N. M. (2019). Naturally-occurring invasive urothelial carcinoma in dogs, a unique model to drive advances in managing muscle invasive bladder cancer in humans. *Frontiers in Oncology, 9*, 1493. https://doi.org/10.3389/fonc.2019.01493

Kol, A., Arzi, B., Athanasiou, K. A., Farmer, D. L., Nolta, J. A., Rebhun, R. B., ... Borjesson, D. L. (2015). Companion animals: Translational scientist's new best friends. *Science Translational Medicine, 7*(308), 308ps321. https://doi.org/10.1126/scitranslmed.aaa9116

van der Kolk, J. H., Endimiani, A., Graubner, C., Gerber, V., & Perreten, V. (2019). Acinetobacter in veterinary medicine, with an emphasis on Acinetobacter baumannii. *Journal of Global Antimicrobial Resistance, 16*, 59–71. https://doi.org/10.1016/j.jgar.2018.08.011

Krause-Parello, C. A. (2012). Pet ownership and older women: The relationships among loneliness, pet attachment support, human social support, and depressed mood. *Geriatric Nursing, 33*(3), 194–203. https://doi.org/10.1016/j.gerinurse.2011.12.005

Krittanawong, C., Kumar, A., Wang, Z., Jneid, H., Virani, S. S., & Levine, G. N. (2020). Pet ownership and cardiovascular health in the US general population. *American Journal of Cardiology, 125*(8), 1158–1161. https://doi.org/10.1016/j.amjcard.2020.01.030

Kumar, V., Abbas, A. K., Aster, J. C., & Perkins, J. A. (2018). *Robbins basic pathology* (10th ed.). Philadelphia, Pennsylvania: Elsevier.

Lane, H. B. Z., & Shannon, D. W. (2013). When reading gets ruff: Canine-assisted reading programs. *The Reading Teacher, 67*, 87–95. https://doi.org/10.1002/TRTR.1204

LeBlanc, A. K., & Mazcko, C. N. (2020). Improving human cancer therapy through the evaluation of pet dogs. *Nature Reviews Cancer*. https://doi.org/10.1038/s41568-020-0297-3

Lee, A. C., Schantz, P. M., Kazacos, K. R., Montgomery, S. P., & Bowman, D. D. (2010). Epidemiologic and zoonotic aspects of ascarid infections in dogs and cats. *Trends in Parasitology, 26*(4), 155–161. https://doi.org/10.1016/j.pt.2010.01.002

Leite-Martins, L., Mahu, M. I., Costa, A. L., Bessa, L. J., Vaz-Pires, P., Loureiro, L., ... Martins da Costa, P. (2015). Prevalence of antimicrobial resistance in faecal enterococci from vet-visiting pets and assessment of risk factors. *Veterinary Record, 176*(26), 674. https://doi.org/10.1136/vr.102888

Leite-Martins, L., Meireles, D., Bessa, L. J., Mendes, A., de Matos, A. J., & da Costa, P. M. (2014). Spread of multidrug-resistant *Enterococcus faecalis* within the household setting. *Microbial Drug Resistance, 20*(5), 501–507. https://doi.org/10.1089/mdr.2013.0217

Leonard, E. K., Pearl, D. L., Finley, R. L., Janecko, N., Reid-Smith, R. J., Peregrine, A. S., & Weese, J. S. (2012). Comparison of antimicrobial resistance patterns of Salmonella spp. and *Escherichia coli* recovered from pet dogs from volunteer households in Ontario (2005-06). *Journal of Antimicrobial Chemotherapy, 67*(1), 174–181. https://doi.org/10.1093/jac/dkr430

Levine, J. M., Levine, G. J., Porter, B. F., Topp, K., & Noble-Haeusslein, L. J. (2011). Naturally occurring disk herniation in dogs: An opportunity for pre-clinical spinal cord injury research. *Journal of Neurotrauma, 28*(4), 675–688. https://doi.org/10.1089/neu.2010.1645

Lindfors, A., van Hage-Hamsten, M., Rietz, H., Wickman, M., & Nordvall, S. L. (1999). Influence of interaction of environmental risk factors and sensitization in young asthmatic children. *Journal of Allergy and Clinical Immunology, 104*(4 Pt 1), 755–762. https://doi.org/10.1016/s0091-6749(99)70284-8

Lippi, G., & Plebani, M. (2019). Diabetes alert dogs: A narrative critical overview. *Clinical Chemistry and Laboratory Medicine, 57*(4), 452–458. https://doi.org/10.1515/cclm-2018-0842

Little, S. E. (2010). Ehrlichiosis and anaplasmosis in dogs and cats. *Veterinary Clinics of North America Small Animal Practice, 40*(6), 1121–1140. https://doi.org/10.1016/j.cvsm.2010.07.004

Littman, M. P., Gerber, B., Goldstein, R. E., Labato, M. A., Lappin, M. R., & Moore, G. E. (2018). ACVIM consensus update on Lyme borreliosis in dogs and cats. *Journal of Veterinary Internal Medicine, 32*(3), 887–903. https://doi.org/10.1111/jvim.15085

Macy, J., & Horvath, T. L. (2017). Comparative medicine: An inclusive crossover discipline. *Yale Journal of Biology and Medicine, 90*(3), 493–498. Retrieved from https://www.ncbi.nlm.nih.gov/pubmed/28955187.

McGreevy, P., Thomson, P., Dhand, N. K., Raubenheimer, D., Masters, S., Mansfield, C. S., ... Hammond, J. (2017). VetCompass Australia: A national big data collection system for veterinary science. *Animals (Basel), 7*(10). https://doi.org/10.3390/ani7100074

Mareze, M., Benitez, A. D. N., Brandao, A. P. D., Pinto-Ferreira, F., Miura, A. C., Martins, F. D. C., ... Navarro, I. T. (2019). Socioeconomic vulnerability associated to *Toxoplasma gondii* exposure in southern Brazil. *PLoS One, 14*(2), e0212375. https://doi.org/10.1371/journal.pone.0212375

Mein, G., & Grant, R. (2018). A cross-sectional exploratory analysis between pet ownership, sleep, exercise, health and neighbourhood perceptions: The Whitehall II cohort study. *BMC Geriatrics, 18*(1), 176. https://doi.org/10.1186/s12877-018-0867-3

Messenger, A. M., Barnes, A. N., & Gray, G. C. (2014). Reverse zoonotic disease transmission (zooanthroponosis): A systematic review of seldom-documented human biological threats to animals. *PLoS One, 9*(2), e89055. https://doi.org/10.1371/journal.pone.0089055

Milberger, S. M., Davis, R. M., & Holm, A. L. (2009). Pet owners' attitudes and behaviours related to smoking and second-hand smoke: A pilot study. *Tobacco Control, 18*(2), 156–158. https://doi.org/10.1136/tc.2008.028282

Miyake, K., Kito, K., Kotemori, A., Sasaki, K., Yamamoto, J., Otagiri, Y., ... Ishihara, J. (2020). Association between pet ownership and obesity: A systematic review and meta-analysis. *International Journal of Environmental Research and Public Health, 17*(10). https://doi.org/10.3390/ijerph17103498

Moreira, A. D. S., Baptista, C. T., Brasil, C. L., Valente, J. S. S., Bruhn, F. R. P., & Pereira, D. I. B. (2018). Risk factors and infection due to Cryptosporidium spp. in dogs and cats in southern Rio Grande do Sul. *Revista Brasileira de Parasitologia Veterinaria, 27*(1), 113–118. https://doi.org/10.1590/S1984-296120180012

Mossello, E., Ridolfi, A., Mello, A. M., Lorenzini, G., Mugnai, F., Piccini, C., ... Marchionni, N. (2011). Animal-assisted activity and emotional status of patients with Alzheimer's disease in day care. *International Psychogeriatrics, 23*(6), 899–905. https://doi.org/10.1017/S1041610211000226

Nagasawa, M., Kikusui, T., Onaka, T., & Ohta, M. (2009). Dog's gaze at its owner increases owner's urinary oxytocin during social interaction. *Hormones and Behavior, 55*(3), 434–441. https://doi.org/10.1016/j.yhbeh.2008.12.002

Nakajima, Y. (2017). Comparing the effect of animal-rearing education in Japan with conventional animal-assisted education. *Frontiers in Veterinary Science, 4*, 85. https://doi.org/10.3389/fvets.2017.00085

National Geographic Society. (2020). *Domestication*. Retrieved from www.nationalgeographic.org/encyclopedia/domestication/#:~:text=Powered%20by-,Domestication%20is%20the%20process%20of%20adapting%20wild%20plants%20and%20animals,and%20cared%20for%20by%20humans.

Neubauer, H., Sprague, L. D., Scholz, H., & Hensel, A. (2001). [Yersinia enterocolitica infections: 1. Impact on animal health]. *Berliner und Munchener Tierarztliche Wochenschrift, 114*(1–2), 8–12. Retrieved from: https://www.ncbi.nlm.nih.gov/pubmed/11225501.

Nogueras, M. M., Pons, I., Pla, J., Ortuno, A., Miret, J., Sanfeliu, I., & Segura, F. (2013). The role of dogs in the eco-epidemiology of Rickettsia typhi, etiological agent of Murine typhus. *Veterinary Microbiology, 163*(1–2), 97–102. https://doi.org/10.1016/j.vetmic.2012.11.043

O'Haire, M. E., Guerin, N. A., & Kirkham, A. C. (2015). Animal-assisted intervention for trauma: A systematic literature review. *Frontiers in Psychology, 6*, 1121. https://doi.org/10.3389/fpsyg.2015.01121

OIE. (2020). *OIE-Listed diseases, infections and infestations in force in 2020*. Retrieved from https://www.oie.int/animal-health-in-the-world/oie-listed-diseases-2020/.

Oliveira, D., & Franco, M. H. P. (2015). Luto pela perda do animal. In G. Casselato (Ed.), *O Resgate da Empatia: Suporte psicológico ao luto não reconhecido*. São Paulo: Summus.

Overgaauw, P. A., van Zutphen, L., Hoek, D., Yaya, F. O., Roelfsema, J., Pinelli, E., … Kortbeek, L. M. (2009). Zoonotic parasites in fecal samples and Fur from dogs and cats in The Netherlands. *Veterinary Parasitology, 163*(1–2), 115–122. https://doi.org/10.1016/j.vetpar.2009.03.044

Overgaauw, P. A. M., Vinke, C. M., Hagen, M., & Lipman, L. J. A. (2020). A one health perspective on the human-companion animal relationship with emphasis on zoonotic aspects. *International Journal of Environmental Research and Public Health, 17*(11). https://doi.org/10.3390/ijerph17113789

Ownby, D. R., Johnson, C. C., & Peterson, E. L. (2002). Exposure to dogs and cats in the first year of life and risk of allergic sensitization at 6 to 7 years of age. *JAMA, 288*(8), 963–972. https://doi.org/10.1001/jama.288.8.963

Partyka, M. L., Bond, R. F., Farrar, J., Falco, A., Cassens, B., Cruse, A., & Atwill, E. R. (2014). Quantifying the sensitivity of scent detection dogs to identify fecal contamination on raw produce. *Journal of Food Protection, 77*(1), 6–14. https://doi.org/10.4315/0362-028X.JFP-13-249

Pasmans, F., Bogaerts, S., Braeckman, J., Cunningham, A. A., Hellebuyck, T., Griffiths, R. A., … Martel, A. (2017). Future of keeping pet reptiles and amphibians: Towards integrating animal welfare, human health and environmental sustainability. *Veterinary Record, 181*(17), 450. https://doi.org/10.1136/vr.104296

Pastorinho, M. R., & Sousa, A. C. A. (2020). *Pets as sentinels, forecasters and promoters of human health*. Cham: Springer.

Pastorinho, M. R., & Sousa, A. C. (2020a). Pets as sentinels of human exposure to neurotoxic metals. In S. A. Pastorinho M (Ed.), *Pets as sentinels, forecasters and promoters of human health*. Cham: Springer.

Pastorinho, M. R., & Sousa, A. C. (2020b). *Pets as sentinels, forecasters and promoters of human health*. Cham: Springer.

Pawelczyk, O., Asman, M., & Solarz, K. (2019). The molecular detection of Anaplasma phagocytophilum and Rickettsia spp. in cat and dog fleas collected from companion animals. *Folia Parasitologica, 66*. https://doi.org/10.14411/fp.2019.020

Pennisi, M. G., Egberink, H., Hartmann, K., Lloret, A., Addie, D., Belak, S., … Horzinek, M. C. (2013). *Yersinia pestis* infection in cats: ABCD guidelines on prevention and management. *Journal of Feline Medicine and Surgery, 15*(7), 582–584. https://doi.org/10.1177/1098612X13489218

Petersson, E., & Athlin, S. (2017). Cat-bite-induced Francisella tularensis infection with a false-positive serological reaction for Bartonella quintana. *JMM Case Reports, 4*(2), e005071. https://doi.org/10.1099/jmmcr.0.005071

Pinello, K. C., Niza-Ribeiro, J., Fonseca, L., & de Matos, A. J. (2019). Incidence, characteristics and geographical distributions of canine and human non-Hodgkin's lymphoma in the Porto region (North West Portugal). *Veterinary Journal, 245*, 70–76. https://doi.org/10.1016/j.tvjl.2019.01.003

Pinho, S. S., Carvalho, S., Cabral, J., Reis, C. A., & Gartner, F. (2012). Canine tumors: A spontaneous animal model of human carcinogenesis. *Translational Research, 159*(3), 165–172. https://doi.org/10.1016/j.trsl.2011.11.005

Poma, G., Malarvannan, G., & Covaci, A. (2020). Pets as sentinels of indoor contamination. In S. A. Pastorinho M (Ed.), *Pets as sentinels, forecasters and promoters of human health*. Cham: Springer.

Priestnall, S. L., Wiinberg, B., Spohr, A., Neuhaus, B., Kuffer, M., Wiedmann, M., & Simpson, K. W. (2004). Evaluation of "*Helicobacter heilmannii*" subtypes in the gastric mucosas of cats and dogs. *Journal of Clinical Microbiology, 42*(5), 2144–2151. https://doi.org/10.1128/jcm.42.5.2144-2151.2004

The power of data. *Veterinary Record, 187*(Suppl. 1), (2020), 8. https://doi.org/10.1136/vr.m4369

Purewal, R., Christley, R., Kordas, K., Joinson, C., Meints, K., Gee, N., & Westgarth, C. (2017). Companion animals and child/adolescent development: A systematic review of the evidence. *International Journal of Environmental Research and Public Health, 14*(3). https://doi.org/10.3390/ijerph14030234

Radford, A. D., Noble, P. J., Coyne, K. P., Gaskell, R. M., Jones, P. H., Bryan, J. G., … Dawson, S. (2011). Antibacterial prescribing patterns in small animal veterinary practice identified via SAVSNET: The small animal veterinary surveillance network. *Veterinary Record, 169*(12), 310. https://doi.org/10.1136/vr.d5062

Redding, L. E., Kelly, B. J., Stefanovski, D., Lautenbach, J. K., Tolomeo, P., Cressman, L., … Lautenbach, E. (2020). Pet ownership protects against recurrence of clostridioides difficile infection. *Open Forum Infectious Diseases, 7*(1), ofz541. https://doi.org/10.1093/ofid/ofz541

Reeve, C. (2018). Biomedical scent detection dogs: Would they pass as a health technology? *Pet Behaviour Science, 6*. https://doi.org/10.21071/pbs.v0i6.10785

Reif, J. S. (2011). Animal sentinels for environmental and public health. *Public Health Reports, 126*(Suppl. 1), 50–57. https://doi.org/10.1177/00333549111260S108

Reif, J. S. (2011). Animal sentinels for environmental and public health. *Public Health Reports, 126*(Suppl. 1), 50–57. https://doi.org/10.1177/00333549111260S108

Reisbig, A. M. J., Hafen, M., Jr., Siqueira Drake, A. A., Girard, D., & Breunig, Z. B. (2017). Companion animal death. *Omega, 75*(2), 124–150. https://doi.org/10.1177/0030222815612607

Rial-Berriel, C., Henríquez-Hernández, L. A., & Luzardo, O. P. (2020). Role of pet dogs and cats as sentinels of human exposure to polycyclic aromatic hydrocarbons. In S. A. Pastorinho M (Ed.), *Pets as sentinels, forecasters and promoters of human health*. Cham: Springer.

Riedler, J., Eder, W., Oberfeld, G., & Schreuer, M. (2000). Austrian children living on a farm have less hay fever, asthma and allergic sensitization. *Clinical & Experimental Allergy, 30*(2), 194–200. https://doi.org/10.1046/j.1365-2222.2000.00799.x

Rijks, J. M., Cito, F., Cunningham, A. A., Rantsios, A. T., & Giovannini, A. (2016). Disease risk assessments involving companion animals: An overview for 15 selected pathogens taking a European perspective. *Journal of Comparative Pathology, 155*(1 Suppl. 1), S75–S97. https://doi.org/10.1016/j.jcpa.2015.08.003

Romero, J. E., Cepeda, J., Quinn, P., & Underwood, S. C. (2018). Prisoner rehabilitation through animal-assisted activities in Argentina: The Huellas de Esperanza prison dog programme. *Revue scientifique et technique, 37*(1), 171–180. https://doi.org/10.20506/rst.37.1.2750

Safra, N., Bassuk, A. G., Ferguson, P. J., Aguilar, M., Coulson, R. L., Thomas, N., … Bannasch, D. L. (2013). Genome-wide association mapping in dogs enables identification of the homeobox gene, NKX2-8, as a genetic component of neural tube defects in humans. *PLoS Genetics, 9*(7), e1003646. https://doi.org/10.1371/journal.pgen.1003646

Samir, A., Abdel-Moein, K. A., & Zaher, H. M. (2020). Emergence of penicillin-macrolide-resistant Streptococcus pyogenes among pet animals: An ongoing public health threat. *Comparative Immunology, Microbiology and Infectious Disease, 68*, 101390. https://doi.org/10.1016/j.cimid.2019.101390

Sanchez-Vizcaino, F., Jones, P. H., Menacere, T., Heayns, B., Wardeh, M., Newman, J., … McConnell, K. (2015). Small animal disease surveillance. *Veterinary Record, 177*(23), 591–594. https://doi.org/10.1136/vr.h6174

Sato, R., Fujiwara, T., Kino, S., Nawa, N., & Kawachi, I. (2019). Pet ownership and children's emotional expression: Propensity score-matched analysis of longitudinal data from Japan. *International Journal of Environmental Research and Public Health, 16*(5). https://doi.org/10.3390/ijerph16050758

Schmidt, P. L. (2009). Companion animals as sentinels for public health. *Veterinary Clinics of North America Small Animal Practice, 39*(2), 241–250. https://doi.org/10.1016/j.cvsm.2008.10.010

Schmidt, P. L. (2009). Companion animals as sentinels for public health. *Veterinary Clinics of North America: Small Animal Practice, 39*(2), 241–250. https://doi.org/10.1016/j.cvsm.2008.10.010

Schuller, S., Francey, T., Hartmann, K., Hugonnard, M., Kohn, B., Nally, J. E., & Sykes, J. (2015). European consensus statement on leptospirosis in dogs and cats. *Journal of Small Animal Practice, 56*(3), 159–179. https://doi.org/10.1111/jsap.12328

Schuppli, C. A., Fraser, D., & Bacon, H. J. (2014). Welfare of non-traditional pets. *Revue scientifique et technique, 33*(1), 221–231. https://doi.org/10.20506/rst.33.1.2287

Seidman, S. M. (2001). *The pet surplus : What every dog and cat owner can due to help reduce it. United States: S.M.* Seidman: Xlibris Corp. distributor.

Seo, I. S., Lee, H. G., Koo, B., Koh, C. S., Park, H. Y., Im, C., & Shin, H. C. (2018). Cross detection for odor of metabolic waste between breast and colorectal cancer using canine olfaction. *PLoS One, 13*(2), e0192629. https://doi.org/10.1371/journal.pone.0192629

Shan Neo, J. P., & Tan, B. H. (2017). The use of animals as a surveillance tool for monitoring environmental health hazards, human health hazards and bioterrorism. *Veterinary Microbiology, 203*, 40–48. https://doi.org/10.1016/j.vetmic.2017.02.007

Shan Neo, J. P., & Tan, B. H. (2017). The use of animals as a surveillance tool for monitoring environmental health hazards, human health hazards and bioterrorism. *Veterinary Microbiology, 203*, 40–48. https://doi.org/10.1016/j.vetmic.2017.02.007

Sharpley, C., Veronese, N., Smith, L., Lopez-Sanchez, G. F., Bitsika, V., Demurtas, J., ... Jackson, S. E. (2020). Pet ownership and symptoms of depression: A prospective study of older adults. *Journal of Affective Disorders, 264*, 35–39. https://doi.org/10.1016/j.jad.2019.11.134

Shirasu, M., & Touhara, K. (2011). The scent of disease: Volatile organic compounds of the human body related to disease and disorder. *Journal of Biochemistry, 150*(3), 257–266. https://doi.org/10.1093/jb/mvr090

Siegel, J. M. (1990). Stressful life events and use of physician services among the elderly: The moderating role of pet ownership. *Journal of Personality and Social Psychology, 58*(6), 1081–1086. https://doi.org/10.1037//0022-3514.58.6.1081

Silva, K.,L. M. (2020). Companion animals and human health: On the need for a comprehensive research agenda toward clinical implementation. In S. A. Pastorinho M (Ed.), *Pets as sentinels, forecasters and promoters of human health*. Cham: Springer.

Smith, M., King, C., Davis, M., Dickson, A., Park, J., Smith, F., ... Flowers, P. (2018). Pet owner and vet interactions: Exploring the drivers of AMR. *Antimicrobial Resistance and Infection Control, 7*, 46. https://doi.org/10.1186/s13756-018-0341-1

Sobo, E. J., Eng, B., & Kassity-Krich, N. (2006). Canine visitation (pet) therapy: Pilot data on decreases in child pain perception. *Journal of Holistic Nursing, 24*(1), 51–57. https://doi.org/10.1177/0898010105280112

Society of Companion Animals Studies. (2019). *Animal assisted interventions: SCAS code of practice for the UK*. Retrieved from https://hub.careinspectorate.com/media/3857/animal-assisted-interventionsscas-code-of-practice-for-the-uk.pdf.

Sonoda, H., Kohnoe, S., Yamazato, T., Satoh, Y., Morizono, G., Shikata, K., ... Maehara, Y. (2011). Colorectal cancer screening with odour material by canine scent detection. *Gut, 60*(6), 814–819. https://doi.org/10.1136/gut.2010.218305

Sousa, L. L. F., Alencar, C. H. M., Almeida, A. M. P., & Cavalcanti, L. P. G. (2017). Seroprevalence and spatial distribution dynamics of *Yersinia pestis* antibodies in dogs and cats from plague foci in the State of Ceara, Northeastern Brazil. *Revista da Sociedade Brasileira de Medicina Tropical, 50*(6), 769–776. https://doi.org/10.1590/0037-8682-0278-2017

Sterneberg-van der Maaten, T., Turner, D., Van Tilburg, J., & Vaarten, J. (2016). Benefits and risks for people and livestock of keeping companion animals: Searching for a healthy balance. *Journal of Comparative Pathology, 155*(1 Suppl. 1), S8–S17. https://doi.org/10.1016/j.jcpa.2015.06.007

Stratton, J., Tagliaro, E., Weaver, J., Sherman, D. M., Carron, M., Di Giacinto, A., ... Caya, F. (2019). Performance of veterinary services pathway evolution and one health aspects. *Revue scientifique et technique, 38*(1), 291–302. https://doi.org/10.20506/rst.38.1.2961

Stroud, C., Dmitriev, I., Kashentseva, E., Bryan, J. N., Curiel, D. T., Rindt, H., ... Harman, R. J. (2016). A One Health overview, facilitating advances in comparative medicine and translational research. *Clinical and Translational Medicine, 5*(Suppl. 1), 26. https://doi.org/10.1186/s40169-016-0107-4

Takashima, G. K., & Day, M. J. (2014). Setting the One Health agenda and the human-companion animal bond. *International Journal of Environmental Research and Public Health, 11*(11), 11110–11120. https://doi.org/10.3390/ijerph111111110

Takeuchi, T., D'Itri, F. M., Fischer, P. V., Annett, C. S., & Okabe, M. (1977). The outbreak of Minamata disease (methyl mercury poisoning) in cats on Northwestern Ontario reserves. *Environmental Research, 13*(2), 215–228. Retrieved from https://www.ncbi.nlm.nih.gov/pubmed/862594.

Taylor, M. T., McCready, J., Broukhanski, G., Kirpalaney, S., Lutz, H., & Powis, J. (2018). Using dog scent detection as a point-of-care tool to identify toxigenic *Clostridium difficile* in stool. *Open Forum Infectious Diseases, 5*(8), ofy179. https://doi.org/10.1093/ofid/ofy179

The International Association of Human-Animal Interaction Organizations (IAHAIO). (2014). *The IAHIO definitions for animal assisted intervention and guidelines for wellness of animals involved.* Retrieved from http://www.iahaio.org/new/fileuploads/9313IAHAIO%20WHITE%20PAPER%20TASK%20FORCE%20-%20FINAL%20REPORT.pdf.

Tun, H. M., Konya, T., Takaro, T. K., Brook, J. R., Chari, R., Field, C. J., ... Investigators, C. S. (2017). Exposure to household furry pets influences the gut microbiota of infant at 3-4 months following various birth scenarios. *Microbiome, 5*(1), 40. https://doi.org/10.1186/s40168-017-0254-x

Tzivian, L., Friger, M., & Kushnir, T. (2015). Associations between stress and quality of life: Differences between owners keeping a living dog or losing a dog by euthanasia. *PLoS One, 10*(3), e0121081. https://doi.org/10.1371/journal.pone.0121081

Vegue Parra, E., Hernandez Garre, J. M., & Echevarria Perez, P. (2021). Benefits of dog-assisted therapy in patients with dementia residing in aged care centers in Spain. *International Journal of Environmental Research Public Health, 18*(4). https://doi.org/10.3390/ijerph18041471

Vincze, S., Stamm, I., Kopp, P. A., Hermes, J., Adlhoch, C., Semmler, T., ... Walther, B. (2014). Alarming proportions of methicillin-resistant *Staphylococcus aureus* (MRSA) in wound samples from companion animals, Germany 2010-2012. *PLoS One, 9*(1), e85656. https://doi.org/10.1371/journal.pone.0085656

Walther, B., Hermes, J., Cuny, C., Wieler, L. H., Vincze, S., Abou Elnaga, Y., ... Lubke-Becker, A. (2012). Sharing more than friendship–nasal colonization with coagulase-positive staphylococci (CPS) and co-habitation aspects of dogs and their owners. *PLoS One, 7*(4), e35197. https://doi.org/10.1371/journal.pone.0035197

Walther, B., Tedin, K., & Lubke-Becker, A. (2017). Multidrug-resistant opportunistic pathogens challenging veterinary infection control. *Veterinary Microbiology, 200*, 71–78. https://doi.org/10.1016/j.vetmic.2016.05.017

Ward, A., Arola, N., Bohnert, A., & Lieb, R. (2017). Social-emotional adjustment and pet ownership among adolescents with autism spectrum disorder. *Journal of Communication Disorders, 65*, 35–42. https://doi.org/10.1016/j.jcomdis.2017.01.002

Weiss, J., & Jones, B. (2020). Using cats as sentinels for human indoor exposure to organic contaminants and potential effects on the thyroid hormone system. In S. A. Pastorinho M (Ed.), *Pets as sentinels, forecasters and promoters of human health.* Cham: Springer.

WHO. (2019). *Mental disorders.* Retrieved from https://www.who.int/news-room/fact-sheets/detail/mental-disorders.

WHO. *Zoonoses. Health topics.* Retrieved from https://www.who.int/topics/zoonoses/en/.

Wilkes, M. S., Conrad, P. A., & Winer, J. N. (2019). One health-one education: Medical and veterinary inter-professional training. *Journal of Veterinary Medical Education, 46*(1), 14–20. https://doi.org/10.3138/jvme.1116-171r

Williams, D. L. (2018). Optimising tear replacement rheology in canine keratoconjunctivitis sicca. *Eye (London), 32*(2), 195–199. https://doi.org/10.1038/eye.2017.272

Wood, L., Giles-Corti, B., & Bulsara, M. (2005). The pet connection: Pets as a conduit for social capital? *Social Science & Medicine, 61*(6), 1159–1173. https://doi.org/10.1016/j.socscimed.2005.01.017

Yeh, T. L., Lei, W. T., Liu, S. J., & Chien, K. L. (2019). A modest protective association between pet ownership and cardiovascular diseases: A systematic review and meta-analysis. *PLoS One, 14*(5), e0216231. https://doi.org/10.1371/journal.pone.0216231

Yousfi, M., Touati, A., Mairi, A., Brasme, L., Gharout-Sait, A., Guillard, T., & De Champs, C. (2016). Emergence of carbapenemase-producing *Escherichia coli* isolated from companion animals in Algeria. *Microbial Drug Resistance, 22*(4), 342–346. https://doi.org/10.1089/mdr.2015.0196

Zajkowska, J., Krol, M., Falkowski, D., Syed, N., & Kamienska, A. (2016). Capnocytophaga canimorsus - an underestimated danger after dog or cat bite - review of literature. *Przeglad Epidemiologiczny, 70*(2), 289–295. Retrieved from: https://www.ncbi.nlm.nih.gov/pubmed/27837588.

Zellner, B., & Huntley, J. F. (2019). Ticks and tularemia: Do we know what we don't know? *Frontiers in Cellular and Infection Microbiology, 9*, 146. https://doi.org/10.3389/fcimb.2019.00146

Zierenberg-Ripoll, A., Pollard, R. E., Stewart, S. L., Allstadt, S. D., Barrett, L. E., Gillem, J. M., & Skorupski, K. A. (2018). Association between environmental factors including second-hand smoke and primary lung cancer in dogs. *Journal of Small Animal Practice, 59*(6), 343–349. https://doi.org/10.1111/jsap.12778

Chapter 5

Food and water security and safety for an ever-expanding human population

João Niza-Ribeiro[a,b,c]
[a]*Departamento de Estudo de Populações, Vet-OncoNet, ICBAS, Instituto de Ciências Biomédicas Abel Salazar, Universidade do Porto, Porto, Portugal;* [b]*EPIUnit - Instituto de Saúde Pública, Universidade do Porto, Porto, Portugal;* [c]*Laboratório para a Investigação Integrativa e Translacional em Saúde Populacional (ITR), Porto, Portugal*

1. Introduction

In a changing world there are opportunities, challenges, and threats. Ending hunger and malnutrition, improving food security and food safety while preserving the health of people and of our planet is one major endeavor of our times. The human health depends on planetary health. Planetary health involves humans, ecosystems, and environment. Ensuring food security and food safety is the second of seventeen Global Goals (SDG - Sustainable Development Goals) adopted by United Nations Member States in 2015 (United Nations, 2020). Producing safe, nutritious, and sustainable food to eight billion souls is a massive challenge to the economy and the ecosystems of the planet. There is a clear tension and competition between the accomplishment of this goal and other Global Goals like the 13th (Climate Action), the 14th (Life Below Water) and the 15th (Life on Land), to mention just a few. Innovative solutions are needed to solve these complex interactions.

Food security is a concept focusing on the populations food's availability, quality and sustainability and strongly related with the Global Food System (GFS) (Parsons et al., 2019) which is responsible for feeding the human world population. Indeed, the GFS is not a formal or hierarchically organized structure, it is rather a complex system of independent, interacting actors trying to cope with the market demand and supply, struggling for price and the control a supply chains, at the same time competing for customers. The regulation of this GFS and the control of food safety can be considered satisfactory, especially among high-income countries, but much remains to be achieved. Important deficiencies remain associated with the way in which the

One Health. https://doi.org/10.1016/B978-0-12-822794-7.00003-4

GFS operates (Godfray et al., 2011). Food security and human health are influenced by market and marketing drivers. Food security deficiencies still exists, as almost one billion human beings still suffer from hunger whereas almost another one billion of people suffers from malnutrition and obesity (FAO, IFAD, UNICEF, WFP, & WHO, 2021). Some other conspicuous inefficiencies of the GFS are the negative impacts on climate and environment and the burden of food waste.

Food safety is a concept focusing on sound and safe food production, transformation, distribution and consumption. Food safety is considered as a right of the citizens and consumers around the world. *People have the right to expect the food they eat to be safe and suitable for consumption* (Codex Alimentarius Commission, FAO, & WHO, 2013). Disparities in food safety at global level are enormous, with developed high-income countries exhibiting better performance in human health indicators (Havelaar et al., 2015). Efforts and programs to improve food safety around the globe are however being carried under the umbrella of WHO and FAO and several NGOs.

Political instability and conflicts at regional at local levels strongly compromise food safety and food security (Läderach, Pacillo, Thornton, Osorio, & Smith, 2021). Organized food safety systems, like the ones existing in the European Union (EU) or in North America, which combines involvement and responsibility from stakeholders and from the governmental authorities have proved to be efficient in reducing the burden of disease associated with food consumption.

Climate changes, wars and political instability across the globe, massive exodus of populations, the impact of climate change on traditional agro-systems like many in Africa, South or Central America are associated with hunger and malnutrition among almost one billion people and to population exodus, challenging the Second Global Goal (Läderach et al., 2021). Resilient, sustainable agro-food systems and politically stable institutions are necessary to revert these problems (Godfray et al., 2011).

The One World One Health concept was formally established in 2004[1] by the conservationist community and it deserved attention from the international agencies and developed countries which funded collaborative agreements to the fight for Transboundary Animal Diseases (TAD) and potential pandemic diseases like influenza, foot and mouth disease or rabies (FAO, OIE, & WHO, 2010); recently (Gruetzmacher et al., 2021) these principles were revisited, in 2018 in Berlin, and their importance to achieve the goal of a healthy world for humans, animals and life in general across the planet was stressed. Food production is based on land and water use hence OH concerns and emergences relate directly or indirectly with food production and food supply.

1. http://www.oneworldonehealth.org/.

The recent COVID 19 pandemic due to the SARS-CoV 2 emergence demonstrates that the worst fears involving the emergence of a human serious infectious of zoonotic origin were realistic (United Nations & FAO, 2020). The expected global impact in human lives loss and in the world, economy is devastating and it exceeds previous forecasts from the World Bank for a hypothetical Influenza Pandemic.

Our purpose is to carry on a review of food security and food safety in the context of one health approach. We will start by addressing the concepts and fundamentals of each concept and proceed with a comprehensive assessment of the interactions between these concepts to create a global landscape for food security and food systems, current global situation, trends, and challenges to end with an overview of food security and food system policies.

2. Food security, food safety and one health

Ensuring food security and food safety is the second of seventeen Global Goals (Sustainable Development Goals) adopted by United Nations Member States in 2015 (United Nations, 2020). However, producing food to a growing population of eight billion souls is a massive challenge to the planet and there is a clear tension and competition between the accomplishment of this goal (second Zero Hunger) and other Global Goals in particular the sixth (Clean Water and Sanitation), 12th (Responsible Consumption and Production), 13th (Climate Action), 14th (Life Below Water) and the 15th (Life on Land). Innovative solutions are needed to solve these complex interactions (Gill, 2018; Hawkes & Parsons, 2019; Sonnino, 2020).

A complex set a definitions and concepts is needed to fully understand the relations between food security, food safety and one health.

3. Food security

In November 13, 1996 the world's leaders reached a formal agreement, that every people in the world had the right to have food security (FAO, 1996). The definition used in the agreement is that "Food security exists when all people, at all times, have physical and economic access to sufficient, safe and nutritious food that meets their dietary needs and food preferences for an active and healthy life" (World Food Summit, 1996) (FAO, 2006). Four main dimensions emerge from this definition: 1) physical *availability*, 2) economical and physical *access*, 3) food *utilization* (nutritional availability) and 4) the *stability* of supply of previous dimensions. Food security (FS) is achieved when all dimensions are met simultaneously. Two new dimensions were added later (HLPE, 2020) to this concept: 5) *sustainability*, linked to the long-term ability of food systems to not produce negative environmental, social and economic impacts and 6) *agency*, associated with people empowerment to choose, It results also from the previous definition that Food Insecurity is the lack of FS.

Food Insecurity, in the form of either hunger or malnutrition, can be chronic or transitory in nature and strategies to address this issue need to take into account both the causes and the effects of the problem (Devereux, 2006). Options to tackle severe food insecurity, include targeted direct feeding programs, food-for-work programs and income transfer programs (Kostas Stamoulis, 2003). Conflicts, climate variability and negative dynamics in countries or regional economies, recently aggravated by Covid-19 pandemics are currently the main forces behind food security problems worldwide (FAO et al., 2021).

Food insecurity happens asymmetrically between countries across the world and high-income countries face different problems, *v.g.* poor access or utilization, than low or middle-income countries which face problems of restricted food availability or poor access. Asia and Africa are the continents more afflicted by the problem (FAO et al., 2021). It also should be kept in mind that *hunger* and *poverty* are closely associated in a feedback negative loop and simultaneously attacking both problems is often considered the best of the approaches (EC/FAO, 2008) as addressed in depth by Amartya-Sen (Sen, 1981).

One of the faces of food insecurity is *malnutrition*. Malnutrition is considered a double burden problem in developing and developed countries (Menon & Peñalvo, 2019). The developmental, economic, social, and medical impacts of the global burden of malnutrition are serious and lasting, for individuals and their families, for communities and for countries (WHO, 2021). Definition of malnutrition refers to deficiencies, excesses, or imbalances in a person's intake of energy and/or nutrients. Three groups of conditions are associated to this health burden closely depending from the type of economic development of the each. In low-income and middle-income countries 1) undernutrition and 2) micronutrient-related malnutrition. These conditions are associated with child physical and mental under development, and reduced resistance to infectious diseases. In high and middle-income countries 3) overweight, obesity and diet related non-communicable diseases, which strongly impacts on the reduction of lifetime expectancy and quality and dispose to other comorbidities. Globally 1.9 billion adults are overweight or obese, while 462 million are underweight. Around 45% of deaths among children under 5 years of age are linked to undernutrition (WHO, 2021).

Malnutrition levels remain alarming worldwide and deserves closer attention. The following statistics refer to children under 5 years of age. Stunting (which refers to a child who is too short for the age) affects 144 million children and it is declining all over the world; almost 40% of these children are in Asia and a slight increase in number was noticed in Africa. Wasting (refers to a child who is too thin for its weight) still impacts the lives of 47 million children. Wasted children are at higher risk of death. Southern and Africa have the highest prevalence of wasted children under 5 years of age.

Overweight (refers to a child who is too heavy for its weight). This is a growing problem affecting almost 40 million kids worldwide; all regions in the world, except Europe, are experiencing an increase of prevalence. Southeastern Asia and Northern America are the only sub-regions which had a significant increase in the number of overweight children since 2000. Not surprisingly, 64% of stunted and 75% of wasted children live in lower or middle-income countries, while 41% of the overweight ones live in high or middle-income countries (United Nations Children's Fund (UNICEF), 2020).

The SDG were facing a trend of positive progress worldwide, since 2014, with indicators of malnutrition and hunger improving in a slow but consistent manner (FAO et al., 2021). With the emergence of Covid-19 there was a sudden but steep inversion of these tendencies and the optimism in reaching the 2030 goals on food security was severely compromised. Without new efforts focusing on humanitarian assistance, social protection systems and on building sustainable and resilient food systems there will be objective difficulties in solving food security problems (United Nations & FAO, 2020).

But the emergence of new menaces of transitory food insecurity in well developed economies was perceived beyond doubt by the European Commission given the recent, and still ongoing, COVID-19 crises (European Commission, 2021). Unforeseen events like impressive shortages over the supply chain, abrupt changes in consumer behavior and habits, losses of certain markets were consecutive to the pandemic onset. The disruptive potential of unprecedented threats happening recurringly every year as droughts, floods, forest fires, biodiversity loss and new and the food chain faces increasing pests, was perceived as a driver for contingency policies focusing on responses to crises and emergencies (European Commission, 2021).

Food systems can be powerful drivers to ensure sustainable food security. These will be addressed deeper in the section of GFS. It will become clear, further down in the text, the critical importance of global food trade in contemporary world to ensure food security. I the last section, FS policies will be discriminated in detail.

Food waste is another remarkable problem worldwide. Food waste and food loss represent occurs along the food chains at production, transformation, distribution, and consumption. Estimates in EU are that 88 million tonnes yearly are wasted or lost. Policies addressed to stakeholders, consumers and decision makers need to be developed to reduce this burden.

4. Food safety

In the context of food security, the application of food safety principles to food production chains contributes decisively to the supply of safe and nutritious food. The focus of food safety is the food chain, from "farm to fork" (Jouve, Stringer, & Baird-Parker, 1998). The food chain can be defined as a series of phases starting with primary production of raw materials and feed, followed by production of vegetables or animal food, transformation, distribution, retail and consumption (Choffnes, Relman, Olsen, Hutton, & Mack, 2012; HLPE, 2017). The main objective of food safety is to protect the health and lives of human population from risks and hazards present in food (Jouve, 1998).

4.1 Fundamentals, scope, and principles

The fundamental principle underlying food safety is that "*People have the right to expect the food they eat to be safe and suitable for consumption*" (Codex Alimentarius Commission et al., 2013). The Codex Alimentarius Commission (CAC) is the world reference organization for standard setting and technical assistance in food trade as well as for assistance in trade disputes between countries under the World Trade Organization (WTO) agreements. Food safety is determinant to protect human health and human lives, to ensure economic prosperity, market access of agriculture and food products, and contributes to ensure safe tourism across the globe.

The principles and tools of food safety are established by the worldwide reference on food hygiene texts, the *Food Hygiene Basic Texts* of the *Codex Alimentarius Commission* (Codex Alimentarius Commission et al., 2013). In 1998, Jean Louis Jouve published a reference paper (Jouve, 1998) reviewing and proposing the guiding principles and Architecture of food safety. The principles were: 1) science-based and 2) flexible food law; 3) the mandatory use risk assessment and 4) adoption of proportionate legal requirements; 5) the adoption of preventive and proactive strategies, 6) with the involvement of all stakeholders without relaxing the identification and attribution of responsibilities. The architecture lied in four pillars building: 1) formulation of food safety and food safety objectives; 2) stakeholders assignment of responsibilities for adequate adoption and implementation of food safety systems; 3) strong and efficient food control and inspection activities; 4) evaluation. These were adopted by the European Commission "*White Paper on Food Safety*" (Commission of the European Communities, 2000) and later included in the Regulation 178/2002,[2] the European current Food Law.

4.1.1 The agreement on sanitary and phytosanitary measures (SPS)[3]

It will become clear, further down in the text, the crucial importance of global food trade in contemporary world to achieve food security. At the basis of feed and food flows across the countries and regions of the globe there is the SPS

2. https://eur-lex.europa.eu/legal-content/EN/ALL/?uri=celex%3A32002R0178.
3. The WTO: https://www.wto.org/index.htm; Urugay Rond Negotiations: https://www.wto.org/english/docs_e/legal_e/legal_e.htm#finalact; The SPS Agrement: https://www.wto.org/english/docs_e/legal_e/15sps_01_e.htm.

Agreement (WTO, 1995). The *Agreement on the Application of Sanitary and Phytosanitary Measures* (the "SPS Agreement") entered into force with the establishment of the World Trade Organization on January 1, 1995. It concerns the application of food safety and animal and plant health regulations. The SPS Agreement are part of Uruguay Round negotiations, signed at the Marrakesh ministerial meeting in April 1994. Along with taxes and technical barriers – product standards - to trade (BTT) the SPS Agreement establishes the current basis for world food trade. The CAC is the reference organism to WTO in matters of food safety and standards.

The fundamental trade principle behind these measures is that "*no Member* [of the WTO] *should be prevented from adopting or enforcing measures necessary to protect human, animal or plant life or health*", but the adoption of such measures shall be "*not applied in a manner which would constitute a means of arbitrary or unjustifiable discrimination between Members where the same conditions prevail or a disguised restriction on international trade*" and, therefore, this measures should not constitute unjustified barriers to trade. Several principles adopted in this agreement like harmonization, Equivalence, Risk Assessment, Regionalization, Control Inspection and Approval Procedures are the core of today's food animal risk assessment and management across the globe.

4.1.2 The scope of food safety

The scope of food safety involves several domains of food. Several definitions are available, but we follow the one defined by the EU food law, "food", or "foodstuff", means any substance or product, whether processed, partially processed or unprocessed, intended to be, or reasonably expected to be ingested by humans, including "drink" (European Union, 2002). Food safety definition, *is the assurance that food does not harm consumer's health when prepared and/or eaten according to its intended use* (Codex Alimentarius Commission et al., 2013). However, other dimensions can be considered under the framework of food safety: after the European Food Law, food safety includes protecting consumer's interests (including from fraud and adulteration) and transparency (European Union, 2002), food loss and food waste (FAO, 2017). These last addressed under section Future of Agriculture.

Technically safe food is neither injurious nor unfit for consumption. Suitable food means that there is the *assurance that food is acceptable for human consumption according to its intended use*. This means that food must be clean and properly managed and stored and free of human health hazards (Codex Alimentarius Commission et al., 2013). Hazards in food can be of biological, chemical, or physical nature and all phases of the food chain must be covered by hygiene preventive and control systems measures or activities (Codex Alimentarius Commission et al., 2013; Choffnes et al., 2012).

4.2 The burden and epidemiology of food safety

Food borne hazards are biological, chemical, or physical agents or conditions present in food that brings the risk of human health to unacceptable levels (Codex Alimentarius Commission et al., 2013). General detailed review of important FB Hazards and risks associated can be found in recent reports from WHO (WHO Europe, 2017; WHO & FERG, 2015).

According to the World Health Organization estimates (WHO, 2019), the global burden of diarrheal disease in 2019 was 79.3 million DALY[4] (1029 DALYs per 100,000 population). Diarrhea ranked third in the same rank in 2000, with 161.0 million (2566 DALYs/100,000). This is a 2.5-fold reduction in importance and represents already a remarkable improvement in food safety assurance to the populations across the world and demonstrates that the global effort in the reduction of food borne disease is being effective. There is still a long way to go, in the potential for improvement, since the burden estimated for 2010 in Europe and North America was around 40 DALY/100,000 against 1270 in some African regions, 31-fold (Havelaar et al., 2015). South Asia regions rank in the middle: around 700 DALY/100,000 persons. This 31-fold difference in risk may be justified by factors associated with public health infrastructure, human health systems and veterinary services in place at country level. But the food safety approach adopted in high income countries, remains the foundation of sustainable food safety and unequivocally account for the success in achieving a low burden of food-born disease (FBD).

Food borne disease agents (hazards) includes a vast list (WHO & FERG, 2015). Global estimates, provided by the Foodborne Disease Burden Epidemiology Reference Group (FERG) contribute to assess relative importance of diarrheal disease, caused by bacteria or viruses. *Campylobacter* spp., *Escherichia coli*, Shiga-toxin *E. coli*, and non-typhoidal *Salmonella* are associated with 84% of disease and 59% of death cases of FBD. Invasive bacteria or viruses like *Listeria monocytogenes*, *Salmonella enterica* Typhi are associated with 4% of illness and 39% casualties. Enteric protozoa such as *Cryptosporidium,* and *Giardia* are responsible for 10% illness and 2% death while invasive parasites like *Ascaris* spp., Cestodes cause 2% disease and 0.2% death. Toxins such as Aflatoxins and Cassava are the main chemicals hazards worldwide; they are not so relevant in the European Union and North America regions, given the food safety measures in place, aimed at preventing contaminated food to access the market (European Union, 2020).

4. DALY: Disability adjusted Life Years. According to World Health Organization, one DALY represents the loss of the equivalent of one year of full health. DALYs for a disease or health condition are the sum of the years of life lost to due to premature mortality (YLLs) and the years lived with a disability (YLDs) due to prevalent cases of the disease or health condition in a population.

In the case of EU, high-income countries region, diseases like *Campylobacter* spp. (aprox. 60 cases/100,000 persons) and non-typhoidal *Salmonella* (aprox. 21/100,000), and Shiga-toxin *E. coli* (aprox. 2/100,000) are responsible for the highest levels or disease incidence but have low fatality rates, 0.03%, 0.19% and 0.22%, respectively. Diseases like *L. monocytogenes* (<1/100,000), on the contrary, have low incidence but high hospitalization with fatality rate of 97% (EFSA & ECDC, 2019). In the EU the success in control has some degree of success variability. Agents like *Brucella melitentis* prevailing in some limited territories, whereas *Campylobacter*, VTEC/ETEC[5] or *L. monocytogenes* the success in control of human incidence is far from being complete (EFSA & ECDC, 2019). Concerning *Mycobacteria bovis* this has been generally well controlled, but reemergence is seen in certain territories associated with endemic infection that has become endemic in wildlife worldwide (Palmer, 2013). In Europe, badgers in the United Kingdom and Ireland, wild boars and cervids in Iberian Peninsula and with the three mentioned wild sources France (Reveillaud et al., 2018). Non-typhoidal *Salmonella* in poultry and turkeys has been highly reduced in the EU after the onset of control programs in the first decade of XXI century, cutting by half the human burden of salmonella in humans in the EU (EFSA & ECDC, 2019).

In the USA it is estimated that, annually, there are circa 40 million cases of illness, of which 59% are caused by viruses, 39% by bacteria, and 2% by parasites. Noroviruses are the more frequent (58%), followed by non-typhoidal *Salmonella* spp. (11%), *C. perfringens* (10%, more frequent than in Europe), and *Campylobacter* spp. (9%, which is by far less important than in Europe) (Scallan et al., 2011). Differences are seen in FBD attributable causes between EU and the USA.

The majority of FBD incidence happens as *individual cases*. But *outbreaks* are important either because often a high number of persons are involved from one single source of exposure, but also because outbreak investigation provide very relevant information to assess the epidemiology of exposure, allowing for the characterization of risk factors of FBD. The study of outbreaks provides information about risk factors. Based on systematic reporting of outbreaks it is estimated that the mean outbreak size in the EU is 9–10 persons, with 10% rate of hospitalization and 0.1% case fatality rate. In the EU, the occurrence of outbreaks has remained stable in a ten-year horizon: in 2009, 5550 and in 2019, 5146 (EFSA & ECDC, 2011, 2019) while in the USA the number of outbreaks recorded yearly appear to be rising (CDC, 2021). *Salmonella* non-*Typhimurium* spp., norovirus, bacterial toxins, and *Campylobacter* spp. are the most frequent agents involved in outbreaks in the EU. Types of food more frequently associated are eggs and egg products (20%), fish and fishery products (16%) and mixed foods (13%). Households (40.5%), restaurants and

5. VTEC: verocytotoxin-producing *Escherichia coli*/ETEC: Enterotoxigenic *Escherichia coli*.

similar (27.6) and canteens or catering (13.4) are the more frequent involved places (EFSA & ECDC, 2019). Results from USA are somewhat different but essentially FB outbreaks have similar characteristics to the ones in the EU.[6]

Estimates of costs associated to FBD show that death tool represents the highest stake of the burden (83%) followed by medical costs (12%) and productivity loss (8%). *Salmonella* (24%), *Toxoplasma* (21%), *Listeria* (18%), *Norovirus* (15%) and *Campylobacter* (12%), together, account for 90% of the total burden (Sandra Hoffmann, Bryan Maculloch, & Batz, 2015).

4.3 Organization of food safety

Food safety holds on a large diversity of activities and sciences which we will attempt to summarize in a few paragraphs. As proposed by Jouve (1998) four pillars are fundamental to organize food safety: (1) policies, (2) systems and stakeholders responsibilities, (3) official authorities control and inspection and (4) monitoring and evaluation tools.

4.3.1 Policies and legislation

The EU approach towards food safety is based in the following principles fundamental: (1) risk analysis, (2) precautionary principle, (3) protection of consumer's interests, (4) transparency in the legislative framework and risk communication. The legal framework is composed by Regulations[7] and derived pieces of legislation (European Union, 2002). The evolution from directives towards regulations, occurred in 2002, represented a significant step in the general and homogeneous enforcement of the food law across EU.

Usually there is a Food Safety Authority at country or region level. In European Union the Health and Food Safety (DG Santé) holds legislative responsibilities and every member state has its own authority being, all together being responsible for risk management.[8] European Food Safety Authority (EFSA), along with European Medicines Agency[9] (EMA) and European Chemicals Agency (ECHA) responsible for risk assessment and risk communication. In the USA the Food and Drug Administration along with 15 other agencies sharing oversight responsibilities in the food safety system, although the two primary agencies are the US Department of Agriculture

6. https://www.cdc.gov/foodsafety/cdc-and-food-safety.html.
7. European food safety legislation relies currently in regulations, delegated, implementing acts and decisions which are pieces of legislation of mandatory application at the Member States level without the need for transposition into national legislation.
8. Food safety risk management: https://ec.europa.eu/food/index_en.
9. Risk assessment and communication: Food, animal health and welfare: ahttps://www.efsa.europa.eu/en; Food Medicines: https://www.ema.europa.eu/en; Chemicals: https://echa.europa.eu/.

(USDA) Food Safety Inspection Service has the federal role of legislation. These are two examples. Organization of food safety worldwide differs a lot be the main principles apply.

Codex Alimentarius Commission (CAC), a commission under the patronage of WHO and FAO provide the guidance globally accepted. The *Codex Alimentarius* is a collection of internationally recognized standards, codes of practice, guidelines, and other recommendations published by the FAO relating to food, food production, food labeling and food safety.[10]

4.3.2 Risk assessment, food safety objectives and policy options

The risk assessment is the scientific foundation of risk analysis and consists of four steps: *hazard identification* (exploring the hazard and the disease), *hazard characterization* (assessing and modeling dose and response relationships), *hazard exposure* (quantitative assessment of introduction, dynamics and fate of the hazard in processes and in food; vias of exposure of people) and *risk characterization.* Risk characterization intendeds to produce a quantitative or qualitative estimate the risk that certain hazards, in certain foods, poses to the health and/or life of populations (FAO & WHO, 2006).

The selection of risk management options links the assessment process to the risk management, through the adoption of an Appropriate Level of Protection (ALOP). The ALOP definition is a cornerstone in defining the food safety policies and protect the human, animal, and plant health in one country or region. The level of protection deemed appropriate by the member (country) establishing a sanitary or phytosanitary measure to protect human, animal or plant life or health within its territory (WTO, 1995). Several options are available when deciding the policy options to the ALOP: examples are the "*zero risk approach*" adopted in the case of BSE, the "*threshold approach*" often used in chemicals such as antimicrobial residues or pesticides, or others such as ALARA,[11] benefit-cost, and precautionary approaches (FAO & WHO, 2006).

Policy tools to put into practice a comprehensive food safety structure include regulatory requirements; guidelines, like good agricultural practices (GAP), good manufacturing practices (GMP) or good hygienic practices (GHP); hazard analysis and critical control point (HACCP) system. Definition of food safety objectives and microbiological criteria for foods is of utmost importance. (Codex Alimentarius Commission et al., 2013). Definition of priority hazards to control, of critical industries, foods, processes, steps of the food chain to be controlled, monitored, inspected, certified is necessary.

---

10. Codex alimentarius: http://www.fao.org/fao-who-codexalimentarius/en/.
11. ALARA: as low as reasonable achievable.

In the European Union, risk assessment, scientific advice and risk communication are performed by EFSA (European Food Safety Authority). This agency is formally independent from the Commission which is responsible for the risk management activities (European Union, 2002).

4.3.2.1 Food safety objectives (FSO) and microbiological criteria

A main instrument of food safety policy are the food safety objectives (FSO) which is the maximum frequency and/or concentration of a hazard, of microbiological or chemical nature, in a food at the time of consumption that provides or contributes to the ALOP (FAO & WHO, 2006). The application of the FSO concept to a food chain allows the identification of the steps, in the chain, where through the application of adequate control measures at the processes the safety is ensured (Gorris, 2005). To achieve effective food chain management, *performance objectives* (PO) need to be applied to specific steps in the food chain (e.g., harvesting, animal production or transformation); once decided those, *process criteria* (a defined quantification in hazard reduction) e.g., 6 log reduction in microbial contamination is specified; to achieve the envisaged reduction correspondent *performance criteria* (measurable physical and/chemical parameters) needs to be defined e.g., the application of 72.5°C during 15 s (FAO & WHO, 2006; Gorris, 2005). FSO and PO can be seen as targets to be met for pathogen and food product combinations (van Schothorst, Zwietering, Ross, Buchanan, & Cole, 2009) and to assess whether food fit these targets, FSO based *microbiological criteria* are needed.

Microbiological criteria (MC) are a risk management tool that indicates the acceptability (in terms of safety or quality) of raw-materials, process, batches of products or consignments (CAC, 1997) and they apply to microorganisms, their toxins, or other markers. MC provides food operators and official authorities with verifiable, objective measurements of food acceptability. The components of an MC should be, at least, the purpose, the food/process/raw material, the food chain point of application, the microorganism concerned and his limits, the analytical methodology, sampling plans and acceptance criteria (CAC, 1997). MC can be used to validate HACCP systems, where they can play the role of critical limits, and processes and to orientate GHP (Gorris, 2005). Ideally MC should be calculated based on FSO and PO (van Schothorst et al., 2009).

4.3.3 Systems and stakeholder responsibilities

Food operators are the responsible to produce, transform, distribute, and sell safe food (Codex Alimentarius Commission et al., 2013; Commission of the European Communities, 2000). Ideally, food companies need state and commit to a food safety police, and should, subsequently develop, implement, assess and maintain integrated food safety systems, properly planned and managed

(Jouve et al., 1998). These can be nominated food safety and food management systems (FSMS). Integration of fundamental tools like GHP, GMP and HACCP if possible with ISO 9000 based quality systems are recommended and proved to be highly effective (Jouve et al., 1998). The elements of a FSMS can be defined as having a *preventive part*: with pre-requisites (GHP and GMP) as well as procedures for traceability, product recall, and communication and an own-check system with procedures base on HACCP (European Commission, 2016).

In the "*Commission notice, on the implementation of food safety management systems covering prerequisite programs (PRPs) and procedures based on the HACCP principles, including the facilitation/flexibility of the implementation in certain food businesses*" (European Commission, 2016) the EU makes a formal approximation to the principle of flexibility to and its relation with European and International Standards. It recommends a risk-based systems development, recognizing the complexity of FSMS should be adjusted to the size and market features of the operators. While medium and big sized companies need to have fully developed systems the traditional, rural, and small sized food operators may need effective but size adjusted systems. The focus of flexibility relay not on relaxing control of CCP but adjusting administrative.

4.3.4 Control by official authorities

The role of official authorities (OA) is essential to ensure that food business operator respect and comply with regulations. Adequate OA activity ensures, not only the consumer protection and safety, but also the market and trade transparency as it allows free movement of certified and safe foods (Commission of the European Communities, 2000). Effective food control systems are needed to accomplish such purposes. Decisions should be based on science, preference on risk-based assessments. This not always happens around the world (Faour-Klingbeil & Todd, 2019). To verify food operators compliance, OA should plan food safety control and inspections in a permanent and comprehensive basis. OA also need to be prepared to combat criminal behaviors, food fraud and food adulteration.

Building blocks of efficient food systems include following domains: (1) food law and regulations; (2) food control management; (3) Inspection services; (4) laboratory system involved either in food monitoring and in epidemiological data production; (5) education and training; (6) communication and information sharing (CAC/WHO, 2003). Controls and inspections need to be carried by competent and well-trained personnel; should be transparent, carried on in a risk-based frequency. Financing and facilities shall be adequate to the requirements and size of the system.

4.3.4.1 Control by official authorities – the case of EU Official Control (OC) system

To illustrate the complexity of a comprehensive food safety control system, we offer brief overview into the European food control system,[12] which is based on Regulation 2017/625. This Regulation is served by 16 Delegated Acts and 17 Implementing acts rooting each one from different articles of the mother Regulation 2017/625. This illustrates *per se* the complexity of the food control legislation.

The competencies of the OC are split between the MS and the Commission. Member States shall perform OC on a regular, transparent, preferably unannounced from, and with a frequency based on risk and historical record of the operator's reliability and compliance. Each Member State shall identify its own competent authority (CA), which shall perform its regular activities based on pluriannual control plans. Delegation of inspection and control is possible but regulated, and responsibility of controls remains always of the CA. Three level system of laboratories support the activity: EU reference laboratories, MS reference laboratories and official control laboratories. Sampling and tests shall follow ISO standards; a cascade system for test selection is foreseen when there are no ISO standards for such tests. Traceability of animals, feed and food is enforced, and FSMS are inspected. Integrated food safety control systems includes Border Control Posts settled at all borders of the EU and controls of food production, transformation and internal food trade. Mechanisms for cooperation and assistance between MS are available. Crisis and emergency measures are foreseen to account for food safety problems involving more than one MS and requiring organized and coordinated intervention.

Competencies of Commission includes planned and permanent controls of OC performed by MS, controls at OC from third countries exporting to the EU, definition of conditions for entry into the EU of animals, plants, food, and feed. Continuous training of the staff serving competent authorities is carried at de EU level by the Commission. The management of the common information systems is one the most important tasks. Systems like RASFF (Rapid Alert System for Feed And Food) and TRACES (Trade And Control Expert System), have been currently integrated in IMSOC (Information Management System for Official Controls) offers support to OC activities and allows its harmonization.

4.4 The animal health and welfare as contribution to food safety and to food security

Food safety and food security are strongly dependent from animal health and animal welfare. Animal diseases can be source of zoonotic agents like

12. Official control regulation: https://ec.europa.eu/food/horizontal-topics/official-controls-and-enforcement/legislation-official-controls-old_en.

tuberculosis, brucellosis, cysticercosis or echinococcosis, or rift valley fever, among many other. For relevant zoonosis associated with animal production it is necessary to establish programs for control, eradication, and disease freedom demonstration. Also, relevant because they imply heavy economic losses and sometimes menace the viability of production systems, although not being zoonosis, are pandemic diseases and transboundary animal diseases like avian influenza (AI), foot and mouth disease (FMD), trypanosomiasis or African swine fever (ASF). Animal welfare practices and legislation contribute decisively to improve animal health and to reduce disease and the reverse is also true: animal health systems and animal health plans improve animal welfare as they contribute to reduce disease and animal disease induced pain and stress. OIE standards[13] are the reference for all these elements and OIE's Veterinary Service evaluation tool (PVS) is an important element for the improvement and harmonization of veterinary services worldwide. The European Union also follows, for decades know, sound and systematic policies on animal health[14] and animal welfare.[15]

5. Water safety and security

As zero hunger, clean water and sanitation are one of the 17 Sustainable Development Goals of the United Nations. However, 780 million individuals still lack access to potable drinking water and 2.5 billion to sanitation (WHO, 2017). Water is essential for human survival to replenish daily losses and guarantee adequate amounts for physiological processes. It is also required for daily activities, such as cooking and personal hygiene. The daily minimum drinking water is estimated in 5.3 L $person^{-1}$ day^{-1} (estimated for the highest need group — lactating women with moderate activity in moderately high temperatures) and the total daily minimum for drinking, cooking, personal hygiene, food hygiene, and hand and face washing is estimated in 20 L $person^{-1}$ day^{-1} (but still insufficient for bathing and laundry) (WHO, 2020). The amount of water used depends on accessibility (time and distance of the water source), but also on the availability through time, reliability, and price. Restricted access to water also increases disease susceptibility by limiting access to hygiene practices (e.g., handwashing). Despite being an essential resource for human survival, water has been widely degraded. Population growth, rapid urbanization, intensive agriculture and industrialization, and lack of wastewater treatment threaten both water security and safety. Agriculture, highly dependent on water, comprises 70% of water use for irrigation, often depleting aquifers, and being a source of water contamination with pesticides, fertilizers, and livestock effluents (OECD, 2021).

13. OIE standards: https://www.oie.int/en/what-we-do/standards/.
14. https://ec.europa.eu/food/animals/animal-health_en.
15. https://ec.europa.eu/food/animals/animal-welfare_en.

In 2020, safe sanitation had only reached 54% of the global population, with 494 million people still practicing open defecation (UNICEF, 2019). Fecal contamination of drinking water is still common. Notably, John Snow traced back a waterborne cholera outbreak in London to the Broad Street pump, which was later found to be dug close to an old cesspit leaking fecal bacterial, founding modern epidemiology. Thus, not only water resources are becoming scarcer, but also becoming increasingly contaminated (both chemically and biologically), contrasting with the growing demand to serve an increasing human population and unrestrained economic growth. Climate change exacerbates the problem, with increasing the incidence of flooding and droughts, deteriorating water quality, and affecting the distribution and availability of rainfall, snowmelt, river flows and groundwater. Sustainable management is required to improve susceptible communities' health and long-term access to water resources.

5.1 Water safety

Water can directly affect health by being a source of harmful agents. Water contaminants can be divided in physical (e.g., sediments), chemicals (e.g., pesticides), radiological (e.g., uranium), and biological (e.g., bacterial), with the last being one of the most relevant for human health. Yearly, diarrhea causes 1.7 billion cases of childhood disease and 525,000 deaths of children under five, mostly spread by faeces-contaminated water with the most common agents being *Rotavirus* and *E. coli* (WHO, 2017). Water-related infections can be divided in four main categories: (i) water-washed infections (from person-to-person due to poor hygiene); (ii) water-based infections (related to parasites with a life cycle in the water); (iii) waterborne infections (infections caused by pathogens in water); and (iv) water-related insect vectors (infections transmitted by water breeding insects) (O'Brien & Xagoraraki, 2019). Many waterborne diseases originate from fecal contamination, with the most important pathways being the ingestion of water followed by recreational exposure through swimming. These pathogens can reach waterbodies through the release of untreated sewage, wastewater overflow during intense rainfalls, those surviving wastewater treatment, and from livestock of wildlife waste. A disease can have multiple transmission routes, as a water-based infection can also be transmitted as person-to-person, foodborne transmission, or through the contamination of crops, and waterbodies can become contaminated by excreta (Griffiths, 2017). An ecological perspective, or even a One Health approach, can benefit the understanding of the complex relationships of waterborne diseases, including some of zoonotic nature. Another consideration is the breeding of insect vectors in waterbodies, such as malaria mosquitoes. For pathogens with water-related pathways, surveillance of human and animal populations can be conducted in waterbodies under a One Health approach (accounting also for zoonotic occurrences) (O'Brien & Xagoraraki, 2019).

Besides waterbodies, wastewater-based surveillance is also useful to monitor human populations. For instance, COVID-19 trends can be monitored through the presence of the agent (SARS-CoV-2) in wastewater, following a similar trend as cases in the region and detecting as low as 3.2 cases per 100,000 people (Tomasino et al., 2021).

Chemical contamination of water is another problem, especially when considering the long-term exposure to low doses of complex mixtures of anthropogenic contaminants. The exposome has emerged to express the measure of all exposures of an individual in a lifetime and their impacts on health. While 80% of global municipal wastewater is released without treatment and industry releases thousands of compounds to the environment, from metals to solvents, agriculture remains the main pollutant due to the shear amount of wastewater produced from irrigation and livestock excreta (FAO, 2018). Agriculture is responsible for the release of organic matter, agrochemicals (e.g., pesticides, chemical fertilizers), and veterinary medicines (e.g., antibiotics, hormones). Nitrates are the most common contaminant released by agriculture which may lead to eutrophication and blue baby syndrome. Eutrophication is a process when there is a rapid enrichment of a waterbody with nutrients (e.g., nitrates), leading to an overabundance of algae and plants, which eventually decompose consuming O_2 and releasing CO_2, degrading inland and coastal waters. High levels of nitrates in water can lead to methemoglobinemia (hemoglobin F^{2+} is converted to Fe^{3+} and uncapable of carrying O_2, leading to cyanosis), also known as the blue baby syndrome, having direct adverse effects on human health. Agricultural pollution threatens 38% of European Union's water bodies. But these are not the only threats. Pharmaceuticals (or their metabolization or degradation products, some still biologically active), personal care products, surfactants and other substances can survive wastewater treatment and reach ecosystems. Persistent contaminants tend to accumulate in the environment (e.g., in sediments) or in organism (e.g., in fats), increasing their concentrations and effects over time. Water contamination may also arise during distribution. For instance, plastic bottles can release chemicals to water (e.g., additives, such as bisphenol A, an endocrine disruptor which mimics hormones even in very low concentrations). Therefore, the European Union has very strict rules regard food contact materials (including water bottles) (Food Contact Materials, n.d.). The Flint water crisis, in Flint, Michigan, USA, began in 2014 when the city decided to switch its water supply from Detroit's system to the Flint River to reduce costs. This led to high levels of lead in drinking water from untreated river water corroding aged lead pipes, an outbreak of Legionnaire's disease, and contamination with coliform bacteria, with severe consequences to the health of residents. In an attempt to solve the problem, high chlorine levels were added to the water, resulting in the additional increase in total trihalomethanes (carcinogenic compounds). The population has so far guaranteed the providing of bottled water and substitution of service lines, which has been a slow process. Similar problems have occurred in China, with bacterial contamination of piped drinking water lessened by the tradition of boiling and consuming hot water, highlighting the effects of cultural norms.

5.2 The importance of water

Global freshwater resources are estimated in the order of 43,750 km^3 $year^{-1}$, at continental level, following an inequal geographical distribution being mainly concentrated in America (45%) (Mancosu et al., 2015). These inequalities are even greater when comparing water resources per inhabitant, with 24,000 m^3 $year^{-1}$ provided by the American continent and 3400 m^3 $year^{-1}$ by Asia. While ground and surface water (e.g., rivers) are renewable resources, deep aquifers are considered non-renewable resources since they do not have significant replenish rates on human time scales. Water can also be subdivided in blue water, liquid water above and below ground (e.g., rivers, groundwater), and green water, in the soil and available to plants. The role of water in human survival goes beyond its direct use as drinking water. The hydrological cycle is interdependent on the conservation of the surrounding ecosystems and provide crucial ecosystem services. Ecosystem services provided by freshwater ecosystems include: (1) drinking water; (2) non-drinking water (e.g., for agriculture); (3) fisheries and aquaculture; (4) raw materials (e.g., algae for fertilizers, wood for energy, sand gravel); (5) water purification; (6) air quality and local climate regulation; (7) erosion prevention and flood protection (e.g., through vegetation); (8) maintaining populations and habitats; (9) pest and disease control; (10) soil formation and composition; (11) carbon sequestration; (12) recreation (e.g., swimming); (13) intellectual or aesthetic appreciation; (14) spiritual or symbolic appreciation; and (15) abiotic energy sources (e.g., hydropower generation) (Grizzetti, Lanzanova, Liquete, Reynaud, & Cardoso, 2016). Therefore, food security and energy production are also interlinked with water. For instance, hydropower accounted for 60% of global renewable energy production in 2018 (4200 TWh), comprising around 16% of the world electricity production (TheWorldBank, 2020). Energy production in hydroelectric plants prevents the burning of fossil fuels, contributing to the mitigation of climate change, one of the largest threats to humanity. While hydroelectric plants provide benefits, such as clean energy, flood prevention, and water storage, they also have large ecological impacts, such as flooding of upstream habitats, reducing water flow, and creating a barrier for the migration of aquatic species. Thus, the construction of new hydroelectric plants must be assessed through environmental impact studies, which always require compromises. Food production is also highly dependent on water resources. Similarly, changes in land use can alter plant evapotranspiration leading to precipitation changes. Therefore, water security and food security are closely interlinked.

5.3 Improving water management

Water, sanitation, and hygiene (WASH) are key public health principles, especially for populations in low- and middle-income countries. Interventions are usually complex as many WASH services rely on self-supply solutions such as private wells or pit latrines. Implementation of WASH principles may include the introduction of technology and infrastructure (e.g., toilets, taps, soap, hardware), practices (behaviors), principles (e.g., sanitation coverage), programs (e.g., village water and sanitation committees), and policies (e.g., subsidies) (Haque & Freeman, 2021). Practices to achieve safe drinking water include microbial treatment of contaminated surface waters (e.g., chlorination), coverage of shallow wells and springs to avoid contamination, flushing tanks (diverting the first flush) and sand filtration of rainwater, and avoiding bathing and excretion in water sources (Griffiths, 2017). Similarly, open defecation must be avoided by making use of latrines (pits with a robust cover, ideally impermeable), ventilated improved pit latrines, septic systems or, ideally, sewage systems connected to treatment plants (Griffiths, 2017). Human waste can the be use as a crop fertilizer, after appropriate treatment, cycling nutrients. However, the growing presence of persistent contaminants (e.g., pharmaceuticals) in human waste may lead to the contamination of agricultural lands. Besides water and sanitation, access to hygiene practices could greatly improve the health of underprivileged communities. Handwashing could reduce diarrheal disease-associated deaths by 50%, reduce the risk of respiratory infections by 16%, and reduce the occurrence of foodborne outbreaks, among others. However, in 2017, only 60% of the global population had access to a handwashing with soap and water at home (UNICEF, 2019). Therefore, efforts must be made to improve hand, face, and body hygiene by improving the availability of water and soap, as well as conducting awareness campaigns. WASH strategies can be summarized in six processes, namely planning, educating, financing, restructuring, quality management (i.e., monitoring and maintenance), and attending to policy context (Haque & Freeman, 2021). However, all involve water security requiring proper water management.

The Integrated Water Resources Management (IWRW) is a long-term process (not a one-shot approach) which promotes the coordinated management (across borders) of water and related resources, to maximize economic and social welfare in an equitable and sustainable manner (The Global Water Partnership, 2020). This is especially difficult considering the temporal and spatial variability and unpredictability of water resources. As water resources are finite, their overuse in one sector (e.g., agriculture) may cause shortages in others (e.g., household consumption). Considering the regional specificities of water management, the IWRW presents guiding principles which can then be adapted to local condition. In 1992, the Dublin Principles were defined as: "1.

Fresh water is a finite and vulnerable resource, essential to sustain life, development, and the environment. 2. Water development and management should be based on a participatory approach, involving users, planners, and policymakers at all levels. 3. Women play a central part in the provision, management and safeguarding of water. 4. Water has an economic value in all its competing uses and should be recognized as an economic good." These principles can be summarized as the equitable, efficient, and sustainable management of water resources as part of ecosystem, society, and economy. Similar to the 3R's, these ideas are summarized in the 3E's: economic efficiency, social equity, and environmental and ecological sustainability. For policymaking and planning, this means accounting for the uses and the people's needs of water, listening to stakeholders (including women and the poor), considering the interrelationship between macroeconomic policies and water management, making decision at the local and basin levels without conflicting with national objectives, and incorporating strategies in broader social, economic, and environmental goals. This framework at a watershed level can be complemented by broader considerations, such as the virtual water trade (i.e., the hidden flow of water in products of services) or foreign direct investments, flows through the atmosphere, surface, and subsurface waters, transfers through physical infrastructure, and altered land-atmosphere-ocean interactions (e.g., saltwater encroachment) (Zipper et al., 2020). In the European Union, the Water Framework Directive addresses aspects of water management attempting to achieve a good ecological and chemical status at a watershed scale. In practice, an effective access to healthy water includes chemical and biological quality, quantity available (considering distance and cost), access (distance, socioeconomic, cultural aspects), reliability, costs (both as tariff or as time and health), and ease of management (e.g., users in rural setting of developing countries are involved in operation, maintenance, and management) (Hunter et al., 2010).

Data can help understand the contribution of each system to water management and dependency on external actors. The water footprint is the total volume of freshwater consumed to produce goods and services or by people (Mancosu et al., 2015). The largest water footprint of the average consumer is attributed to cereal products (27%), followed by meat (22%) and milk products (7%). Therefore, food importers (especially cereals) can compensate local water shortages. Total renewable water resources (TRWR), defined by FAO, helps understand the dependency from neighboring countries, being the sum of internal renewable water resources (IRWR) and external renewable water resources (ERWR), corresponding to the yearly amount of water available to a country (Mancosu et al., 2015). Water resources (surface water and groundwater) generated from precipitation within a country or catchments is IRWR. Surface water can also replenish groundwater through seepage in the riverbed, and aquifers contribute to base flow by discharging into rivers (especially in dry periods). ERWR are water resources entering from upstream countries through rivers (external surface water) or aquifers (external groundwater resources). The ratio between ERWR and TRWR expresses de dependency of a country on external water resources. These indicators help understand the use and distribution of water resources, aiding their proper management.

Other aspects must be considered regarding water management. Water safety can be improved through water and wastewater treatment. Water treatment is comprised of processes which improve the quality of the water, usually making it safer for drinking. The most drinking water treatments for surface waters include: (1) coagulation and flocculation, with the neutralization of negatively charged dissolved particles forming larger particles (floc); (2) sedimentation, settling of the floc to the bottom; (3) filtration, further removing dissolved particles, agents (e.g., bacteria), and chemicals; (4) disinfection, by adding a disinfectant (e.g., chlorine chloramine) to remove remaining organism from the water (CDC, 2015). Other solutions are available to low-income communities which lack access to water treatment infrastructures, including the use of disinfectant tablets (i.e., sodium hypochlorite tablets), solar disinfection (i.e., leveraging the solar heat and UV radiation), ceramic filters (i.e., porous ceramic filters capable of removing microorganisms and particles), and biosand filters (i.e., slowly filtering water through sand and gravel of different granulometry). Conversely, a high-cost solution to water scarcity is the desalinization of seawater, usually by reverse osmosis, with a high energy cost. One factor to take into account is the acceptability of the water resources, as strong taste and odors arising from iron content of groundwater or chlorination can hinder use despite being safer (Hunter et al., 2010). Wastewater treatment intends to safely dispose of effluents originating from human activities. The main types of wastewater treatment can be divided as: (1) pre-treatment, consisting on the removal of coarse solids (e.g., grit removal); (2) primary treatment, sedimentation to the bottom and skimming of floating materials; (3) secondary treatment, removing residual organics and suspended solids, generally by aerobic biologic treatment (e.g., activated sludge); (4) tertiary treatment, removing remaining inorganic compounds using coagulants (e.g., Alum to remove phosphorus) and microorganisms through the use of disinfection technologies (e.g., chlorine, ultraviolet irradiation) (FAO, 2021). The resulting solids can be landfilled, incinerated, or used as fertilizer, but concerns are being raised over the persistence and introduction contaminants in agricultural lands. Indeed, water and wastewater systems are not always effective in removing persistent anthropogenic contaminants. Treated wastewater effluents (or reclaimed water) can vary in quality, from quality standards for use in agriculture to drinkable water. Generally, treated effluents are released back to the environment, in freshwater bodies or directly in the ocean. Wastewater treatment can also involve other solutions, such as sewage lagoons or effluent ponds, in which nutrients and microorganisms are slowly treated in a properly engineered natural-based system. High-cost strategies are often unavailable to low-income countries, which lack sewage systems and piping. In these cases, the solution may rely on proper management of communities' septic tanks and pit latrines, such as impermeabilization and frequent emptying by the authorities. Practices on human and animal waste management can also help improve the health of communities. Agriculture as the major water consumer should also contribute to a more sustainable use (Mancosu et al., 2015). This includes monitoring and computing how much water is needed by crops with regards to climate conditions and soils to satisfy agricultural demand and then improve the efficiency of the irrigation application. Other improvements may include irrigation scheduling in terms of timing and quantity of water, crop management (e.g., planting date), and cultivar (e.g., drought-tolerant cultivars). In pair with sewage management, reclaimed water (i.e., treated effluent) and greywater (i.e., untreated wastewater uncontaminated with feces) can be used for irrigation, to avoid the use of valuable drinking water resources. However, the indiscriminate use of greywater may carry risks regarding soil contamination, and these systems require investment in a separate piping system.

6. One health

6.1 Brief introduction

One Heath is a recent concept anchored in old principles and a long history of human medical and veterinary sciences (Evans & Leighton, 2014). In 2004 the Wildlife Conservation Society held a meeting entitled "Building Interdisciplinary Bridges to Health in a Globalized World" in the Rockefeller Foundation (Manhattan) in which consolidated the term "*One World One Health* ™." From this meeting emerged the Twelve Manhattan Principles. According to these principles, the health of the humankind, the animal health and environment health are inextricably linked, and are interdependent. "*To win the disease battles of the 21st Century while ensuring the biological integrity of the Earth for future generations requires interdisciplinary and cross-sectoral approaches to disease prevention, surveillance, monitoring, control and mitigation as well as to environmental conservation more broadly*".[16] As emphasised by Treadwell (2008),[17] the convergence of forces at overlapping boundaries between human, veterinarian and environment domains are responsible zoonotic disease emerging, re-emergence or persistence.

Later, the World Health Organization (WHO), Food and Agriculture Organization (FAO) and World Animal Health Organization (OIE) developed a conjoint initiative, the "*Tripartite Concept Note*", endorsed by the World

16. https://oneworldonehealth.wcs.org/About-Us/Mission/The-Manhattan-Principles.aspx.
17. At the Institute of Medicine/National Research Council Workshop on Sustainable Global Capacity for Surveillance and Response to Emerging Zoonoses.

Bank (FAO-OIE-WHO, 2010). After this agreement the vision of a "world capable of preventing, detecting, containing, eliminating and responding to animal and public health risks attributable to zoonosis and animal diseases with an impact on food security" (FAO et al., 2010), in line with the Manhattan Principles was recognized as the best response to health problems like Pandemic Avian Influenza, Transboundary animal diseases like Rinderpest or Rift Valley Fever, Antimicrobial Resistance fighting, just to mention a few. Food security and safety were at the heart of the vision, but also human health (World Bank, 2010).

More recently, the Berlin Principles revisited, enlarged, and clarified the Manhattan principles (Gruetzmacher et al., 2021). Non-communicable diseases became clear part of the mandate of One Health, expanding the vision adopted earlier: "*Today broad consensus exists that health entails more than the absence of infectious disease; it must incorporate socio-economic, political, evolutionary, and environmental factors while considering individual attributes and behaviors. To address the myriad of health challenges of the 21st century while ensuring the biological integrity of the planet for current and future generations, we need to strengthen existing interdisciplinary and cross-sectoral approaches that address not only disease prevention (communicable and non-communicable), surveillance, monitoring, control, and mitigation but also biodiversity conservation*".

The recent pandemic of COVID-19, caused by SARS CoV-2, proved the judgment underlying OH principles to be right, behind the frontier of doubt.

6.2 Characteristics of one health approaches

The vision of "*a world capable of preventing, detecting, containing, eliminating, and responding to animal and public health risks attributable to zoonoses and animal diseases with an impact on food security through multisectoral cooperation and strong partnerships*" (FAO-OIE-WHO, 2010) requires a multisectoral, multidisciplinary and system-based approach (Rüegg, Häsler, & Zinsstag, 2018; World Bank, 2018). Given the multiplicity of factors involved the three domains, and their complex interactions, contributing to emergence, re-emergence or maintenance of zoonotic diseases the implementation of OH methodologies a reorganization of the way governments respond is necessary (World Bank, 2010). The question is not anymore "*what is my job*? Or *what do I have to do*?" rather "*what needs to be done*" (World Bank, 2010). Operationalization of OH principles into practice requires the adoption a new organizational paradigms, new initiatives and multisectoral coordination (World Bank, 2018) along with knowledge inputs from many sciences and techniques (Rüegg et al., 2018).

One Health approach requires systemic and adaptative reasoning. The operationalization of OH approaches requires, the formal involvement and commitment of concerned countries or regions, governments, institutions, stakeholders, often the civil society (FAO-OIE-WHO, 2010). Recommendation is that "Multisectoral OH Coordination Mechanisms" (MCM) should be created at country level to coordinate OH activities with adequate leadership and technical functions, dynamic and fruitful collaboration, communication and coordination across all sectors (WHO-FAO-OIE, 2019). Possible OH activities are (1) emergency preparedness, (2) surveillance and information sharing regarding zoonotic diseases, (3) coordinated investigation and response, (4) joint risk assessment of disease threats, (5) risk reduction and risk communication (WHO-FAO-OIE, 2019; World Bank, 2018). Often the different stages in these operations are prevention, detection, response and recover. Financing and resourcing, community and stakeholder engagement (World Bank, 2010, 2018) are fundamental activities. Operations include problem identification, mapping stakeholders, disease prioritization, selection of control alternatives, implementation and monitorization (Rüegg et al., 2018).

6.3 Contribution of one health to food security and safety

The adoption of One Health methodologies is essential to reduce or avoid the burden and the economic and social costs of typical OH issues (World Bank, 2010, 2018). Unnecessary heavy burden in human lives and economic losses is often endured before such approaches are adopted, as reported (Ibrahim et al., 2021; Zinsstag, Schelling, Wyss, & Mahamat, 2005). The following are, a few, good examples of the necessity of joining efforts in medical, veterinary and agronomic domains: (1) Salmonellosis reduction by 50% in human prevalence in Europe, after control programs implementation in poultry and turkey (Balleste-Delpierre & Vila Estape, 2016); (2) Q-Fever and unexpected resurgence in the Netherlands, causing human outbreaks after infection of few goat herds (Schneeberger, Wintenberger, van der Hoek, & Stahl, 2014); (3) Brucellosis and Q-Fever also a problem better addressed combining veterinarian and public health programs Ethiopia (Ibrahim et al., 2021).

The economic burden can be national, regional, or global and includes direct and indirect costs associated with human disease and death; animal disease and death with consequent production losses, loss of market for animal products (due to trade restrictions and bans); infrastructure, and health system; loss a tourism market (Rüegg et al., 2018; World Bank, 2010). Typical OH issues, associated with food security or food safety includes zoonoses, vector borne or food-borne diseases, and antimicrobial resistance. A few examples of cost from global problems are BSE (Global, 2003–07 $11 billion), H5N1 (Asia, 2003; exceeded $20 billion), SARS (2003, Asia and Canada, exceeded $ 40 billion). The global societal costs of de COVID-19 world pandemic, currently ongoing, have been estimated in approximately 6% of world's GDP in 2020, circa $ 4 trillion.[18]

Examples of relevant areas for action are the fighting against Antimicrobial Resistance, Action on the control of Neglected Zoonosis (in particular Rabies, Antrax and Brucella), Food Safety (see above), and Vector-borne Diseases (such as malaria, leishmaniosis, yellow fever or dengue) (World Bank, 2018). Next two relevant cases are presented.

6.3.1 The INFOSAN (International Food Safety Authorities Network)

INFOSAN is an important example of a global information change network complying objectives and activities mentioned above. The INFOSAN it is a global network of national food safety authorities and was created in 2004 after a conjoint initiative of the WHO and FAO. The main purpose of "*Halt the international spread of contaminated food, prevent foodborne disease outbreaks, and strengthen food safety systems globally to reduce the burden of foodborne diseases*" (WHO-FAO, 2019). The objectives are: (1) Promote the rapid exchange of information during food safety incidents, happening mostly in incidents and crisis; (2) Share information and resources on important food safety related issues of global interest, mostly in peace time; (3) Promote partnerships and collaboration between national agencies, meaning stimulation of information and data exchange among national authorities between countries, and between networks; (4) Help countries strengthen their capacity to manage food safety emergencies, an objective directed mainly to lower income countries (WHO-FAO, 2019).

6.3.2 Antimicrobial resistance emergence

Antimicrobial Resistance (AMR) is a global health and eco-health growing problem (ECDC/EMEA, 2009; O'neill, 2016; OCED, 2016). It results from the extensive utilization of antimicrobials to treat human and animal infections (World Bank, 2017). The vast industrial capacity to produce pharmaceutical drugs, the globalized economy and the increasing in expansion of health systems the world, explain a permanent increasing in demand and explain the permanent growth in the global consumption of AM (European Commission, 2017; World Bank, 2017). The negative impacts are a permanent increase in AMR with potential catastrophic increase of human death due to AMR problems (O'neill, 2016). Food production results, in part, from animal production which utilizes AM, and AMR access humans also through the food system: food and drinking water. The health burden of antimicrobial resistance in Europe was recently estimated in 33 thousand attributable deaths and 874 thousand DALY (Cassini et al., 2019). Action against this problem is needs to

18. https://www.statista.com/topics/6139/covid-19-impact-on-the-global-economy/.

be global and coordinated (European Commission, 2017; WHO, 2015). The international agencies involved in this problem, WHO (WHO, 2015), OIE (OIE, 2016), FAO, produced documents with global strategies to address and contain RAM increase. The strategies of all these organizations focus on some or all of the following actions to (1) reduce the amount of AM utilization by reducing infection, increase biosecurity and implementation of best practices; (2) seek alternatives to fight infections like an increase in the use of vaccines; (3) develop new drugs to overcome resistance emergence; (4) produce evidence on consumption and on AMR dynamics, through the implementation of surveillance systems and information sharing and (5) investigation on epidemiology and mitigation strategies.

The EU developed a three pillar strategy designated "A Europe One Health Action Plan against Antimicrobial Resistance"; the pillars are (1) making the EU a best practice region (in medicine and veterinary medicine activity), (2) boost on research and innovation (tests, vaccines, new molecules) and (3) shaping the global agenda (European Commission, 2017). This action plan endorses a One Health approach since extends the call for integrated and concerted action all the relevant stakeholders from human health to animal production and small animal clinic.

6.3.3 Corona virus – COVID-19

The COVID-19 crisis threatens the food security and nutrition of millions of people, many of whom were already suffering. A large global food emergency is looming. In the longer term, we face possible disruptions to the functioning of food systems, with severe consequences for health and nutrition. (United Nations & FAO, 2020).

6.3.4 In summary

"*Globalization of the food supply has created conditions favorable for emergence, reemergence, and spread of food-borne pathogens and has compounded the challenge of anticipating, detecting, and effectively responding to food-borne threats to health*" (Choffnes et al., 2012).

The One health approach brings a *new Vision* for (1) solving complex dilemmas and wicked problems, requiring (2) adopts an integrated and holistic adaptative systems approach, (3) requires collaborative work across science disciplines, professions, institutions, (4) recognizes that all relevant stakeholders for a problem solution – governments, operators, academy, consumers – needs to be involved, (5) changes the focus of attention from human infection to sources and ways infection accesses humans, and (6) work on human, animal, vegetal and environmental health domains, survey and collect data on these domains and share it.

6.4 Water as an essential one health issue

Water security and safety is dependent on a transdisciplinary approach characteristic of the One Health approach. The availability of water is dependent on its use, which is contingent on economic activities and governmental planning, but also on land use and conservation, which can alter the hydrological cycle. These policies are also relevant at an international level since watershed often cross borders. Resilient ecosystems and biodiversity survival are also dependent on the quantity and quality of water availably. For instance, the loss of ponds (small, isolated freshwater bodies) led to the loss of essential habitat for many species, including threatened amphibians, dragonflies, and aquatic plants (Rannap et al., 2009). Agriculture is closely interlinked with water, being dependent on water consumption but also being responsible for water depletion and pollution. Therefore, food and water security and safety are both interlinked. Water safety also depends on proper practices regarding drinking water (e.g., boiling water) and human waste (e.g., pit latrines), highly influenced by cultural norms. Therefore, awareness campaigns can help prevent high risk behaviors. Access to basic hygiene must also be improved to prevent water-washed infections. Another aspect of water management, typical of One Health, is the involvement of various stakeholders, working at all levels of society. Water resources are often managed at a local level, often with high involvement of local communities, which must be considered by authorities during planning. Knowledge sharing, resource sharing, and infrastructure building are long-term objectives to provide proper water and wastewater treatment and could be endorsed by developed countries. Finally, governments must address unrestrained economic growth based on pillaging natural resources, such as water, and the release of high amounts of contaminants, with immeasurable adverse effects on human populations and ecosystems. All these problems would be better addressed under a One Health framework, involving an interdisciplinary team working at a local level and addressing the needs of the population and stakeholders, while preserving the environment.

7. The food system: challenges and trends in the anthropocene epoch

7.1 Introductory concepts

Briefly addressing concepts of the Anthropocene epoch, complexity and complex adaptative systems is relevant to understand food systems as well as their drivers, problems, challenges and solutions.

7.1.1 The anthropocene and the Earth System

According to some authors the humanity entered the Anthropocene age in the eighteenth century. Development of engines using fossil fuels contributed to noticeable anthropogenic emission of carbon dioxide and methane (Crutzen, 2002). The science and technological achievements of the last centuries gave humanity an unprecedented domain over the planet and the effects of human activities have already changed the face of the earth in many aspects in an irreversible way, with negative effects of the human activity and presence over the planet are perceived and measured (Steffen et al., 2005). *Steffen* et al. conceptualized three periods in the Holocene epoch: Pre-Anthropocene Events starting 12,000 years ago, the Industrial Era Stage I (1800−945) and the Great Acceleration Stage II (1945−2015), the last showing abrupt significant increased shifts in CO_2, NO_x, CH_4 in air, water, and land. The Stewardship of the Earth System, Stage III started after 2015 and the ending is currently going on (Will Steffen, Crutzen, & McNeill, 2007). The positive effects of science technology and economy are that wellbeing and quality of life of the world population have globally increased. The negative side effects are profound alterations in Earth System balances and interrelations (Steffen et al., 2005). The response of the society, the human institutions and the economy has proven to be able to overcome the challenges and unbalances when regarding solutions to specific problems like chlorofluorocarbons and the ozone layer, the acid rains and water nitrification at regional level (Will Steffen et al., 2007). Even the capacity to respond quickly and efficiently to global menaces helped in overcoming the potential catastrophic effects, like the Covid-19 pandemic[19] provide some optimism regarding the human society capacity to respond to catastrophic menaces.

The conjoint food production systems on Earth form the GFS (Global Food System) (Parsons et al., 2019). The GFS, in the Anthropocene, at the same time allows to nurture the human population and endangers it viability as one of the forces that is shaping the surface of the planet and the use of resources. Solutions shall be comprehensive and ideally operate in the safe operating space for food systems (Willett et al., 2019).

Food insecurity remains a problem in many regions of the globe (FAO et al., 2021). Disparities remain and the Global Development Goals for 2030, clearly witness the main challenges for the humankind. The pressure over the world's natural resources is excessive and unsustainable, the globe is interconnected, and the unregulated systems created by man have an enormous

19. https://www.who.int/emergencies/diseases/novel-coronavirus-2019/interactive-timeline?gclid=CjwKCAjw9aiIBhA1EiwAJ_GTSm7Wtgzsh0lM59o6h5JL9KUIYmwzZtINB7ntudO9YwWsX5YoMTk6cRoCE0oQAvD_BwE#event-115.

influence over the natural systems creating complex dynamics unbalances which are very difficult to manage and compensate or even to predict.

The climatic changes are strongly influenced by the food production system and, at the same time, they strongly influence agriculture efficiency and resilience, menacing the local or regional agricultural production in many parts of the world. Globally the economic processes, driven by market, social, political, and capital forces are dependent from the energy production industry which strongly influences climate change. Agricultural outputs are vital to feed the world's population. Agriculture as the source of food to sustain the humankind remains strongly rooted on the planet, and it is deeply influenced by all the extremely complex systems of forces of geological, geographical, climatic and societal nature. The challenges to ensure that food security and food safety are achieved are complex and difficult to solve.

7.1.2 Complexity and complex adaptative systems

Understanding the concept of complexity is relevant to enter the problems, trends and challenges associated with the Food System. A comprehensive and commonly accepted definition of complexity[20] is difficult to find but is generally accepted that complex systems operate under an unknown order and their outcomes are difficult or even impossible to predict. At least four dimensions exists in complex systems: (1) they are formed of many components (systems), operating at different scales, from which interaction new functions of higher level emerge; (2) the response of the system is non-linear, hence responses are impossible to predict accurately; (3) they form a network, and relationships and flows between the elements determine their respective importance and influence on the system; (4) they have autonomy and show adaptation, there is non-centralized coordination; these systems experience evolution in response to inner or outer system forces. Complexity has specific meanings associated with several scientific fields (Ercsey-Ravasz, Toroczkai, Lakner, & Baranyi, 2012; Kurtz & Snowden, 2003; Levin, 1998).

Examples of complex adaptive systems can be ecosystems within the biosphere (Steffen et al., 2005); global, regional, country level economy and the interactions between all levels.[21] The Global Food System which is inextricably associated to Food Security is definitively a complex system (Gill, 2018). Its elements are the food chains, the stakeholders and people involved, its drivers and outcomes and the interrelations between these elements. The interplay of the five dimensions influencing the drivers, stakeholders and relations in food chains already identified: economy, politics and policies,

20. https://www.youtube.com/watch?v=vp8v2Udd_PM (complex system). What is a Complex System?
21. http://oecdinsights.org/?s=NAEC+complexity (OCED).

environment, society, and health creates a highly delicate, unstable, and complex system of which the activity outcomes have desired but often also undesired effects.

There is a clearly inextricably tension between some SDG, after defining the achievement of food security in the world, negative consequences will probably impact on land, water, biodiversity, climate which are also the target of other SDG goals. Systems thinking approach has been proposed to address these problems and helping in solution disclosure (Zhang, 2018).

7.2 Global food system supply (GFS)

Several definitions are found, being the following a general and inclusive one: *the food system, is the interconnected system of everything and everybody that influences, and is influenced by, the activities involved in bringing food from farm to fork and beyond* (*Parsons et al., 2019*). Another operative definition is "*The food system incorporates all elements and activities that relate to the production, processing, distribution, preparation and consumption of food, as well as its disposal. This includes the environment, people, processes, infrastructure, institutions and the effects of their activities on our society, economy, landscape and climate*" (Gill, 2018). In a seminal paper Erikson (2008) discussed existing types of food systems (traditional or modern), their activities (production, processing and packaging, distribution and retail, consumption), and their outcomes including the several dimensions of food security (*see above*) but extending the reflection to include social welfare and economics dimensions and environmental impacts (Ericksen, 2008). In this review we will use the term Global Food System (GFS) (Choffnes et al., 2012), as applying to all food production systems which feeds the world, through trade from local to global and vice versa thus considering food systems in its globality.

Food fluxes across the world are complex and are continuously changing and the market drives these flows (Ercsey-Ravasz et al., 2012; Heinrich Böll Foundation, Rosa Luxemburg Foundation, & Friends of the Earth Europe, 2017). The production of food is evenly distributed around the globe and there are countries or regions which are net exporters like the United States, Brazil or Australia and regions which are net importers like China, India, the Middle East (FAO, 2017). The biggest exporters of food – EU and USA – are, simultaneously, the bigger importers; seven countries in the world account for 77% of the world food trade (Ercsey-Ravasz et al., 2012). Trade of food and animal feed across the globe ensures that most of the world population is fed and, currently, both the world trade and population are increasing (Ercsey-Ravasz et al., 2012). The major stake of this trade is responsibility of a few gigantic multinational players which are responsible to make available food access to the consumers though out the world: data suggests that less than 50 multinational companies control half of the world's food trade (Heinrich Böll

Foundation et al., 2017). Closely associated with the huge current global food trade, the risk of introducing and spread food born hazards increased as well immensely (Choffnes et al., 2012).

The *elements* of the GFS can be categorized in four domains: (i) the food production systems and the food chain; (ii) the forces influencing and resulting from these systems interactions; (iii) the stakeholders and the consumers; (iv) the interconnections and fluxes between the previous elements (Ercsey-Ravasz et al., 2012). The *food production systems* are often represented as the food chain. The food chain is the sequence of phases, from "stable to table": from animal feed production to retailors and catering/restaurants or home consumption, requiring animals and animal products production at farm level and going through processing/industry. Transportation and logistics play a central role, from the farms to the final consumption place. Fish, and marine products also have often extensive chains (Godfray et al., 2011).

Five dimensions may be found among the previously mentioned drivers. These drivers are simultaneously driven by the GFS, mutually shaping each other: (i) Politics and policies, (ii) Economics; (iii) Environment, (iv) Heath and (v) Society (Parsons et al., 2019). *Politics* domain is composed by the governance and government organizations, policies and legislation, taxes and subsidies, political parties interventions. *Economics* involves trade, skills and jobs, size, and relationship among the food operators. *Environmental* drivers are composed by water, land, forests available and are influenced by climate and geographical location on earth. *Society* influences de GFS through the population structure, culture and education, gender issues, media and marketing, ways of living. *Health* through the diets and consumption habits or preferences, the nutritional value, the health messages on nutrition and diets coming from the national health authorities.

According to Hueston (Choffnes et al., 2012; Parsons et al., 2019), five major features shape the organization of this system systems, hence the risk for food safety:

(1) *Continuous and dynamic change*, induced basically by the interactions among the "five dimension drivers";
(2) *Panarchy*, is the type of hierarchy currently existing in the market among food suppliers, always competing for supremacy and pursuing continuous growth, this resulting in constant reduction of diversity. On the long run this translates in global less resilient systems and on destruction of traditional food systems (FAO, 2017);
(3) *Demand driven economy* - competition for market share, often offering safe food at the lowest price, between the big food players drives the relations between the players in the chain;
(4) *Culture clash* – the enormous differences, among countries and regions, between food preparation tradition and cultures, legislation economic

and environmental ambiance creates marked differences in food safety risks and attribution of responsibility.

(5) *Increasing dominance of private standards* – private standards like GIFS, ISO series, Global GAP, to mention just a few, tend to be ahead of food safety legislation and are a factor of dynamics for the food systems and shape the relations among food operators.

7.3 In summary

Food security depends on the Global Food System's ability to provide food after the defined requirements: affordable, accessible, sustainable, nutritious. "There is no single global food system but rather a multitude of interdependent food systems driven by the diverse needs of different countries and populations" (Parsons et al., 2019). *Globalization of the food supply has served to expand the range of food-borne pathogens as well as to amplify health and economic impacts of a single contamination incident* (Choffnes et al., 2012; Ercsey-Ravasz et al., 2012).

8. Outcomes, trends and challenges of food systems in a globalized world

"Conflict, climate variability and extremes, and economic slowdowns and downturns (now exacerbated by COVID-19 pandemic) are major drivers of food insecurity and malnutrition that continue to increase in both frequency and intensity, and are occurring more frequently in combination (FAO et al., 2021).

Population probable growth up to 10 billion people in 2050 and the corresponding market demand are one key challenge for world sustainability. This will happen in a scenario of profound inequalities, conflicts, economic instability and uncertainty, and profound climate change effects. It is important to understand the structure and the dynamics of these challenges to understand food security and food safety. Relevant trends and challenges are analyzed the present section to provide a deeper insight about forces driving and shaping food security in the XXI century.

8.1 Outcomes of food systems, at present

Hunger and the different forms of malnutrition are presently expanding, associated with conflicts, regional or continental extreme climate events. The quality of diets is highly uneven, causing problems with obesity and malnutrition metabolic and health associated disorders in middle and high-income countries. Food safety remains a problem in a large number of countries, with important impact on disease and death, and there is long way to go until acceptable level of food safety are achieved. In some regions of the world the livelihood of farmers is precarious resulting from diverse factors like uneven power relations in food systems, pests, diseases and climate changes which compromises the resilience of their production systems. External impacts of food systems still remain elevated and many regions of the world, particular where highly intensive industrial agriculture is performed, food systems have crossed the boundaries of sustainability (HLPE, 2020).

8.2 Trends and challenges

Without change, the global food system will continue to degrade the environment and compromise the world's capacity to produce food in the future, as well as contributing to climate change and the destruction of biodiversity (Godfray et al., 2011).

8.2.1 Population related trends: growth and demographics, urbanization, food consumption and nutrition

The *world population* is currently at 8 billion and is still growing. This growth will happen probably until de end of the XXI century. Since the middle of the XX century Europe, North America and Oceania show a steady, even slight decrease in the population. Asia and Africa account for the increase but in annual rate of growth is decreasing. The projection of population on earth by the end of this century is expected to fall between 9.7 and 11.2 billion people. *Urbanization* is a progressive tendency: today 54% of the world's population is urban and by 2050 urban population will represent more than 65%. Urban populations are increasing faster in low- and middle-income countries while remaining relatively stable in high-income ones. Urban populations are less fertile than rural ones and this will reduce the population growth rate. The *population age* will tend to increase in all continents and the share of people older than 65 is raising. In the next decades, the globe will grow older, more populated, and more urban (FAO, 2017).

Migration it is a growing global phenomenon driven by poverty, climate change, political instability, and competition by natural resources. Migration shapes the population demographics either at source or at destination. At the origin, migrants from rural impoverished regions tend to create a gender shift of women remaining in agricultural areas which is an increased burden for food security. At destination the burden on food security among ethnic groups and other groups of destination of these immigrants is also relevant (United Nations, 2015).

Nutrition and *consumption* patterns are linked. Consumption patterns vary greatly across the world. They are linked to tradition and culture, ethics, religion, social determinants. Between 1990 and 2013 the consumption of more nutritious food increased worldwide. The tendency is to be a convergence among the consumption profiles between low- and middle income with those from high-income countries: the consumption of animal proteins will rise in the first group of countries, and it will remain stable but probably will be reduced in the last. The same happens with fruit and vegetables. The tendency is to occur a reduction in the *per capita* cereal consumption. Global food production needs to accommodate the expected increase particularly in consumption of animal protein, fruits and vegetables and supply needs to accompany this demand (FAO, 2017). With economic growth, however, urbanization and globalization, countries often go through a "nutrition transition" (HLPE, 2017) whereby consumption of highly processed foods increase. Consumption of high-energy beverages and snacks, as well as other processed and ultra-processed foods have been on the rise in lower middle-income countries, especially in urban areas. In higher-income countries, the trend has been less stark in recent years, with a plateau or slight decline from a relatively high level of ultra-processed food consumption (Baker & Friel, 2016).

Not all diets have the same impact on greenhouse gases emission (GHG). Income dependent diets are the ones with higher impact, when compared with the Mediterranean type which in turn is higher than the vegetarian. Needless to stress that animal proteins are essential components of nutritional balanced diets. Policies addressing the change in consumption habits in high-income countries need to consider the balance between nutritional needs and the impact diets have on the environment (FAO, 2017).

8.2.2 Trends in economy, food waste, food prices and relation between poverty and food insecurity

Agricultural trade follows closely the global trade trends (FAO, 2017). Agricultural products global trade has tripled between 2000 and 2015. Even though most of food consumed is still produced at the country level. North (USA and Canada), South America (Brazil, Argentina, Uruguay and Paraguay) along with Oceania and Russia are net export regions, while Asia, Middle East and Africa are net importers. Europe is a particular case of an almost neutral, but very active, trade balance (FAO, 2017). Investment in agriculture tend to be relatively low compared with other activities (industry or services), especially in low- and middle-income countries. World food trade is very concentrated, being 7 countries responsible for trading with 77% of all countries in the world (Ercsey-Ravasz et al., 2012). The movement for concentrating food business is, like in any other business: the 50 largest food manufactures hold 50% of market global sales (Heinrich Böll Foundation et al., 2017).

Food waste is not easy to quantify (FAO, 2017). It is estimated that between 25 and 35% of food produced worldwide is wasted, and this occurs in all the stages of the food system, from production to final consumer in a notorious asymmetric way. In high-income countries, where losses can be up to 300 kg/*capita* (HLPE, 2014), consumers are responsible for more than 10% of food waste whereas the food production, packaging and distribution sectors, together, account for approximately another 10%, and the remaining 10% occur in primary production, whereas in low-income countries the relation in 5%, 20%, 10%, respectively. A significant part of this avoidable and policies and measures to respond are needed (Bagherzadeh, Inamura, & Jeong, 2014).

Food prices are a sensitive matter for food security (HLPE, 2017). The volatility of food prices is dependent from many variables linked to demand and offer: the resources available (water and land), the climate changes, the natural disasters, the competition at the market can directly affect the availability of food. The resilience of agricultural systems and they capacity to resist to highly variable food prices (Tendall et al., 2015) and to climatic changes is also very important (Anderson, 2015). The peak in the price of food commodities may jeopardize governmental programs of subsidies to maintain populations feed and low prices may be the reason to collapse of local rural agricultural systems less efficient and less competitive than intensive ones integrated in multinational food companies (FAO, 2017).

There is a *link between poverty, inequality, and food security* (FAO, 2017; FAO et al., 2021). While there has been a global consistent reduction in extreme poverty, in Sub-Saharan Africa this was not the case. Rural areas in some places Africa, Asia and South America are the most exposed. In these areas even if agriculture is key to poverty and hunger alleviation, rural populations have not the means to break the circle of poverty (Olinto, Beegle, & Uematsu, 2013). These populations are highly exposed to extreme climate conditions and political instability which worsen their living condition. In fast growing middle-income countries and in high-income countries within countries inequalities associated with poverty can also be important (IMF et al., 2015), affecting people in rural areas, in particular some regions, ethnic groups, children and women (Olinto et al., 2013).

The food and agricultural markets have known increasing and rapid changes in recent decades with increasing importance of world food trade, growing concentration in agri-food supply chains deeply rooted in heavy financial actors. In this context financial crisis and uncertainties like the ones recently faced (sub-prime in 2007/2008, COVID-19) have profound impact on food security especially in low- and middle-income countries. The "industrialization" and "monetarization" around food systems – land, commodities, distribution chains, food products - increase the uncertainty and reduce access to food (HLPE, 2020).

8.2.3 Political trends, conflicts, and crisis

A *political decrease in state regulation*, at national and international levels, of food trade and property of land other structural food security resources has been seen in last decades in western countries accompanied by a declining trend of public sector investment in agriculture (HLPE, 2020). This void was progressively filled by an aggressive increase of the presence of multinational corporations' intensive capital based (Heinrich Böll Foundation et al., 2017). There is a need for reinforcement in food security governance and the adoption of coherent and comprehensive food and nutrition policies is needed, rebalancing the equilibrium around the stakeholders in the food chain – farm to consumers (Godfray et al., 2011; European Commission, 2021).

Political and military conflicts are increasing worldwide, affecting large millions of the world population in Asia, Africa, and South America, even in some high-income countries. "*The main drivers of conflicts include ethnic and religious differences, discrimination and marginalization, poor governance, limited state capacity, population pressure, rapid urbanization, poverty and youth unemployment*". But agriculture can also be a driver for conflicts which "*include competition for land, water and other natural resources, food insecurity, environmental mismanagement, and government neglect of poor and marginalized areas, such as arid and semi-arid zones essential for livestock-dependent populations and subsistence fishing grounds* (FAO, 2017)". Regions in conflict or permanent crises experience the highest levels of food security problems.

8.2.4 Climate, biodiversity, water and land use, diseases, and pests

Climate change is already affecting every inhabited region across the globe with human influence contributing to many observed changes in weather and climate extremes (IPCC & WGI, 2021). Climate change is already having an impact on food security due to temperature rises, changes in precipitation patterns and an increase in the occurrence of extreme weather events. Agriculture, Forestry, and Other Land Use (AFOLU) GHG emissions account for more than 23% of the anthropogenic contribution. The global food system emissions with up to 36% of anthropogenic emissions (IPCC, 2019).

Climate change comes associated with extreme whether episodes and with climate related disasters. Natural disasters associated with storms and floods (more frequent in Asia, Latin America and Caribbean) or with droughts and extreme temperatures (Africa and Near East) or fires (North America and Australia) become more frequent both in frequency and in magnitude in the last ten years, when compared with levels of the 80's of the twentieth century (FAO, 2017). Climate issues became a vicious cycle with double side effect: the food system activities have an important negative impact on climate chance and climate change with its droughts, floods and storms has very negative impacts in agricultural production particularly in developing countries. Mitigation options are possible, and its adoption is urgently needed, and agricultural production, the need to prepare for change (Godfray et al., 2011).

There are widespread problems with l*and and water use,* e.g., soil loss due to erosion, loss of soil fertility, salination and other forms of degradation; rates of water extraction for irrigation are exceeding rates of replenishment in many places (Charles Godfray et al., 2011). A significant part of the worlds resources are used by the food system: agriculture uses 70% of worlds fresh water and pesticides and antimicrobial residues contaminate water resources; more than 30% of land is currently dedicated to agricultural and animal husbandry utilization (HLPE, 2017).

Biodiversity and ecosystem services are highly impacted by agricultural production, by deforestation driven by agricultural needs and intensive non-sustainable production (Charles Godfray et al., 2011). The thrive for intensification and efficiency in production is putting at risk thousands of species of domestic breeds and crops and their wild relatives globally, reducing the availability of locally highly adapted species (HLPE, 2017). Ecosystems services such as water purification, carbon sequestration, pollination, natural flood protection are at risk (UNESCO & Connor, 2020).

Globalization increases the risks of introduction *of animal diseases and plant pests* (FAO, 2017). Organization of veterinary services is essential to protect animal and plant health, to protect environment, and to maintain an efficient and productive food system.[22] Antimicrobial resistance rise (O'neill, 2016) and animal diseases control, particularly with TAD is critical to maintain sound, and resilient animal production systems. Menaces like ASF, AI or FMD contribute to endanger food security (FAO, 2017). Plant pests are important problems and the control associated with environmental impacts.

8.2.5 Trends in innovation and research

Food systems could be addressed as constituted by three groups pf elements: food supply chains food environments, and consumer behavior (HLPE, 2017). The current trends and drivers shaping the global food security requires the adoption of systems approach to new research and innovation which enables a better understanding of key interactions between a multitude of actors, government levels and processes (production, consumption, distribution) and involving stakeholders is crucial to delivery of transformation (Gill, 2018). Some institutions and governments already recognize the food system needs to be transformed in holistic and comprehensive manner (Sonnino, 2020) and solutions to questions like shifting towards more healthier and sustainable diets or the empowerment of the food system actors needs systemic approach

22. OIE, role of veterinary services: https://www.oie.int/en/what-we-do/global-initiatives/food-safety/role-of-veterinary-services/.

science and evidence. A participatory and transdisciplinary approach is necessary for transformation. The research and innovation shall respond in a way that previous signaled trends can be reversed (Gill, 2018; Zhang, 2018).

8.2.5.1 Towards food supply chains sustainability and resilience

To meet the world's food demand in 2050, food systems will need to increase production by 50% (FAO, 2017). The volatility induced by world markets and lack of governance of the system, and the volatility induced by climate change will be a serious menace to food systems (Anderson, 2015; Tendall et al., 2015). Current yield differences, from agricultural systems, among nations are large. In 2005 leading group of nations were three times higher than lowest group and 1.3 times higher than the middle group. Trajectories of global agricultural development that are directed to greater achievement of the technology improvement and technology transfer frontier would meet 2050 crop demand with much lower environmental impacts than past trends, because strategic intensification that elevates yields of existing croplands of underyielding nations can do can greatly reducing land clearing and GHG emissions. However, yield increases in poorer nations will require significant investments in innovative adaptation of technologies to new soil types, climates, and pests, as well as new infrastructure (Godfray & Muir, 2010; Tilman, Balzer, Hill, & Befort, 2011).

The potential of already existing in knowledge, science, and organization of food supply chains already allows for significant *improvements in productivity efficiency* and efficacy (FAO, 2017). Sustainable intensification, defined as the capacity to produce more out of the same area while reducing environmental impacts (Godfray & Muir, 2010) can make important contributions in low- and middle-income countries farming systems or organic farming (European Commission, 2021) can be one solution for European countries, for instance, by offering solutions adapted for concrete regional or local needs. Farm level measures offer a wide range of possibilities, such as precision farming, improvement on animal or plant genetics; better use and water management, like ecosystem compatible draining systems coupled with irrigation and water harvesting technologies; improving animal health and animal welfare, along with adoption of herd health programs in which animal nutrition efficiency and GHG reduction methodologies; adoption of climate adapted and pest resilient crops, intensification of cover crops utilization (EEA, 2019; Godfray et al., 2011).

8.3 Challenges

The challenges ahead, involving the GFS can be summarized in brief. (1) Achieve global food security and ending hunger and malnutrition; (2) develop resilient and sustainable food systems, able ensure stability of food supplies;

(3) introduce mechanisms to efficiently balance demand and supply and make food affordable, e.g., governance, science and innovation, reducing waste, better nutrition; (4) reorganize food systems processes in a way that make them contribute to climate change mitigation, and (5) also to keep biodiversity and ecosystem services (FAO et al., 2021; Godfray et al., 2011). Global problems require local solutions, because heterogeneity exists, complexity is present.

9. The way ahead – policies and transformation

9.1 Introduction

9.1.1 Problems in decision making within complex environment

From previous exposition, it comes clear that at present, decision makers in Food Security context face complex problems which poses serious challenges. In conflict situations (Läderach et al., 2021) or pandemic emergencies (United Nations & FAO, 2020) chaos often emerge. The Cynefin framework – which recognizes four types of situations: simple, complicated, complex, chaotic - proposed by Kurtz and Snowden (2003) is one effective way of addressing decision making when exposed to chaotic situations. Frequently, the important effort is to recognize chaotic or complex systems and drive complexity into complicated or simple situations. The systems approach, can provide tools for this path (Nguyen, 2018; Zhang, 2018).

Recognizing that the complex problems we are facing requires intergovernmental and multistakeholder, interdisciplinary, multilateral solutions, the preamble of the Berlin Principles states that "*No one group, discipline or sector of society holds enough knowledge and resources to singlehandedly prevent the emergence or resurgence of diseases, while maintaining and improving the health and well-being of all species in today's globalized world*". The promotion of interdisciplinary and transdisciplinary research accompanied by the adoption of "*multidisciplinary and multilateral solutions, while boldly integrating current uncertainties to address the opportunities and challenges ahead*" is essential to respond to novel challenges.

Several contributions have recently been made towards methods for comprehensive and efficient policy design (Hawkes and Parsons, 2019; HLPE, 2017, 2020), in which regard the need for systemic thinking adoption is generally recommended. Several models of the food system have been proposed, some have been referenced in this review: Ericksen in 2008, several times afterwards referenced and modified but still valid (Ericksen, 2008); the High Level Panel of Experts, which offers the *conceptual framework of food systems for diets and nutrition* (HLPE, 2017 and Zhang, 2018), which emphasizes the relevance of the adoption of systemic thinking. Adoption of adaptative systemic thinking is highly recommended since "*one solution does not fit all*" and careful application of principles to each country or region is necessary (Rüegg et al., 2018). The implementation of policies often requires the food system transformation, and this is a hard and complex difficult task to embrace: political leadership and motivation is needed, also to overcome barriers and obstacles (imbalances across systems or conflicts of interests) and elicit the adequate conditions for transformation (HLPE, 2017). Food system transformation is also addressed in the previous references but will be based on the proposal by FAO in "*The State of Food Security and Nutrition in the World 2021*" (FAO et al., 2021).

9.1.2 Evidence of policy impacts

The European Common Agricultural policy (CAP) is a good example of how policy measures, anchored in specific legislation and adequately financed impacts the food system. Directed initially to ensure availability and access of food, i.e., focused on supply chains, it evolved during decades, subject to several reforms. The current reform, even subjected to criticism is much more focused on environmental and climate impacts mitigation than previous ones (Scown, Brady, & Nicholas, 2020). Outcomes of this policy were undoubtedly of great importance to Europe's food supply in terms of unique quality, safety, quantity, and affordability (*agency*). Regarding utilization and sustainability efforts need now to be made to improve these two pillars (SAPEA, 2020).

9.2 Policies for systemic response

To address in a complete and comprehensive manner Food Security and, at the same time, to do so based in sustainable, robust and sustainable way, food policies need to consider health, economy, environment, societal and political issues (FAO et al., 2021; Parsons et al., 2019; United Nations & FAO, 2020). The concept of Food Policy (Hawkes and Parsons 2019) has been tracked down to the beginning of the XX Century but it has evolved since then, from the initial purpose of balancing demand and supply to strategies like the "farm to fork strategy" which provides a comprehensive attempt to efficiently respond to the complex problem of balancing the food supply chain negative impacts with sustainability and consumer needs and preferences (European Commission, 2021).

In building portfolios of policies drivers, challenges, and trends should address the need for radical transformation (in many cases), recognize the complex inter play between food systems and other sector systems, focus on hunger and malnutrition and recognize that different situations require, often, different solutions (Nguyen, 2018; HLPE, 2020).

9.2.1 Food supply chain policies

Such policies should consider the different phases of the chain: production systems and post-harvest practices, market and store access, the nutrient content of foods, food quality and food safety, and the added value chain involved. Conflicts of interest along the chain and balances of power (HLPE, 2017). Impacts on biodiversity, ecological systems involved, on climate, on land use and on water (EEA, 2019). Focus on contribution that production technologies, medicines, novel feed sources, genetics, breeds, seeds, information technologies can introduce to promote sustainability and reduce or mitigate impacts (EFSA & ECDC, 2011). Sustainable intensification, definition of the safe operating space for food the systems involved and establish clear objectives and goals can be beneficial (Willett et al., 2019). Adoption of good practices in the use of chemicals (antimicrobials and pesticides) is relevant (European Commission, 2017; O'neill, 2016; OCED, 2016). Prevention of raw-materials and food waste during production and transformation phases is very relevant in low- and middle-income countries (Bagherzadeh et al., 2014).

9.2.2 Food environment policies

Food environment should contribute to promote better (sustainable, environmentally friendly and healthy) food consumer choices. Policies on social inclusions, removing poverty and inequity barriers, mother with young child particular access are important in affordability; in this regard the COVID-19 pandemic brought to the light existing and new social fragilities (United Nations & FAO, 2020). Food safety infrastructures and operator responsibility. Promotion of healthy diets through taxes and subsidies and discriminatory trade policies. To act upon consumer information in labeling and marketing activities (HLPE, 2017). Act firmly against fraud (European Council, 2019).

9.2.3 Food and nutrition policies

Nutrition policies should be able to promote a healthy and less environmental damaging consumer choices. Education, children and young people, parents and consumers in general is a cornerstone of this policy. The choice of where and what to eat, the food to buy and its provenance, the way of cooking, serving, and storing have nutrition and safety implications. Empower the consumers with knowledge on diets (Willett et al., 2019), hygiene practices in food handling and, also on product origin. The particular care with wasting avoidance is of enormous impact in high-income countries (EFSA & ECDC, 2019; HLPE, 2014).

The WHO response to malnutrition was adopted by the World Health Assembly in 2004 after resolution WHA57.17 in a document entitled "WHO Global Strategy on Diet, Physical Activity and Health", (WHO, 2004). There is a clear link between heathy diets, adequate levels of daily physical activity, tobacco control and long-life health, recognized by the WHO. "Unhealthy diets and physical inactivity are thus among the leading causes of the major noncommunicable diseases (NCDs), including cardiovascular disease, type 2 diabetes and certain types of cancer, and contribute substantially to the global burden of disease, death and disability" (WHO, 2004). The "Politic Declaration on Noncommunicable Disease" recognized, after almost 10 years of the publication of the Global Strategy that NCD remained a major threat and to public health and that food and nutrition polices should be implemented world-wide, by member states (WHO, 2012). These should address, consumer choices and multisectoral and multistakeholder domains – agriculture, industry, transports, and food retailers.

9.2.4 Research and innovation

Research and innovation are one pillar of food system policies. It should contribute to increase the efficiency, resilience and sustainability of food supply chain, and to create better environment (Gill, 2018). To overcome the challenges ahead, to understand the systems, understand consumer behavior and stakeholders interests, to discover new medicines, new technologies, just to mention a few issues, R&I is an essential step for success (Godfray et al., 2011; HLPE, 2020). Complexity of the issue requires adoption of flexible and new frameworks for research calls (Sonnino, 2020). A better understanding of key interactions between a multitude of actors, government levels and processes (production, consumption, distribution) and involving stakeholders is crucial to delivery of transformation (Gill, 2018). Research areas like circularity of economy, knowledge in nutrition, climate is important.

9.2.5 Food security policies

Until recently food security policies were considered guaranteed in economically developed countries, supported by a set of comprehensive agricultural, social and economic policies, these countries have ordinarily in place. Consequently, no particular concern was raised regarding food security policies, but recent emergency of COVID-19 pandemic dramatically changed this perception and demonstrated the need for contingency measures and policies. On the other hand, in less developed countries or in countries facing military conflicts these policies are mostly endorsed by International Organizations.

At a global level, international organizations such as UN and dependent organizations like FAO, UNICEF or the WFP,[23] are committed to fight hunger and ensure food security. After COVID-19 onset a document issued by the UN offered a global view about the conceptual structure to respond to food security crises. These develop around three axes: (1) recentre the focus of actions in critical targeted groups, make food production marketing and distributions essential services anywhere, monitoring systems and provide liquidity and financial inclusion (2) strength social protection systems tailoring them to make nutrition as one major delivery, through a series of actions concerted

23. World Food Program - https://www.wfp.org/.

under three domains: social protection system, food system, health system; (3) focus on the transformation of food systems to make them better adapted with nature and climate, more inclusive, and resilient (United Nations & FAO, 2020). Malnutrition is a particular incidence of Food Insecurity. Policies (Menon & Peñalvo, 2019).

The importance and necessity of create policies, at country or region level, that ensures food security was perceived beyond doubt by the European Commission given the recent, and still ongoing, COVID-19 crises (European Commission, 2021). Unforeseen events like impressive shortages over the supply chain, abrupt changes in consumer behavior and habits, losses of certain markets were consecutive to the pandemic onset. The disruptive potential of unprecedented threats happening recurringly every year as droughts, floods, forest fires, biodiversity loss and new and the food chain faces increasing pests, was perceived as a driver for new policies. As a response a contingency plan for ensuring food supply and food security will be developed in the European Union along with common response mechanisms to crises affecting food systems (European Union, 2002).

9.3 One vision: from moving towards food systems transformation

A key challenge that precludes successful transformation of food systems is that existing national, regional and governmental policies, strategies, investments and legislation are compartmentalized (FAO et al., 2021), another key is that multinational corporations and funds, which play currently a main role in food security, pursue their own legitimate interests which may not coincide with public food security interests (Heinrich Böll Foundation et al., 2017). For example, climate mitigation policies, public agricultural policies, restrictions to world trade agreements, social and health care systems are discussed and negotiated independently.

9.3.1 The pathway for food system transformation

In the document recently published "*The state of food security and nutrition in the world*" the leading international agencies under the patronage of the UN (FAO, UNICEF, IFAD, WFP) make the case for a methodology for transformation of food systems (FAO et al., 2021). Six pathways are proposed, to build resilient and sustainable food systems, oriented to food security, improved nutrition, and access to affordable diets for all, requiring adoption of cross-sectoral policies. One or more these pathways can be used to resolve the situation in a particular country: (1) in conflict affected areas: consider the integration of humanitarian, development and peacebuilding policies; (2) reinforce the climate resilience across all food systems; (3) consider the introduction of social and health policies focused on those vulnerable to economic adversity; (4) tackle poverty and structural inequalities; (5) consider lowering the cost of nutritious foods; (6) act upon consumer behaviors to induce the adoption of healthier and environmental friendly nutrition habits (FAO et al., 2021). The pathway method for food system transformation requires the formulation and implementation of portfolios of cross-sectoral integrated policies, involving the relevant elements of the three domains of food systems – food supply chain, food environment, consumer choices (HLPE, 2017). The portfolio of policies is developed after a structured situation analysis. The implementation shall be monitored, evaluated and accountability measures should be in place (FAO et al., 2021). Several case studies are added in the document.

9.3.2 Farm to fork strategy: the European Union response for change

Another example of a comprehensive approach towards a food system transformation is the "farm to fork policy (F2F)" (European Commission, 2021) adopted in 2021 by the EU in the context of the "Green Deal" initiative undertake by the Commission, the Parliament and the Council. The F2F strategy covers the three domains of food systems (HLPE, 2017). It aims to "*build a food supply chain that works for consumers, producers, climate and the environment*" developing a comprehensive set of measures directed to a sustainable food production, food transformation, distribution, selling and marketing, and to food waste reduction; food security is also envisaged. The F2F strategy put a strong emphasis on the "*transition measures*" through a set of measures directed to the reinforcement of research, innovation, technology, and correlated investments; it also focus on skills, data, and knowledge sharing. The commitment with "*global transition*" which is a group of measures with international incidence such as international cooperation, measures restricting the importation of products associated with global deforestation, or meat from countries not having antimicrobial use properly regulated, completes the strategy. Financing the strategy and the identification of a portfolio of legislative acts necessary to complete the existing regulations and directives is discriminated in the strategy.

This strategy is the result of several consultations with stakeholders, research on domains like environment and climate, trade, consumer domain, common agricultural policy (CAP), among others. Specific financial resources will be allocated to the implementation of the policies which are supported on sectoral Regulations and Directives.

References

Anderson, M. D. (2015). The role of knowledge in building food security resilience across food system domains. *Journal of Environmental Studies and Sciences, 5*(4), 543–559. https://doi.org/10.1007/s13412-015-0311-3

Bagherzadeh, M., Inamura, M., & Jeong, H. (2014). *Food waste along the food chain*, 18156797. Paris: OECD.

Baker, P., & Friel, S. (2016). Food systems transformations, ultra-processed food markets and the nutrition transition in Asia. *Global Health, 12*(1), 80. https://doi.org/10.1186/s12992-016-0223-3

Balleste-Delpierre, C., & Vila Estape, J. (2016). Why are we still detecting food-related *Salmonella* outbreaks in Spain? *Enfermedades Infecciosas y Microbiología Clínica, 34*(9), 541–543. https://doi.org/10.1016/j.eimc.2016.08.001

CAC. (1997). *Principles and guidelines for the establishment and application of microbiological criteria related to foods. Rome.*

CAC/WHO. (2003). *Assuring food safety and quality: Guidelines for strengthening national food control systems* (Vol. 76). Rome: Italy.

Cassini, A., Högberg, L. D., Plachouras, D., Quattrocchi, A., Hoxha, A., Simonsen, G. S., … Hopkins, S. (2019). Attributable deaths and disability-adjusted life-years caused by infections with antibiotic-resistant bacteria in the EU and the European economic area in 2015: A population-level modelling analysis. *The Lancet Infectious Diseases, 19*(1), 56–66. https://doi.org/10.1016/s1473-3099(18)30605-4

CDC. (2015). *Water treatment.* https://www.cdc.gov/healthywater/drinking/public/water_treatment.html.

CDC. (2021). In CDC (Ed.), *CDC and food safety.*

Charles, H., Godfray, J., Beddington, J. R., Crute, I. R., Haddad, L., Lawrence, D., James, F., Muir, J. P., Robinson, S., Thomas, S. M., & Toulmin, C. (2010). Food security: The challenge of feeding 9 billion people. *Science, 327.*

Choffnes, E. R., Relman, D. A., Olsen, L., Hutton, R., & Mack, A. (2012). *Improving food safety through a one health approach.* The National Academies Press.

Codex Alimentarius Commission, FAO, & WHO. (2013). *Food hygiene - basic texts. Rome, Italy.*

Commission of the European Communities. (2000). *White paper on food safety. Brussels.*

Crutzen, P. J. (2002). Geology of mankind. *Nature, 415.*

Devereux, S. (2006). Distinguishing between chronic and transitory food insecurity in emergency needs assessments. In *Rome: World food programme, emergency Needs assessment branch (ODAN).*

EC/FAO, F. S. P. (2008). *An introduction to the basic concepts of food security.*

ECDC/EMEA. (2009). *The bacterial challenge: Time to react – A call to narrow the gap between multidrug-resistant bacteria in the EU and the development of new antibacterial agents.*

EEA. (2019). *Climate change adaptation in the agriculture sector in Europe.*

EFSA, ECDC. (2011). The European Union summary report on trends and sources of zoonoses, zoonotic agents and food-borne outbreaks in 2009. *EFSA Journal, 9*(3). https://doi.org/10.2903/j.efsa.2011.2090

EFSA, ECDC. (2019). The European union one health 2018 zoonoses report. *EFSA Journal, 17*(12), e05926. https://doi.org/10.2903/j.efsa.2019.5926

Ercsey-Ravasz, M., Toroczkai, Z., Lakner, Z., & Baranyi, J. (2012). Complexity of the international agro-food trade network and its impact on food safety. *PLoS One, 7*(5), e37810. https://doi.org/10.1371/journal.pone.0037810

Ericksen, P. J. (2008). Conceptualizing food systems for global environmental change research. *Global Environmental Change, 18*(1), 234–245. https://doi.org/10.1016/j.gloenvcha.2007.09.002

European Union. (2002). *Regulation (EC) No 178/2002 of the European parliament and of the council, of 28 January 2002 laying down the general principles and requirements of food law, establishing the European Food Safety Authorityand laying down procedures in matters of food safety. Regulation (Ec) No 178/2002 C.F.R.*

European Commission. (2016). *Commission Notice, on the implementation of food safety management systems covering prerequisite programs (PRPs) and procedures based on the HACCP principles, including the facilitation/flexibility of the implementation in certain food businesses. (C 278/1).*

European Commission. (2017). *A European one health against antimicrobial resistance (AMR).* European Commission Retrieved from https://ec.europa.eu/health/sites/default/files/antimicrobial_resistance/docs/amr_2017_action-plan.pdf.

European Commission. (2021). *Farm to fork strategy.* Retrieved from https://ec.europa.eu/food/horizontal-topics/farm-fork-strategy_en.

European Council. (2019). *Next steps how to better tackle and deter fraudulent practices in the agrifood chain - Council Conclusions (16 December 2019). Regulation (Ec) No 178/2002 of the European Parliament and of the Council, of 28 January 2002 laying down the general principles and requirements of food law, establishing the European Food Safety Authority and laying down procedures in matters of food safety, Regulation (Ec) No 178/2002 C.F.R. (2002).*

European Union. (2020). *RASFF — the rapid Alert system for food and feed — annual report 2019.*

Evans, B. R., & Leighton, F. A. (2014). A history of one health. *Revue scientifique et technique (International Office of Epizootics), 33*(2), 413−420.

FAO. (1996). *Rome declaration on world food security.* Retrieved from http://www.fao.org/3/w3613e/w3613e00.htm.

FAO. (2006). *Food security.* Retrieved from https://reliefweb.int/sites/reliefweb.int/files/resources/pdf_Food_Security_Cocept_Note.pdf.

FAO (Ed.). (2017). *The future of food and agriculture — trends and challenges.*

FAO. (2018). *More people, more food, worse water?.* http://www.fao.org/3/ca0146en/ca0146en.pdf.

FAO. (2021). *Wastewater treatment.* http://www.fao.org/3/t0551e/t0551e05.htm.

FAO-OIE-WHO. (2010). The FAO-OIE-WHO Collaboration. Sharing responsibilities and coordinating global activities to address health risks at the animal-human-ecosystems interfaces. *A Tripartite Concept Note.*

FAO, IFAD, UNICEF, WFP, & WHO. (2021). *The state of food security and nutrition in the world 2021.*

FAO, OIE, & WHO. (2010). *The FAO-OIE-WHO Collaboration. Sharing responsibilities and coordinating global activities to address health risks at the animal-human-ecosystems interfaces.*

Faour-Klingbeil, D., & Todd, E. C. D. (2019). Prevention and control of foodborne diseases in middle-East North African countries: Review of national control systems. *International Journal of Environmental Research and Public Health, 17*(1). https://doi.org/10.3390/ijerph17010070

FAO, & WHO. (2006). *Food safety risk analysis - a guide for national food safety authorities.* Rome: FAO Comunications Division.

Food contact materials. (n.d.). EFSA. Retrieved June 8, 2021, from https://www.efsa.europa.eu/en/topics/topic/food-contact-materials.

Gill, M., den Boer, A. C. L., Kok, K. P. W., Breda, J., Cahill, J., Callenius, C., … Broerse, J. E. W. (2018). *Brief 4 A systems approach to research and innovation for food system transformation.* Published by FIT4FOOD2030 https://fit4food2030.eu/eu-think-tank-policy-brief/.

Godfray, C., Crute, I., Haddad, L., Lawrence, D., Muir, J., Pretty, J., … Toulmin, C. (2011). *Foresight. The future of food and farming. Executive summary.*

Gorris, L. G. M. (2005). Food safety objective: An integral part of food chain management. *Food Control, 16*(9), 801−809. https://doi.org/10.1016/j.foodcont.2004.10.020

Griffiths, J. K. (2017). Waterborne diseases. In *International encyclopedia of Public Health* (pp. 388−401). Elsevier. https://doi.org/10.1016/B978-0-12-803678-5.00490-2

Grizzetti, B., Lanzanova, D., Liquete, C., Reynaud, A., & Cardoso, A. C. (2016). Assessing water ecosystem services for water resource management. *Environmental Science & Policy, 61*, 194−203. https://doi.org/10.1016/j.envsci.2016.04.008

Gruetzmacher, K., Karesh, W. B., Amuasi, J. H., Arshad, A., Farlow, A., Gabrysch, S., … Walzer, C. (2021). The Berlin principles on one health - bridging global health and conservation. *The Science of the Total Environment, 764*, 142919. https://doi.org/10.1016/j.scitotenv.2020.142919

Haque, S. S., & Freeman, M. C. (2021). The applications of implementation science in Water, Sanitation, and Hygiene (WASH) research and practice. *Environmental Health Perspectives, 129*(6), 065002. https://doi.org/10.1289/EHP7762

Havelaar, A. H., Kirk, M. D., Torgerson, P. R., Gibb, H. J., Hald, T., Lake, R. J., … World Health Organization Foodborne Disease Burden Epidemiology Reference, G. (2015). World health organization global estimates and regional comparisons of the burden of foodborne disease in 2010. *PLoS Medicine, 12*(12), e1001923. https://doi.org/10.1371/journal.pmed.1001923

Hawkes, C., & Parsons, K. (2019). *Brief 1: Tackling food systems challenges: The role of food policy*. London: Centre for Food Policy.

Heinrich Böll Foundation, Rosa Luxemburg Foundation, Friends of the Earth Europe. (2017). *Agrifoodatlas- Facts and figures about the corporations that-control what we eat*.

HLPE. (2014). *Food losses and waste in the context of sustainable food systems. A report by the high level Panel of Experts on food security and nutrition of the committee on world food security, Rome*.

HLPE. (2017). *Nutrition and food systems. A report by the high level Panel of Experts on food security and nutrition of the committee on world food security, Rome*.

HLPE. (2020). *Food security and nutrition: Building a global narrative towards 2030. A report by the high*.

Hunter, P. R., MacDonald, A. M., & Carter, R. C. (2010). Water supply and health. *PLoS Medicine, 7*(11), e1000361. https://doi.org/10.1371/journal.pmed.1000361

Ibrahim, M., Schelling, E., Zinsstag, J., Hattendorf, J., Andargie, E., & Tschopp, R. (2021). Seroprevalence of brucellosis, Q-fever and Rift Valley fever in humans and livestock in Somali Region, Ethiopia. *PLoS Neglected Tropical Diseases, 15*(1), e0008100. https://doi.org/10.1371/journal.pntd.0008100

IMF, Fund, I. M., Dabla-Norris, E., Kochhar, K., Ricka, F., Suphaphiphat, N., & Tsounta, E. (2015). *Causes and consequences of income inequality: A global perspective*.

IPCC. (2019). *Special report on climate change, desertification, land degradation, sustainable land management, food security, and greenhouse gas fluxes in terrestrial ecosystems*. Summary for Policymakers.

IPCC, & WGI (Eds.). (2021). *Climate change 2021: The Physical science basis*. Summary for Policymakers.

Jouve, J. L. (1998). Principles of food safety legislation. *Food Control, 9*, 75−81.

Jouve, J. L., Stringer, M. F., & Baird-Parker, D. A. C. (1998). *Food safety management tools*.

Kostas Stamoulis, A. Z. (2003). *A conceptual framework for national agricultural, rural development, and food security strategies and policies*. ESA Working Paper No. 03-17.

Kurtz, C. F., & Snowden, D. J. (2003). IBM complexity and complication - new dynamics of strategy. *IBM Systems Journal, 42*(3).

Läderach, P., Pacillo, G., Thornton, P., Osorio, D., & Smith, D. (2021). Food systems for peace and security in a climate crisis. *The Lancet Planetary Health, 5*(5), e249–e250. https://doi.org/10.1016/s2542-5196(21)00056-5

Level Panel of Experts on food security and nutrition of the committee on world food security, Rome.

Levin, S. A. (1998). Ecosystems and the biosphere as complex adaptive systems. *Ecosystems, 1*, 431–436.

Mancosu, N., Snyder, R., Kyriakakis, G., & Spano, D. (2015). Water scarcity and future challenges for food production. *Water, 7*(12), 975–992. https://doi.org/10.3390/w7030975

Menon, S., & Peñalvo, J. L. (2019). Actions targeting the double burden of malnutrition: A scoping review. *Nutrients, 12*(1). https://doi.org/10.3390/nu12010081

Nguyen, H. (2018). In FAO (Ed.), *Sustainable food systems - concept and framework.*

O'Brien, E., & Xagoraraki, I. (2019). A water-focused one-health approach for early detection and prevention of viral outbreaks. *One Health, 7*, 100094. https://doi.org/10.1016/j.onehlt.2019.100094

OCED. (2016). *Antimicrobial resistance - policy insights OCED.*

OECD. (2021). *Water and agriculture.* https://www.oecd.org/agriculture/topics/water-and-agriculture/.

OIE. (2016). *The OIE strategy on antimicrobial resistance and the prudent use of antimicrobials.*

Olinto, P., Beegle, K., Sobrado, C. E., & Uematsu, H. (2013). *The State of the poor: Where are the poor, where is extreme poverty harder to end, and what is the current profile of the world's poor.* Washington, DC: World Bank.

O'neill, C. B. J. (2016). *The review on antimicrobial resistance - tackling drug-resistant infections globally: Final report and recommendations. Wellcome trust.* HM Government.

Palmer, M. V. (2013). Mycobacterium bovis: Characteristics of wildlife reservoir hosts. *Transboundary and Emerging Disesases, 60*(Suppl. 1), 1–13. https://doi.org/10.1111/tbed.12115

Parsons, K., Hawkes, C., & Wells, R. (2019). *Brief 2: Understanding the food system: Why it matters for food policy. London.*

Rannap, R., Lõhmus, A., & Briggs, L. (2009). Restoring ponds for amphibians: A success story. *Hydrobiologia, 634*(1), 87–95. https://doi.org/10.1007/s10750-009-9884-8

Reveillaud, E., Desvaux, S., Boschiroli, M. L., Hars, J., Faure, E., Fediaevsky, A., … Richomme, C. (2018). Infection of wildlife by *Mycobacterium bovis* in France assessment through a national surveillance system, Sylvatub. *Frontiers in Vertinary Science, 5*, 262. https://doi.org/10.3389/fvets.2018.00262

Rüegg, S. R., Häsler, B., & Zinsstag, J. (2018). *Integrated approaches to health. A handbook for the evaluation of One Health.* The Netherlands: Wageningen Academic Publishers.

Sandra, H., Bryan, M., & Batz, M. (2015). *Economic burden of major foodborne illnesses acquired in the United States. EIB-140.* U.S. Department of Agriculture,.

SAPEA. (2020). *Science advice for policy by European academies. A sustainable food system for the European union. Berlin.* https://doi.org/10.26356/sustainablefood

Scallan, E., Hoekstra, R. M., Angulo, F. J., Tauxe, R. V., Widdowson, M. A., Roy, S. L., … Griffin, P. M. (2011). Foodborne illness acquired in the United States–major pathogens. *Emerging Infectious Diseases, 17*(1), 7–15. https://doi.org/10.3201/eid1701.P11101

Schneeberger, P. M., Wintenberger, C., van der Hoek, W., & Stahl, J. P. (2014). Q fever in The Netherlands - 2007-2010: What we learned from the largest outbreak ever. *Medecine et Maladies Infectieuses, 44*(8), 339–353. https://doi.org/10.1016/j.medmal.2014.02.006

van Schothorst, M., Zwietering, M. H., Ross, T., Buchanan, R. L., & Cole, M. B. (2009). Relating microbiological criteria to food safety objectives and performance objectives. *Food Control, 20*(11), 967–979. https://doi.org/10.1016/j.foodcont.2008.11.005

Scown, M. W., Brady, M. V., & Nicholas, K. A. (2020). Billions in Misspent EU agricultural subsidies could support the sustainable development goals. *One Earth, 3*(2), 237–250. https://doi.org/10.1016/j.oneear.2020.07.011

Sen, A. (1981). *Poverty and famines, an essay on entitlement and DeprivationAmartya-Sen | 1981.pdf.* New York: Oxford University Press.

Sonnino, R., Csallenius, C., Lähteenmäki, L., Breda, J., Cahill, J., Caron, P., … Gill, M. (2020). *Brief-3 research and innovation supporting the farm to fork strategy of the European commission.* https://doi.org/10.13140/RG.2.2.10891.23840

Steffen, W., Sanderson, A., Tyson, P. D., Jäger, J., Matson, P. A., Moore, B., III, … Wasson, R. J. (2005). *Global change and the earth system - a planet under pressure (2nd ed. Vol. Ch. 3 - the Anthropocene Era: How humans are changing the earth system).* Berlin Heidelberg New York: Springer.

Tendall, D. M., Joerin, J., Kopainsky, B., Edwards, P., Shreck, A., Le, Q. B., … Six, J. (2015). Food system resilience: Defining the concept. *Global Food Security, 6*, 17–23. https://doi.org/10.1016/j.gfs.2015.08.001

The Global Water Partnership. (2020). *The need for an integrated approach.* https://www.gwp.org/en/About/why/the-need-for-an-integrated-approach/.

The World Bank. (2020). *Operation and maintenance strategies for hydropower: Handbook for practitioners and decision makers.* https://www.worldbank.org/en/topic/energy/publication/operation-and-maintenance-strategies-for-hydropower.

Tilman, D., Balzer, C., Hill, J., & Befort, B. L. (2011). Global food demand and the sustainable intensification of agriculture. *Proceedings of the National Academy of Sciences of the United States of America, 108*(50), 20260–20264. https://doi.org/10.1073/pnas.1116437108

Tomasino, M. P., Semedo, M., Vieira e Moreira, P., Ferraz, E., Rocha, A., Carvalho, M. F., … Mucha, A. P. (2021). SARS-CoV-2 RNA detected in urban wastewater from Porto, Portugal: Method optimization and continuous 25-week monitoring. *The Science of the Total Environment, 792*, 148467. https://doi.org/10.1016/j.scitotenv.2021.148467

A Tripartite Concept Note. Retrieved from http://www.fao.org/ag/againfo/home/en/news_archive/AGA_in_action/2013_Tripartite_partnership_at_the_human-animal-ecosystem_interface.html.

UNESCO, & Connor, R. (2020). In U. Nations (Ed.), *The united Nations world water development report 2020. Water and climate change. Executive summary.* UNESCO World Water Assessment Programme.

UNICEF. (2019). *Progress on household drinking water, sanitation and hygiene I 2000-2017.* https://data.unicef.org/resources/progress-on-household-drinking-water-sanitation-and-hygiene-2000-2020/.

United Nations. (2015). Trends in International Migrant stock: The 2015 revision. In *Department of Economic and Social Affairs of the United Nations POP/DB/MIG/Stock/Rev.2015.*

United Nations. (2020). *Sustainable development goals.* Retrieved from https://www.undp.org/content/undp/en/home/sustainable-development-goals.html.

United Nations Children's Fund (UNICEF), WHO, & International Bank for Reconstruction and Development/The World Bank. (2020). *Levels and trends in child malnutrition: Key findings of the 2020 edition of the joint child malnutrition estimates in Geneva.*

United Nations, & FAO. (2020). *Policy brief: The impact of Covid-19 on food security and nutrition.* Retrieved from https://www.un.org/sites/un2.un.org/files/sg_policy_brief_on_covid_impact_on_food_security.pdf.

World Food Summit. (1996). https://www.fao.org/3/w3613e/w3613e00.htm.

WHO. (2004). *WHO global strategy on diet, physical activity and health.*

WHO. (2012). *Political declaration of the high-level meeting of the general assembly on the prevention and control of non-communicable diseases. (Sixty-sixth session, general assembly).*

WHO. (2015). *WHO Strategic Plan Global action plan on antimicrobial resistance.*

WHO. (2017). *Diarrhoeal disease.* https://www.who.int/en/news-room/fact-sheets/detail/diarrhoeal-disease.

WHO. (2019). *Global health estimates 2019 (DALY): Disease burden by cause, age, sex, by country and by region, 2000-2019.* Retrieved from https://www.who.int/healthinfo/global_burden_disease/estimates/en/index1.html.

WHO. (2020). *Domestic water quantity, service level and health.* https://www.who.int/publications/i/item/9789240015241.

WHO. (2021). *Malnutrition.* Retrieved from https://www.who.int/news-room/fact-sheets/detail/malnutrition.

WHO Europe. (2017). The burden OF foodborne diseases. In *THE WHO EUROPEAN REGION. UN city,Marmorvej 51, DK-2100 Copenhagen Ø, Denmark.*

WHO-FAO. (2019). *Global INFOSAN strategic plan 2020-2025.*

WHO-FAO-OIE. (2019). *Taking a multisectoral, one health approach: A tripartite guide to addressing zoonotic diseases in countries.*

WHO, & FERG. (2015). *WHO estimates OF the global burden OF foodborne diseases.*

Will, S., Crutzen, P. J., & McNeill, J. R. (2007). The Anthropocene: Are humans now overwhelming the great forces of nature? *Ambio, 36*(8), 614–621.

Willett, W., Rockström, J., Loken, B., Springmann, M., Lang, T., Vermeulen, S., ... Murray, C. J. L. (2019). Food in the Anthropocene: The EAT–Lancet commission on healthy diets from sustainable food systems. *The Lancet, 393*(10170), 447–492. https://doi.org/10.1016/s0140-6736(18)31788-4

World Bank. (2010). *People, pathogens, and our planet. Volume 1: Towards a one health approach for controlling zoonotic diseases.* Vol. Report No. 50833-GLB.

World Bank. (2017). *Drug resistant infections. A threat to our economic future.* Executive Summary.

World Bank. (2018). *One health. Operational framework for strengthening human, animal, and environmental public health systems at their interface.*

WTO. (1995). *Agreement on sanitary and phytosanitary measures.*

Zhang, W., Gowdy, J., Bassi, A. M., Santamaria, M., DeClerck, F., Adegboyega, A., ... Wood, S. L. R. (2018). Systems thinking: An approach for understanding 'eco-agri-food systems. In *TEEB for agriculture & food: Scientific and economic foundations.* Geneva: UN Environment.

Zinsstag, J., Schelling, E., Wyss, K., & Mahamat, M. B. (2005). Potential of cooperation between human and animal health to strengthen health systems. *The Lancet, 366*(9503), 2142–2145. https://doi.org/10.1016/s0140-6736(05)67731-8

Zipper, S. C., Jaramillo, F., Wang-Erlandsson, L., Cornell, S. E., Gleeson, T., Porkka, M., ... Gordon, L. (2020). Integrating the water planetary boundary with water management from local to global scales. *Earth's Future, 8*(2). https://doi.org/10.1029/2019EF001377

Chapter 6

The influence of social and economic environment on health

Diogo Guedes Vidal[a], Gisela Marta Oliveira[a], Manuela Pontes[a], Rui Leandro Maia[a] and Maria Pia Ferraz[b,c,d]

[a]*UFP Energy, Environment and Health Research Unit (FP-ENAS), University Fernando Pessoa, Porto, Portugal;* [b]*Departamento de Engenharia Metalúrgica e de Materiais, Faculdade de Engenharia da Universidade do Porto, Porto, Portugal;* [c]*i3S — Instituto de Investigação e Inovação em Saúde, Universidade do Porto, Porto, Portugal;* [d]*INEB — Instituto de Engenharia Biomédica, Universidade do Porto, Porto, Portugal*

1. Introduction

One Health approach recognizes that interconnections among humans, animals, and environment is a key issue in public health and wellbeing, that stimulates interdisciplinary collaborations to develop a more holistic understand and effective action against public health threats (Stenvinkel, 2020). The One Health strategy to develop knowledge on human and environmental health by working at the local, regional, national, and global levels involves a network of international and national organizations and multidisciplinary teams (including physicians, veterinarians, ecologists, and many others professionals). The aim is to achieve better health outcomes by monitoring and restraining threats to public health (Alonso Aguirre et al., 2019; Machalaba et al., 2018) and to learn how diseases spread among people, animals (domesticated or wild), and their shared ecosystems (World Health Organization (WHO) & Nations (FAO) and World Organisation for Animal Health (OIE), 2019).

Several scientific evidences (Commission on Social Determinants of Health, 2008; Costa, Santana, Dimitroulopoulou, & Burstrom, 2019; Graham & White, 2016; Mackenbach et al., 2017) support the fact that human health is no longer considered only as a consequence of individual biological causes but is strongly influenced by environmental, cultural, social, economic and political factors. Human health is dependent on life conditions which has a social, cultural, and economic context that is shaped by several environmental

One Health. https://doi.org/10.1016/B978-0-12-822794-7.00005-8

pressure factors such as air, water, or food quality (World Health Organization (WHO), 2017). Among other, factors affecting health include sanitation conditions, housing quality, environmental quality, access to healthcare and education, work conditions, and safety (Bambra et al., 2010; Mitsakou et al., 2018).

Considered by the World Health Organization as one of the current major challenges (Krech, 2011), the concerns on this topic are not new. In the nineteen century, three key pioneers Rudolf Virchow, Robert Koch, and Oswaldo Cruz (Raviglione & Krech, 2011), have devoted attention to how social and economic conditions influenced tuberculosis. In a work of Chisholm (1949), new insights on the causes of diseases are discussed concluding that the death rate from tuberculosis was a mirror of the socioeconomic status. In 1978, at the International Conference on Primary Health Care, the Declaration of Alma-Ata (World Health Organization, 1978) was published urging for a global commitment to support health for all at all ages and remarking that health condition is not only the absence of disease or infirmity but, is, above all, a human right. This document also stresses the need to develop strategies to reduce the disparities in health outcomes between developed and underdeveloped nations. In 1986 the Ottawa Charter for Health Promotion (World Health Organization, 1986) and subsequent documents highlighted the importance of including health issues across all policies, to prevent disease and promote health.

This chapter discusses the multifaceted interaction between social and economic conditions and health since individual's health is a consequence of life conditions which are influenced by the access to healthcare services, green spaces, secure food, sanitary conditions, exposure to pollution, risky behaviors and addictions, among other factors.

2. How social, economic and environmental conditions constraint health outcomes

Inequities in health are systematic, socially produced, that means modifiable and unfair (Whitehead & Dahlgren, 2007). The systematic pattern of health inequities is expressed when health outcomes are different across socioeconomic groups (Whitehead & Dahlgren, 2007). Throughout the life course, individuals may be exposed to situations that can aggravate health conditions. For example, growing in a poor and disadvantaged community can cause privations of safe and quality food and pharmaceuticals. In these low income conditions, individuals have less opportunities to achieve better living conditions (Committee on Assessing Interactions Among Social Behavioral, Hernandez, & Blaze, 2006). This can be called as a cumulative multi-level of exposure which represents the most challenging dimension in the health equity promotion. Several studies evidenced that most common indicators of health outcomes, namely mortality and morbidity, worsen as social position declines

(Bambra et al., 2010; Oliveira et al., 2019; Phelan, Link, & Tehranifar, 2010; Vidal, Pontes, Barreira, Oliveira, & Maia, 2018). Worldwide, specifically in regions without adequate health services and healthcare infrastructures, which experience more vulnerability to poverty and deprivation, health inequalities are aggravated (Arruda, Maia, & Alves, 2018; Barreto, 2017; Doetsch, Pilot, Santana, & Krafft, 2017; Oliveira et al., 2019; Vidal, Pontes, et al., 2018), translated in the raise of amenable deaths (expressed by the percentage of deaths that could be avoided if timely and effective health care were provided (Charlton, Hartley, Silver, & Holland, 1983; Davis, 2014; Gianino, Lenzi, Fantini, Ricciardi, & Damiani, 2017; Rutstein et al., 1976).

Individual lifestyles, i.e. diet, physical activity, alcohol, cigarette, other drug use, hand washing, are health determinants that are developed and constructed by individuals living environment, thus have a strong influence on a person's global wellbeing (Graham & White, 2016; Hillger, 2008). Unhealthy lifestyles are attributed to the increase of several diseases like obesity (Cha et al., 2015), cardiovascular diseases (Gaziano, 2017), and even some cancer types (Katzke, Kaaks, & Kühn, 2015). Being aware that lifestyles are far from being an individual choice implies having in mind that the opportunities to fully achieve proper life conditions and good health are not equal to everyone. In deprived communities, children may become more vulnerable, with lower chances to be well nourished, which may lead to the development of non-communicable diseases (Allen et al., 2017; World Health Organization, 2011). There are multiple harmful lifestyles/behaviors, being the most commonly indicated poor nutrition, low physical activity levels, and substance abuse. When combined with economic, social and cultural environment, lifestyles affect the individuals' health status in a determinant way (Foster et al., 2018), since they are more difficult to modify (the harmful ones). Thus, social and fiscal policies that reduce poverty are needed alongside with public health and individual-level interventions to address a wider range of lifestyle factors in poor regions.

Communicable diseases, triggered by poverty, poor sanitation, and hygiene, were the major problem concerning mortality. However, non-communicable diseases, more evident since industrialization and increased urbanization, are another major burden of disease caused by changes in lifestyles, through an accelerated, urban and modern life conditions (Oliveira, Vidal, & Ferraz, 2020). According to the World Health Organization (World Health Organization, 2011), nearly 30 % of deaths related to non-communicable diseases happen in low-income countries, whereas in high-income countries the proportion is only 13 %.

2.1 Nature in urban spaces – an (in)equal issue

The presence of green spaces in urban settings have a positive influence in many dimensions, namely in the urban environment and in people health status, in its physical, social, and mental domains (Ridgley et al., 2020; Veras et al., 2020; Vidal, Barros, & Maia, 2020). The Millennium Ecosystem Assessment publication (Millennium Ecosystem Assessment, 2005), has assessed the consequences of ecosystem change for human wellbeing and the Common International Classification of Ecosystem Services (CICES) (Haines-Young & Potschin, 2018) have identified the final services that link to the goods and benefits that are valuable to people. These services are divided into three types: (i) provisioning services, covering nutritional and non-nutritional material and energetic outputs, such as food, raw materials, freshwater and medical resources; (ii) regulation and maintenance services, concerning mediation and moderation of the environment that affects human health, safety and comfort; and (iii) cultural services that relate to all the characteristics of elements of nature that provide opportunities for people to derive cultural goods or benefits, such as recreation and mental/physical health, tourism, aesthetic appreciation and inspiration for culture, art, and design and spiritual experience and sense of place.

The health benefits provided by green spaces are well documented in several reports (University of Leeds, 2015; World Health Organization, 2017a, 2017b). These benefits are far from being fully understood. The current state of the art refers to its positive influence in the improvement of birth outcomes (Dadvand et al., 2012; Frumkin et al., 2017), in the prevention of cancer, cardiovascular (Gidlow et al., 2016; Song et al., 2013, 2014; Song, Ikei, Igarashi, Takagaki, & Miyazaki, 2015) and respiratory diseases (Cavaleiro Rufo, Paciência, et al., 2020; Cavaleiro Rufo, Ribeiro, Paciência, Delgado, & Moreira, 2020; Squillacioti, Bellisario, Levra, Piccioni, & Bono, 2019), in the improvement of mental health (Gubbels et al., 2016; Mayer, Frantz, Bruehlman-Senecal, & Dolliver, 2008; Tyrväinen et al., 2014) and, consequently, globally decreasing mortality (Crouse et al., 2017; Takano, Nakamura, & Watanabe, 2002; Villeneuve et al., 2012).

The recognition of green spaces importance in urban settlements is expressed in the 2030 Agenda for Sustainable Development (United Nations, 2015). In SDG 11 – Make cities and human settlements inclusive, safe, resilient, and sustainable – the presence of nature in the urban space is visible in the target to provide universal access to safe, inclusive and accessible green and public spaces. Several studies (Hoffimann, Barros, & Ribeiro, 2017; Mears & Brindley, 2019; Vidal, Fernandes, Viterbo, Barros, & Maia, 2020, 2021; World Health Organization, 2012; Wüstemann et al., 2017) have documented that green spaces distribution is not fair, thus aggravating environmental injustice. Green spaces distribution is associated with socioeconomic, ethnic, and cultural background of its potential users (Dai, 2011; Ferguson, Roberts, McEachan, & Dallimer, 2018; Mears, Brindley, Maheswaran, & Jorgensen, 2019). This implies that green spaces provision needs to be based on the principle of equity, ensuring public access for all regardless of an individual's residential location, socioeconomic background, or ethnicity/race (Hoffimann et al., 2017; Schlosberg, 2007; Vidal, Maia, Vilaça, Barros, & Oliveira, 2018).

As Wood & DeClerck (2015) stated, economically deprived groups are those that most directly depend on access to green spaces ecosystems and their services. In fact, the most recent worldwide reports on this topic confirms an association between human development index (HDI) (United Nations Development Programme, 2019) and the share of green spaces availability (World Cities Culture Forum, 2018). Table 6.1 presents, at the top, the three cities with the highest share of public green spaces available and, at the bottom, the three cities with the lowest share of public green spaces available.

This information should be analyzed taking into account the city population density and HDI which highlights the inequality dimension of this issue. At the top of the list are high-income cities, such as Oslo, Singapore, and Sydney. Oslo is the city with the lowest population density and is also which have the biggest share of public green space (68.0%) and is at the top HDI ranking. On the other hand, at the bottom of the list are cities with low HDI, where disparities between rich and poor are demarked (Focus Taiwan News Channel, 2017) and present the lowest values of share of public green space. Through this data, it is observed a possible association between the city HDI and the availability of urban green spaces.

World Health Organization (World Health Organization, 2017a) suggests that socio-ecological benefits of green spaces are more evident and expressive in urban spaces than in others – environmental inequalities –, namely because of its quality and ecosystem services provision. Everyone can benefit from urban green spaces interventions, but they can be of particular relevance for socially disadvantaged or underserved community groups, which often have the least access to high-quality green spaces. Alongside, some recent studies

TABLE 6.1 Comparison between the human development index and the share of public green space available.

City	Population density (no. Inhabitants/km^2)	HDI[a]	Share of public green space[b] (% of city area)
Oslo	1645	0.954	68.0
Singapore	8358	0.935	47.0
Sydney	2037	0.938	46.0
Bogotá	4310	0.761	4.9
Taipei	9918	0.758	3.4
Istanbul	2523	0.806	2.2

[a] *HDI varies between 0 and 1, and the closer to one, the higher the human development level of the country. Source: Human Development Reports website http://hdr.undp.org/en/content/human-development-index-hdi.*
[b] *Source: World Cities Forum website http://www.worldcitiesculturefoum.com/data/of-public-green-space-parks-and-gardens.*

have revealed that provision of green spaces in low-income communities can act as a protective factor in mothers with low education levels and low income (Cusack, Larkin, Carozza, & Hystad, 2017) and in deprived communities residential duration (Łaszkiewicz, Kronenberg, & Marcińczak, 2018). These health benefits derived from the ecosystem services provided by urban green spaces in four main mechanisms (Hartig, Mitchell, de Vries, & Frumkin, 2014): contributing to better air and environmental quality, by mitigating extreme temperatures, reducing noise and depletion of air pollutants; enhancing physical activity; improving social connections between its users and residents on the surrounding area; contributing to restoration by reducing stress symptoms, contributing to the wellbeing.

A clear example of how health can be addressed by changing environmental variables is visible in the study developed by Song et al. (2013). This study gives an important contribution to a physiological understanding of human contact with the natural environment. The study comprised seventeen participants that walked in an urban park and in a non-natural environment, such as walking in a busy street. After walking in the urban park a higher parasympathetic nervous activity and lower sympathetic nervous activity was identified in comparison with the non-natural environment. The heart rate was significantly lower when walking in the urban park. Negative feelings such as "tension-anxiety" and "fatigue" were significantly lower. The anxiety dimension score was also significantly lower after walking in the urban park when compared to a non-natural context. As can be seen, environmental variables have a greater influence on people health, suggesting that the contact with nature in urban spaces have positive impacts on health outcomes. If nature in urban spaces, such as green spaces, is not equally distributed means that some groups are not taking advantage of the health benefits provided by them resulting in health and environmental inequalities.

2.2 The social-economic and environmental contexts influence in the process of disease transmission

2.2.1 Humans–animal interaction

Humans-animal interaction has positive impacts on individual quality of life, namely acting as a positive social determinant of health (Mueller, Gee, & Bures, 2018), however, this interaction can lead to the development of serious public health concerns, namely being the trigger of pandemics. Contemporary societies have brought the most diverse animals (even wild and threatened species) to humans habitats, coexisting in the same house, independently of the housing conditions and provision of a healthy environment to those animals. Currently, humans cohabit not only with companion animals, but also with wild ones in a complex and interdependent relationship (World Health Organization, 2020) that could be the source of diseases with unpredictable possible impacts in public health.

Zoonoses (diseases that are transmitted directly or indirectly between animals and humans), as well as antimicrobial resistance pose major risks for public health having significant social and economic impact, especially when transmitted by animals products (such as meat, dairy or eggs that are important food sources or by exotic food items), i.e., via food chain (Alegbeleye, Singleton, & Sant'Ana, 2018; Bintsis, 2018; Koopmans & Duizer, 2004). One Health concept has become more important in recent years because many anthropogenic and environmental driven factors have changed the already known interactions of the triangle humans—animals - environment and, further, uncover new forms of environmental stress caused by extreme climate events which impact on human and ecosystems health. These changes have effects on the human-animals bidirectional transmission of disease, the dissemination of diseases, and have led to the emergence and re-emergence of many diseases (Watts et al., 2017).

The nature of the factors that determine zoonoses transmission is complex and difficult to fully mention. Human population growth combined with economic development has resulted in increased demand for livestock-derived food products, which has led to larger livestock populations, increased production intensity and changes in trade volumes and patterns. This facilitates the evolution and spread of infectious zoonotic pathogens, including those with antibiotic resistance genes (Daszak, Cunningham, & Hyatt, 2000; McMichael, 2004; Palumbi, 2001; Pearce-Duvet, 2006). The environmental impact of the management of human effluent, waste from livestock and aquaculture facilities, and from manufacturers of pharmaceuticals is recognized as a potential hotspot for antimicrobial resistance (Berendonk et al., 2015).

Pigs and poultry have a key role as potential source of new zoonotic diseases among domestic animals due to their large populations which are often kept in high densities with high turnover rates. In the case of influenza virus infections, the similarity in respiratory epithelium receptors between humans and pigs increases the likelihood of cross-infectivity (Greger, 2007). A recent literature review (Buzanovsky, Sanchez-Vazquez, Maia-Elkhoury, & Werneck, 2020) states that zoonoses development can be potentiated by several environmental and socioeconomic factors, which create barriers for the presence of vectors, reservoirs, and parasites. These factors relate to migration, urbanization, loss of biodiversity, deforestation process, sanitation, and income and education level. It has been found that these factors could act as protective factors when improved, such as good sanitation facilities, high income, and high education level (Buzanovsky et al., 2020).

WHO (2006) estimates that 60 % of human infectious diseases is caused by zoonotic pathogens. In several rural areas of the world, poor families directly depend on livestock production and agriculture and, in many cases, these activities are led by women and children, which results in a higher risk of exposure to humans-animal interaction and, consequently, increase in health risks. In the low income regions of the globe, these vulnerable groups represent up to 70 % of the population (Food and Agriculture Organization, 2004). Thus, populations with this type of living conditions not only have a higher risk of contracting disease through humans - animal interaction but are also more vulnerable to suffer from food deprivation or poor nutrition, putting in evidence the cumulative multi-levels of exposure (Committee on Assessing Interactions Among Social Behavioral et al., 2006) and the poverty cycle - set of factors or events by which poverty, once started, is likely to continue unless there is outside intervention. Effective surveillance is needed to control zoonotic diseases and requires a multisectoral collaboration involving human health, veterinary, agricultural, educational, wildlife, environment, and sanitation sectors (Molyneux et al., 2011). European Union surveillance strategy (One Health European Joint Programme, 2019) is focused on selected high-risk areas; however, globalization poses new challenges and broader and flexible actions to detect hazards, reservoirs, vectors, trends, and transmission routes, as well as common approaches, timely data analysis, are necessary.

2.2.2 Antimicrobial resistance and socioeconomic factors

Antimicrobial resistance is a very complex issue that should be integrated into a larger perspective than the biomedical one. Infectious diseases are transversal to all society but affect unevenly different population groups, like poor and marginalized populations which are already more vulnerable. Poverty is a risk factor to contract infectious diseases, due to living conditions such as poor sanitation, low quality of water for human consumption, building conditions, housing overcrowding, and inadequate nutrition, among other factors (Alividza et al., 2018; Molyneux et al., 2011). Socioeconomic factors, such as income and education level and housing conditions, have been directly linked to antimicrobial misuse, which leads to the increase of antimicrobial resistance (Miller-Petrie & Gelband, 2017), contributing to a positive continuous feedback cycle of poverty and the tendency to illness. A global study (Collignon, Beggs, Walsh, Gandra, & Laxminarayan, 2018) demonstrated that GDP per capita, education, infrastructure, public health-care spending, and antibiotic consumption were all inversely correlated with antimicrobial resistance. On the other hand, the same study relates the lack of services infrastructures (like potable water, sanitation, energy), poor governance, and low expenditure in healthcare systems and services, with higher levels of antimicrobial resistance. Antimicrobial resistance is more severe in lower- and upper-middle-income countries (Alvarez-Uria, Gandra, & Laxminarayan, 2016; Klein et al., 2018; Malik & Bhattacharyya, 2019) and another factor contributing to antimicrobial resistance worldwide disparities is the limited capacity for microbiology testing in low income countries (Aiken, Karuri, Wanyoro, & Macleod, 2012). Resources to perform proper microbiological analysis require both trained personnel and equipment to maintain cell cultures in refrigerated conditions.

This is a key aspect to tackle antimicrobial resistance, because in many poor countries electricity continues to be a non-guaranteed service. Around one billion people still do not have access to electricity and about 80 % of this deprived population is concentrated in the Sub-Saharan region of Africa (SDG 7 Technical Advisory Group, 2018).

2.2.3 Energy poverty and air quality: a contemporary social inequality

Alleviation of world energy poverty would result in several benefits for human health by (i) ensuring the provision of electricity to support healthcare systems and services, as in the referred example; (ii) improving thermal comfort through heat, ventilation and air conditioning means; and also by (iii) reducing the use of biomass for cooking which is one of the main causes for bad indoor air quality. Extreme weather events caused by climate change, especially extreme heat and cold waves, are expected to increase in the next years (Johnson, Xie, Kosaka, & Xichen, 2018) inducing a growing need to ensure thermal comfort in housing and services buildings to prevent amenable deaths mainly in vulnerable population groups: those with respiratory or circulatory health conditions, children and the elderly (US Global Change Research Program (USGCRP), 2018). Energy poverty is also a critical issue in human health concerning indoor air quality as almost three billion people do not have access to clean-cooking solutions resorting to biomass and to kerosene. These forms of basic fuels emit particulate matter, volatile organic compounds and black carbon, among other harmful substances, that can be accumulated in indoor households, especially in rustic and poor constructed dwellings (WHO Regional Office for Europe & WHO European Center for Environment and Health, 2010). Women, children, and the elderly are the groups mostly exposed to dangerous levels of indoor air pollution, which results in millions of avoidable deaths each year (World Health Organization, 2018). Besides chemical substances, poor indoor air quality may also be the vector for airborne diseases, transmitted mostly by bioaerosols inhalation (Facciponte et al., 2018). Bioaerosols may transport bacteria, viruses, microalgae, and mold, aggravating already existing respiratory disease or even being the cause for public health outbreaks such as the case of *Legionella* induced pneumonia due to the contamination of refrigerating water and filters of air conditioning systems (Prussin, Schwake, & Marr, 2017).

The potential for fungal diseases to become an emerging threat has been underestimated (Forouzanfar et al., 2016) taking into consideration that the dampness and the consequent presence of mold in indoor households air is a very common problem, especially in poorly ventilated, bad preserved dwellings and basements. As with many other pollutants present in indoor air, mold inhalation aggravates health conditions of vulnerable population groups like children and the elderly, and those with respiratory diseases and allergies

(Cox-Ganser, 2015). However, globally, atmospheric pollution is the most important environmental risk factor to human health, leading to an increased mortality and morbidity, because air pollutants exposure can affect human health in several ways (Landrigan et al., 2018). Many studies have linked the exposure to atmospheric pollution with the aggravation of respiratory, heart circulatory and allergic diseases; and other works found associations air pollution and the intensification of some types of cancer, neurological disorders, and metabolic conditions (Feigin et al., 2016; Gakidou et al., 2017; Manisalidis, Stavropoulou, Stavropoulos, & Bezirtzoglou, 2020; Nolte et al., 2018; WHO Department of Public Health Environmental and Social Determinants of Health (PHE), 2016). In 2016, WHO estimated that air pollution was the cause of more than 3 million deaths every year, with almost 91 % of the world population breathing air with poor quality, which means that pollutants concentration in the air are above the recommended levels defined by WHO Guidelines (World Health Organization, 2016).

2.2.4 Linking food safety with socioeconomic determinants

Foodborne illnesses, with wide regional variations, affect particularly children under the age of five, people living in disadvantaged conditions and immunocompromized people: the elderly, pregnant women, people infected with the human immunodeficiency virus (HIV), transplant recipients, cancer patients, and drug addicts (Murray, Rosenthal, & Pfaller, 2005). Foodborne diseases caused by bacteria (e.g. *Salmonella*, *Listeria monocytogenes*, *Escherichia coli* and *Staphylococcus aureus*), viruses (rotavirus and norovirus) and, to a lesser extent, chemicals, affect one in ten people annually and cause 420,000 deaths, representing about 33 million years of life lost (World Health Organization, 2015).

Morbidity and mortality due to the consumption of risky foods responsible for various diseases is a concern for health authorities and has a significant impact on countries' socioeconomic development. A study aiming to evaluate, in several European countries, microbial survival after cooking in foods such as poultry meat, demonstrated that there is a high risk of contamination with *Salmonella* or *Campylobacter*. The understanding of good practices is more related to the cultural aspects, than to the fulfilment of hygiene and food safety rules. As an example, some consumers use the internal color or texture of the meat to decide on the cooking time, although these approaches do not guarantee the elimination of pathogens (Langsrud et al., 2020). The case studies explain the influence that sociodemographic profile of participants have on production, manipulation, and consumption, with an impact, in general, on what is done and on the understanding of how it is done, functioning as a kind of behavior duplicating devices, which help to understand logics of action, influencing public health (Bourdieu, 1983, 1997). Individuals move in their "social spaces" and act according to standards that govern social relations or, what also happens, integrate, with or without lasting, into "social spaces" that condition forms of action. Although they know that eating certain foods entails risks, they act guided by social standards that derive from the contexts in which they find themselves. The situations of social pressure justify behaviors: someone who is offered a specific food prepared, with enthusiasm and affection, hardly will not taste it under penalty of generating, in any case, feelings of adversity (Veflen, Scholderer, & Langsrud, 2020).

The generational effect is also felt with particular relevance in more developed countries, where the prevalence of unhealthy diets among young people has risen constantly, with an impact on public health: unhealthy eating habits have become established over the past 20 years, consumption of ready to eat foods and fast-food, which translate "hectic" lifestyles and low awareness of health threats (Chin & Mansori, 2019). The same is true for forms of production or food preparation. In an investigation on the occurrence of *Salmonella* spp. in eggs from chickens reared in domestic environments in Portugal and Romania, it has been demonstrated that there is low compliance with safety practices (Ferreira et al., 2020). In a study on hygiene practices before, during, and after food preparation in domestic kitchens, including the analysis of responses to a questionnaire, administered in 10 European countries, to a total of 9966 respondents, and results of microbiological tests on 30 kitchen surfaces, it can be seen that the behaviors are motivated by routines and that these are dependent on factors such as age, the nature of the surfaces used, which make hygiene conditions, characteristics of the environment or the training of stakeholders (Møretrø et al., 2020).

Those individuals most vulnerable to foodborne infections are generally less aware of the risk. In a study aimed at evaluating the knowledge and practices of food safety in pregnant women in Portugal, with a particular focus on listeriosis, it was found that only 12.2 % of the 956 respondents had heard about the infection and that 32.3 % did not change their domestic food preparation and cooking habits during pregnancy (Mateus, Silva, Maia, & Teixeira, 2013). The number of information received influenced the foods avoided outside the home. The greater number of pregnancies corresponded to lower economic resources and less education for these women. A study of family and community health carried out among 494 middle-aged African-Americans shows that, compared to the rest of the population, the low levels of education and income, the disadvantage of the neighborhood in which they live and the social discrimination to which they are exposed are factors which explain their higher risk behaviors concerning diet, exercise, and alcohol consumption (Simons et al., 2020).

Failure to comply with official recommendations, practices associated with cultural references or consumption motivations, which go beyond concerns about hygiene and safety issues, sometimes completely absent, justify that health authorities study and define models and practices to adopt to minimize the negative effects of foodborne illnesses. The "2030 Agenda for Sustainable

Development" (United Nations, 2015), particularly for goals 2, 3, 4 and 12, appears with an extensive set of goals whose achievement, in part, will depend on the transformation of mentalities with effects on practices, between production and consumption, which involve food. One of the challenges facing those responsible for implementing food safety models and practices is to develop interventions that change the power of the rules and that allow people to say no to something they prefer not to eat to reduce the risks of contamination.

Lifestyles dictate patterns of food consumption, which, perhaps, can be changed according to new realities. The results of a retrospective study on the assessment of eating habits with fifty-eight individuals diagnosed with gastric or colorectal cancer show that, when comparing before and after disease, patients tend to adopt, with recommendations from health professionals, healthier habits (Silva, Maia, Teixeira, Santos, & Goncalves, 2009, pp. 362–363). In responding to the phenomenon of consumption of ready to eat foods and fast food, social marketing is an approach that plays a critical role because it can influence individuals to engage in healthier eating behaviors and to promote social change (Chin & Mansori, 2019). The selection of food-contact materials can motivate users to clean and reduce risks, while campaigns should be implemented to lead to a greater adoption of the habit of cleaning surfaces during the entire process of use (Møretrø et al., 2020). An investigation into knowledge and attitudes of food safety carried out with 990 consumers residing in Istanbul, Turkey, demonstrates the need to obtain information, and can take different forms guided by the concern to achieve behavioral changes that reduce risks of diseases (Bolek, 2020).

At a global level the production of food is one of the largest users of water, accounting for more than 80 % of blue water (the set of renewable freshwater resources such as springs, aquifers, rivers, and lakes which are fed by rainwater) (Hoff, 2011). Alongside with food security, consumption of water with poor quality has been a paramount cause for several diseases acting as a transmission vector of microorganisms and of chemical contaminants (Geissen et al., 2015). In addition to water, wastewaters are dissemination means for antibiotics, hormones, and many other pharmaceuticals that became environmentally widespread affecting food chain in both marine and terrestrial habitats (Cheng et al., 2020; Craddock et al., 2020).

The scarcity of water with quality for human consumption is a very important issue in many low income countries, specially where slum dwellers are a common and frequent reality in very populated cities (National Academies of Sciences Engineering Medicine, 2018). Lack of hygiene and proper sanitation infrastructures and services, alongside with the use of unsafe water, altogether account for 3.5 million deaths worldwide, representing 25 % of the premature deaths of children younger than 14 (UNEP & WHO, 2016). Scarcity of freshwater resources may be aggravated as a consequence of climate change effects that include the raise of ocean levels and atmosphere temperatures, which contribute to the occurrence of extreme weather events, the increase of desert regions caused by draughts, floods and wildfires, and consequently the reduction of available arable land (Cramer et al., 2014). All these climate change driven factors add extra pressure on an already critical issue, because the supply of water with quality for human consumption also has a direct impact in food production systems and in food security. In addition, the pressure on regional water resources is increasing from the competition of its multiple uses (ecosystems, agriculture, urban settlements, industry, energy production), as well as population growth and the increase of individual water consumption.

3. Final remarks

Modern medicine tends to emphasize lifestyles as the main explanatory factor for health disparities, explaining, for example, how issues of ethnic adversity, structurally rooted, of a cultural nature, are responsible for the precocious biological decline of minorities with similar effects on disease onset and mortality. These differences can only be addressed by the implementation and development of social policies and programs that promote economic and social equity (Simons et al., 2020).

Research efforts are needed, in particular by carrying out experimental studies, which can identify which communication factors may have a causal effect on the modification of people's risk behaviors and how these factors can influence the processing of information by consumers (Maia, Teixeira, & Mateus, 2019; Mateus et al., 2013). Interested entities, public and private, health or other areas of intervention, can contribute to improvements in food hygiene and safety throughout the entire chain, incorporating them in the development of local, national, regional, and international policies. Improving hygiene and food safety practices is critical to obtaining gains in public health. The adoption of a set of measures, in a holistic dimension, must be considered so that the gradual change of behaviors translates into the reduction of economic and social burdens on health.

The elimination of risks of transmission of foodborne diseases must take place through the implementation of preventive measures, with emphasis on its transversal nature, the role to be played by education in issues of hygiene and food safety. While reducing the incidence of foodborne illnesses, increasing the skills of consumers in handling and consuming safe food, providing knowledge and solutions for best practices for preserving the shelf life of food, especially in consumerist countries, the challenge will be to promote practices that lead to waste reduction (De Laurentiis, Caldeira, & Sala, 2020). Waste is also produced due to consumers' lack of confidence, which is a lack of knowledge about the time and conditions for food preservation.

On the other hand, it is necessary to adopt education plans in hygiene and food safety that result from the continuous articulation between health authorities, official entities and other dynamic agents, such as NGOs, capable of contributing to the dissemination information and good practices that translate into predictive actions to be adopted, simultaneously, by producers and consumers.

There is a growing need to improve human and animal health services to protect global health and food security, therefore there are increasing demands on disease surveillance, emergency response and disease control due to an increase in the emergence of new diseases or the re-emergence of existing diseases, many of which are zoonotic (Harper & Armelagos, 2010; Jones et al., 2008). Factors influencing emergence include human population and behavior changes, increasing livestock production, intensification of production, trade, habitat change, loss of biodiversity, and globalisation.

The United States National Action Plan (The White House, 2015) proposed the strengthening of a "One Health" national surveillance system (for humans, animals and environment) with improved international collaboration and capacity and abroad, systems-based approach to complex problems (Zinsstag, Schelling, Waltner-Toews, & Tanner, 2011). Therefore, this action is suitable for antimicrobial resistance surveillance once it considers some of the main structural factors that influence antimicrobial resistance, namely social, political, material, biological, and economic ones (Kock, 2015). Antimicrobial resistance has been focused only in human health outcomes, neglecting that this is also an ecological problem that results from the interface of human, animal and planetary health, and also with food hygiene and environmental science.

Taking different living conditions into account it becomes clear that promoting the public's health is a challenging mission for the individual itself but first of all for policies, institutions, the global market, and the community. For this reason campaigns and interventions that address health promotion should also focus on cultural and socioeconomic factors. Therefore, actions and interventions in various settings, such as kindergartens, schools, workplaces, or local communities could be realized efficiently by practicing healthy behaviors and using preventive healthcare services.

Intervention strategies are primarily focused within either the food, animal, or medical domains. However, the growing concern about increased antimicrobial resistance requires a stronger stewardship about antimicrobial usage in both humans and animals. The development of robust, evidence-based interventions and guidelines in animals is however not as advanced as in humans.

As the increase in antimicrobial resistance continues and fewer new drugs are being developed, calls for action to avert the impending crisis of a "post antibiotic era" are being heard. To promote effectiveness and economic efficiency, interventions need to be designed from sound evidence gained from surveillance. A broad approach to the evidence gathering surveillance, data analysis, intervention design and evaluation is needed. Given the scale of the problem and the expected socio-economic costs, the additional monetary, social and time investments are likely to be recovered by the resulting benefits, which include quantifiable financial efficiencies and improved human and animal health outcomes.

Acknowledgments

Diogo Guedes Vidal was funded by the Fundação para a Ciência e a Tecnologia, I.P., through the Doctoral Grant SFRH/BD/143238/2019.

References

Aiken, A. M., Karuri, D. M., Wanyoro, A. K., & Macleod, J. (2012). Interventional studies for preventing surgical site infections in sub-Saharan Africa – a systematic review. *International Journal of Surgery, 10*(5), 242–249. https://doi.org/10.1016/j.ijsu.2012.04.004

Alegbeleye, O. O., Singleton, I., & Sant'Ana, A. S. (2018). Sources and contamination routes of microbial pathogens to fresh produce during field cultivation: A review. *Food Microbiology, 73*, 177–208. https://doi.org/10.1016/j.fm.2018.01.003

Alividza, V., Mariano, V., Ahmad, R., Charani, E., Rawson, T. M., Holmes, A. H., & Castro-Sánchez, E. (2018). Investigating the impact of poverty on colonization and infection with drug-resistant organisms in humans: A systematic review. *Infectious Diseases of Poverty, 7(1)*(76). https://doi.org/10.1186/s40249-018-0459-7

Allen, L., Williams, J., Townsend, N., Mikkelsen, B., Roberts, N., Foster, C., & Wickramasinghe, K. (2017). Socioeconomic status and non-communicable disease behavioural risk factors in low-income and lower-middle-income countries: A systematic review. *The Lancet Global Health, 5*(3), e277–e289. https://doi.org/10.1016/S2214-109X(17)30058-X

Alonso Aguirre, A., Basu, N., Kahn, L. H., Morin, X. K., Echaubard, P., Wilcox, B. A., & Beasley, V. R. (2019). Transdisciplinary and social-ecological health frameworks—novel approaches to emerging parasitic and vector-borne diseases. *Parasite Epidemiology and Control, 4*, e00084. https://doi.org/10.1016/j.parepi.2019.e00084

Alvarez-Uria, G., Gandra, S., & Laxminarayan, R. (2016). Poverty and prevalence of antimicrobial resistance in invasive isolates. *International Journal of Infectious Diseases: IJID: Official Publication of the International Society for Infectious Diseases, 52*, 59–61. https://doi.org/10.1016/j.ijid.2016.09.026

Arruda, N. M., Maia, A. G., & Alves, L. C. (2018). Desigualdade no acesso à saúde entre as áreas urbanas e rurais do Brasil: uma decomposição de fatores entre 1998 a 2008. *Cadernos de Saúde Pública, 34*(6), 1–14. https://doi.org/10.1590/0102-311x00213816

Bambra, C., Gibson, M., Sowden, A., Wright, K., Whitehead, M., & Petticrew, M. (2010). Tackling the wider social determinants of health and health inequalities: Evidence from systematic reviews. *Journal of Epidemiology & Community Health*, 284–291. https://doi.org/10.1136/jech.2008.082743

Barreto, M. L. (2017). Desigualdades em saúde: uma perspectiva global. *Ciência & Saúde Coletiva, 22*(7), 2097–2108. https://doi.org/10.1590/1413-81232017227.02742017

Berendonk, T. U., Manaia, C. M., Merlin, C., Fatta-Kassinos, D., Cytryn, E., Walsh, F., … Martinez, J. L. (May 2015). Tackling antibiotic resistance: The environmental framework. Nature reviews. *Microbiology, 13*, 310–317. https://doi.org/10.1038/nrmicro3439

Bintsis, T. (2018). Microbial pollution and food safety. *AIMS Microbiology, 4*(3), 377–396. https://doi.org/10.3934/microbiol.2018.3.377

Bolek, S. (2020). Consumer knowledge, attitudes, and judgments about food safety: A consumer analysis. *Trends in Food Science & Technology, 102*, 242–248. https://doi.org/10.1016/j.tifs.2020.03.009

Bourdieu, P. (1983). *Sociologia. São Paulo: Editora África.*

Bourdieu, P. (1997). *Razões práticas. Sobre a teoria da acção. (Oeiras: Celta Editora).*

Buzanovsky, L. P., Sanchez-Vazquez, M. J., Maia-Elkhoury, A. N. S., & Werneck, G. L. (2020). Major environmental and socioeconomic determinants of cutaneous leishmaniasis in Brazil - a systematic literature review. *Revista Da Sociedade Brasileira de Medicina Tropical, 53.* Retrieved from http://www.scielo.br/scielo.php?script=sci_arttext&pid=S0037-86822020000100202&nrm=iso.

Cavaleiro Rufo, J., Paciência, I. R., Hoffimann, E., Moreira, A. M. A., Barros, H., & Ribeiro, A. I. (2020). The neighbourhood natural environment is associated with asthma in children: A birth cohort study. *Allergy.* https://doi.org/10.1111/all.14493

Cavaleiro Rufo, J., Ribeiro, A. I., Paciência, I., Delgado, L., & Moreira, A. (2020). The influence of species richness in primary school surroundings on children lung function and allergic disease development. *Pediatric Allergy and Immunology, 31*(4), 358–363. https://doi.org/10.1111/pai.13213

Cha, E., Akazawa, M. K., Kim, K. H., Dawkins, C. R., Lerner, H. M., Umpierrez, G., & Dunbar, S. B. (2015). Lifestyle habits and obesity progression in overweight and obese American young adults: Lessons for promoting cardiometabolic health. *Nursing & Health Sciences, 17*(4), 467–475. https://doi.org/10.1111/nhs.12218

Charlton, J. R., Hartley, R. M., Silver, R., & Holland, W. W. (1983). Geographical variation in mortality from conditions amenable to medical intervention in England and Wales. *Lancet (London, England), 1*(8326 Pt 1), 691–696.

Cheng, D., Ngo, H. H., Guo, W., Chang, S. W., Nguyen, D. D., Liu, Y., ... Wei, D. (2020). A critical review on antibiotics and hormones in swine wastewater: Water pollution problems and control approaches. *Journal of Hazardous Materials, 387*(121682). https://doi.org/10.1016/j.jhazmat.2019.121682

Chin, J. H., & Mansori, S. B. (2019). *Social marketing in foods: A review of behavioural change models of healthy eating*. https://doi.org/10.1016/B978-0-08-100596-5.22654-1

Chisholm, B. (1949). Social medicine. *Scientific American, 180*, 11–15. https://doi.org/10.1038/scientificamerican0449-11

Collignon, P., Beggs, J. J., Walsh, T. R., Gandra, S., & Laxminarayan, R. (2018). Anthropological and socioeconomic factors contributing to global antimicrobial resistance: A univariate and multivariable analysis. *The Lancet Planetary Health, 2*(9), e398–e405. https://doi.org/10.1016/S2542-5196(18)30186-4

Commission on Social Determinants of Health. (2008). *Closing the gap in a generation: Health equity through action on the social determinants of health.* Retrieved fromhttps://apps.who.int/iris/bitstream/handle/10665/43943/9789241563703_eng.pdf;jsessionid=A41734C2C69796C98EA28D15E63331DF?sequence=1.

Committee on Assessing Interactions Among Social Behavioral, and Genetic Factors in Health. (2006). The impact of social and cultural environment on health. In L. M. Hernandez, & D. G. Blaze (Eds.), *Genes, behavior, and the social environment: Moving beyond the nature/nurture debate* (pp. 25–43). Retrieved from https://www.ncbi.nlm.nih.gov/books/NBK19929/pdf/Bookshelf_NBK19929.pdf.

Costa, C., Santana, P., Dimitroulopoulou, S., & Burstrom, B. (2019). Population health inequalities across and within European metropolitan areas through the lens of the EURO-HEALTHY population health index. *International Journal of Environmental Research and Public Health, 16*(836), 1–17. https://doi.org/10.3390/ijerph16050836

Cox-Ganser, J. M. (2015). Indoor dampness and mould health effects - ongoing questions on microbial exposures and allergic versus nonallergic mechanisms. *Clinical and Experimental Allergy: Journal of the British Society for Allergy and Clinical Immunology, 45*(10), 1478–1482. https://doi.org/10.1111/cea.12601

Craddock, H. A., Chattopadhyay, S., Rjoub, Y., Rosen, D., Greif, J., Lipchin, C., … Sapkota, A. R. (2020). Antibiotic-resistant *Escherichia coli* and *Klebsiella* spp. in greywater reuse systems and pond water used for agricultural irrigation in the West Bank, Palestinian Territories. *Environmental Research, 188*(109777). https://doi.org/10.1016/j.envres.2020.109777

Cramer, W., Yohe, G. W., Auffhammer, M., Huggel, C., Molau, U., da Silva Dias, M. A. F., … Tibig, L. (2014). Detection and attribution of observed impacts. In C. B. Field, V. R. Barros, D. J. Dokken, K. J. Mach, M. D. Mastrandrea, T. E. Bilir, … L. L. White (Eds.), *Climate change 2014: Impacts, adaptation, and vulnerability. Part A: Global and sectoral aspects. Contribution of working group II to the Fifth assessment report of the Intergovernmental Panel on climate change* (pp. 979–1037). United Kingdom and New York: Cambridge University Press.

Crouse, D. L., Pinault, L., Balram, A., Hystad, P., Peters, P. A., Chen, H., … Villeneuve, P. J. (2017). Urban greenness and mortality in Canada's largest cities: A national cohort study. *The Lancet Planetary Health, 1*(7), e289–e297. https://doi.org/10.1016/S2542-5196(17)30118-3

Cusack, L., Larkin, A., Carozza, S., & Hystad, P. (2017). Associations between residential greenness and birth outcomes across Texas. *Environmental Research, 152*, 88–95. https://doi.org/10.1016/j.envres.2016.10.003

Dadvand, P., Sunyer, J., Basagana, X., Ballester, F., Lertxundi, A., Fernandez-Somoano, A., … Nieuwenhuijsen, M. J. (2012). Surrounding greenness and pregnancy outcomes in four Spanish birth cohorts. *Environmental Health Perspectives, 120*(10), 1481–1487. https://doi.org/10.1289/ehp.1205244

Dai, D. (2011). Racial/ethnic and socioeconomic disparities in urban green space accessibility: Where to intervene? *Landscape and Urban Planning, 102*(4), 234–244. https://doi.org/10.1016/j.landurbplan.2011.05.002

Daszak, P., Cunningham, A. A., & Hyatt, A. D. (2000). Emerging infectious diseases of wildlife—threats to biodiversity and human health. *Science, 287*(5452), 443 LP–449. https://doi.org/10.1126/science.287.5452.443

Davis, D. (2014). *Some preliminary thoughts on inequality and urban space: Looking back, thinking comparatively, heading forward*. Retrieved May 10, 2019, from The Cities Papers: an essay collection from The Decent City initiative website http://citiespapers.ssrc.org/some-preliminary-thoughts-on-inequality-and-urban-space-looking-back-thinking-comparatively-heading-forward/.

De Laurentiis, V., Caldeira, C., & Sala, S. (2020). No time to waste: Assessing the performance of food waste prevention actions. Resources. *Conservation and Recycling, 161*(104946). https://doi.org/10.1016/j.resconrec.2020.104946

Doetsch, J., Pilot, E., Santana, P., & Krafft, T. (2017). Potential barriers in healthcare access of the elderly population influenced by the economic crisis and the troika agreement: A qualitative case study in Lisbon. *Portugal*, 1–17. https://doi.org/10.1186/s12939-017-0679-7

Facciponte, D. N., Bough, M. W., Seidler, D., Carroll, J. L., Ashare, A., Andrew, A. S., ... Stommel, E. W. (2018). Identifying aerosolized cyanobacteria in the human respiratory tract: A proposed mechanism for cyanotoxin-associated diseases. *Science of The Total Environment, 645*, 1003–1013. https://doi.org/10.1016/j.scitotenv.2018.07.226

Feigin, V. L., Roth, G. A., Naghavi, M., Parmar, P., Krishnamurthi, R., Chugh, S., ... Forouzanfar, M. H. (2016). Global burden of stroke and risk factors in 188 countries, during 1990–2013: A systematic analysis for the global burden of disease study 2013. *The Lancet Neurology, 15*(9), 913–924. https://doi.org/10.1016/S1474-4422(16)30073-4

Ferguson, M., Roberts, H. E., McEachan, R. R. C., & Dallimer, M. (2018). Contrasting distributions of urban green infrastructure across social and ethno-racial groups. *Landscape and Urban Planning, 175*, 136–148. https://doi.org/10.1016/j.landurbplan.2018.03.020

Ferreira, V., Cardoso, M. J., Magalhães, R., Maia, R., Neagu, C., Dumitraşcu, L., ... Teixeira, P. (2020). Occurrence of *Salmonella* spp. in eggs from backyard chicken flocks in Portugal and Romania - results of a preliminary study. *Food Control, 113*(107180). https://doi.org/10.1016/j.foodcont.2020.107180

Focus Taiwan News Channel. (2017). *Income gap between rich, poor households remains wide: MOF.* Retrieved June 17, 2019, from http://focustaiwan.tw/news/aeco/201707040022.aspx.

Food and Agriculture Organization. (2004). *The pro-poor livestock policy initiatives: A living from livestock.* Retrieved from http://www.fao.org/docs/up/easypol/572/pplpi-which_poor-to-target_195en.pdf.

Forouzanfar, M. H., Afshin, A., Alexander, L. T., Anderson, H. R., Bhutta, Z. A., Biryukov, S., ... Murray, C. J. L. (2016). Global, regional, and national comparative risk assessment of 79 behavioural, environmental and occupational, and metabolic risks or clusters of risks, 1990–2015: A systematic analysis for the global burden of disease study 2015. *The Lancet, 388*(10053), 1659–1724. https://doi.org/10.1016/S0140-6736(16)31679-8

Foster, H. M. E., Celis-Morales, C. A., Nicholl, B. I., Petermann-Rocha, F., Pell, J. P., Gill, J. M. R., ... Mair, F. S. (2018). The effect of socioeconomic deprivation on the association between an extended measurement of unhealthy lifestyle factors and health outcomes: A prospective analysis of the UK Biobank cohort. *The Lancet Public Health, 3*(12), e576–e585. https://doi.org/10.1016/S2468-2667(18)30200-7

Frumkin, H., Bratman, G. N., Breslow, S. J., Cochran, B., Kahn, P. H. J., Lawler, J. J., ... Wood, S. A. (2017). Nature contact and human health: A Research Agenda. *Environmental Health Perspectives, 125*(7), 75001. https://doi.org/10.1289/EHP1663

Gakidou, E., Afshin, A., Abajobir, A. A., Abate, K. H., Abbafati, C., Abbas, K. M., ... Murray, C. J. L. (2017). Global, regional, and national comparative risk assessment of 84 behavioural, environmental and occupational, and metabolic risks or clusters of risks, 1990–2016: A systematic analysis for the global burden of disease study 2016. *The Lancet, 390*(10100), 1345–1422. https://doi.org/10.1016/S0140-6736(17)32366-8

Gaziano, T. A. (2017). Lifestyle and cardiovascular disease. *Journal of the American College of Cardiology, 69*(9), 1126 LP–1128. https://doi.org/10.1016/j.jacc.2016.12.019

Geissen, V., Mol, H., Klumpp, E., Umlauf, G., Nadal, M., van der Ploeg, M., ... Ritsema, C. J. (2015). Emerging pollutants in the environment: A challenge for water resource management. *International Soil and Water Conservation Research, 3*(1), 57–65. https://doi.org/10.1016/j.iswcr.2015.03.002

Gianino, M. M., Lenzi, J., Fantini, M. P., Ricciardi, W., & Damiani, G. (2017). Declining amenable mortality: A reflection of health care systems? *BMC Health Services Research, 17*(1), 735. https://doi.org/10.1186/s12913-017-2708-z

Gidlow, C. J., Jones, M. V., Hurst, G., Masterson, D., Clark-Carter, D., Tarvainen, M. P., ... Nieuwenhuijsen, M. (2016). Where to put your best foot forward: Psycho-physiological responses to walking in natural and urban environments. *Journal of Environmental Psychology, 45*, 22–29. https://doi.org/10.1016/j.jenvp.2015.11.003

Graham, H., & White, P. C. L. (2016). Social determinants and lifestyles: Integrating environmental and public health perspectives. *Public Health, 141*, 270–278. https://doi.org/10.1016/j.puhe.2016.09.019

Greger, M. (2007). The human/animal interface: Emergence and resurgence of zoonotic infectious diseases. *Critical Reviews in Microbiology, 33*(4), 243–299. https://doi.org/10.1080/10408410701647594

Gubbels, J. S., Kremers, S. P. J., Droomers, M., Hoefnagels, C., Stronks, K., Hosman, C., & de Vries, S. (2016). The impact of greenery on physical activity and mental health of adolescent and adult residents of deprived neighborhoods: A longitudinal study. *Health & Place, 40*, 153–160. https://doi.org/10.1016/j.healthplace.2016.06.002

Haines-Young, R., & Potschin, M. B. (2018). *Common international classification of ecosystem services (CICES) V5.1 and guidance on the application of the revised structure*. In Fabis Consulting Ltd. Retrieved from www.cices.eu.

Harper, K., & Armelagos, G. (2010). The changing disease-scape in the third epidemiological transition. *International Journal of Environmental Research and Public Health, 7*(2), 675–697. https://doi.org/10.3390/ijerph7020675

Hartig, T., Mitchell, R., de Vries, S., & Frumkin, H. (2014). *Nature and Health. Annu. Rev. Public Health, 35*, 207–208. https://doi.org/10.1146/annurev-publhealth-032013-182443

Hillger, C. (2008). Lifestyle and health determinants. In W. Kirch (Ed.), *Encyclopedia of public health* (pp. 854–861). https://doi.org/10.1007/978-1-4020-5614-7_1982

Hoff, H. (2011). *Understanding the Nexus. Background paper for the Bonn 2011 conference: The water, energy and food security Nexus. The water, energy and food security Nexus: Solutions for the green economy.* Bonn.

Hoffimann, E., Barros, H., & Ribeiro, A. I. (2017). Socioeconomic inequalities in green space quality and Accessibility—evidence from a Southern European city. *International Journal of Environmental Research and Public Health*. https://doi.org/10.3390/ijerph14080916

Johnson, N. C., Xie, S., Kosaka, Y., & Xichen, L. (2018). Increasing occurrence of cold and warm extremes during the recent global warming slowdown. *Nature Communications, 9*(1724). https://doi.org/10.1038/s41467-018-04040-y

Jones, K. E., Patel, N. G., Levy, M. A., Storeygard, A., Balk, D., Gittleman, J. L., & Daszak, P. (2008). Global trends in emerging infectious diseases. *Nature, 451*(7181), 990–993. https://doi.org/10.1038/nature06536

Katzke, V. A., Kaaks, R., & Kühn, T. (2015). Lifestyle and cancer risk. *Cancer Journal (Sudbury, Mass.), 21*(2), 104–110. https://doi.org/10.1097/PPO.0000000000000101

Klein, E. Y., Van Boeckel, T. P., Martinez, E. M., Pant, S., Gandra, S., Levin, S. A., ... Laxminarayan, R. (2018). Global increase and geographic convergence in antibiotic consumption between 2000 and 2015. *Proceedings of the National Academy of Sciences, 115*(15), E3463–E3470. https://doi.org/10.1073/pnas.1717295115

Kock, R. (2015). Structural one health-are we there yet? *The Veterinary Record, 176*(6), 140–142. https://doi.org/10.1136/vr.h193

Koopmans, M., & Duizer, E. (2004). Foodborne viruses: An emerging problem. *International Journal of Food Microbiology, 90*(1), 23–41. https://doi.org/10.1016/s0168-1605(03)00169-7

Krech, R. (2011). Social determinants of health: Practical solutions to deal with a well-recognized issue. *Bulletin of the World Health Organization, 89*(10), 703. https://doi.org/10.2471/BLT.11.094870

Landrigan, P. J., Fuller, R., Acosta, N. J. R., Adeyi, O., Arnold, R., Basu, N., … Zhong, M. (2018). The *Lancet* Commission on pollution and health. *The Lancet, 391*(10119), 462–512. https://doi.org/10.1016/S0140-6736(17)32345-0

Langsrud, S., Sørheim, O., Skuland, S. E., Almli, V. L., Jensen, M. R., Grøvlen, M. S., … Møretrø, T. (2020). Cooking chicken at home: Common or recommended approaches to judge doneness may not assure sufficient inactivation of pathogens. *PLoS One, 15*(4), e0230928. https://doi.org/10.1371/journal.pone.0230928. Retrieved from.

Łaszkiewicz, E., Kronenberg, J., & Marcińczak, S. (2018). Attached to or bound to a place? The impact of green space availability on residential duration: The environmental justice perspective. *Ecosystem Services, 30*, 309–317. https://doi.org/10.1016/j.ecoser.2017.10.002

Machalaba, C. C., Salerno, R. H., Barton Behravesh, C., Benigno, S., Berthe, F. C. J., Chungong, S., … Wannous, C. (2018). Institutionalizing one health: From assessment to action. *Health Security, 16*(S1). https://doi.org/10.1089/hs.2018.0064. S37–S43.

Mackenbach, J. P., Hu, Y., Artnik, B., Bopp, M., Costa, G., Kalediene, R., … Nusselder, W. J. (2017). Trends in inequalities in mortality amenable to health care in 17 European countries. *Health Affairs, 36*(6). https://doi.org/10.1377/hlthaff.2016.1674

Maia, R. L., Teixeira, P., & Mateus, T. L. (2019). Risk communication strategies (on listeriosis) for high-risk groups. *Trends in Food Science & Technology, 84*, 68–70. https://doi.org/10.1016/j.tifs.2018.03.006

Malik, B., & Bhattacharyya, S. (2019). Antibiotic drug-resistance as a complex system driven by socio-economic growth and antibiotic misuse. *Scientific Reports, 9*(1), 9788. https://doi.org/10.1038/s41598-019-46078-y

Manisalidis, I., Stavropoulou, E., Stavropoulos, A., & Bezirtzoglou, E. (2020). Environmental and health impacts of air pollution: A review. *Frontiers in Public Health, 8*. https://doi.org/10.3389/fpubh.2020.00014, 14–14.

Mateus, T., Silva, J., Maia, R. L., & Teixeira, P. (2013). Listeriosis during pregnancy: A public health concern. *ISRN Obstetrics and Gynecology, 2013*, 851712. https://doi.org/10.1155/2013/851712

Mayer, F. S., Frantz, C. M., Bruehlman-Senecal, E., & Dolliver, K. (2008). Why is nature Beneficial?: The role of connectedness to nature. *Environment and Behavior, 41*(5), 607–643. https://doi.org/10.1177/0013916508319745

McMichael, A. J. (2004). Environmental and social influences on emerging infectious diseases: Past, present and future. *Philosophical Transactions of the Royal society of London. Series B, Biological Sciences, 359*(1447), 1049–1058. https://doi.org/10.1098/rstb.2004.1480

Mears, M., & Brindley, P. (2019). Measuring urban greenspace distribution equity: The importance of appropriate methodological approaches. *ISPRS International Journal of Geo-Information, 8*(6). https://doi.org/10.3390/ijgi8060286

Mears, M., Brindley, P., Maheswaran, R., & Jorgensen, A. (2019). Understanding the socioeconomic equity of publicly accessible greenspace distribution: The example of Sheffield, UK. *Geoforum, 103*(April), 126–137. https://doi.org/10.1016/j.geoforum.2019.04.016

Millennium Ecosystem Assessment. (2005). *Ecosystems and human well-being. Our human planet*. Washington, D.C.: Island Press.

Miller-Petrie, M., & Gelband, H. (2017). *Socioeconomics, antimicrobial use and antrimicrobial resistance*. Retrieved July 16, 2020, from AMR Control - Overcoming Global Antimicrobial Resistance website http://resistancecontrol.info/2017/socioeconomics-antimicrobial-use-and-antimicrobial-resistance/.

Mitsakou, C., Corman, D., Freitas, Â., Zengarini, N., Schweikart, J., Camprubí, L., ... Santana, P. (2018). Population health inequalities across metropolitan areas: Evidence from the EURO-HEALTHY project. *European Journal of Public Health, 28*(Suppl. 1_4).

Molyneux, D., Hallaj, Z., Keusch, G. T., McManus, D. P., Ngowi, H., Cleaveland, S., ... Kioy, D. (2011). Zoonoses and marginalised infectious diseases of poverty: Where do we stand? *Parasites & Vectors, 4*(106). https://doi.org/10.1186/1756-3305-4-106

Møretrø, T., Martens, L., Teixeira, P., Ferreira, V. B., Maia, R., Maugesten, T., & Langsrud, S. (2020). Is visual motivation for cleaning surfaces in the kitchen consistent with a hygienically clean environment? *Food Control, 111*(107077). https://doi.org/10.1016/j.foodcont.2019.107077

Mueller, M. K., Gee, N. R., & Bures, R. M. (2018). Human-animal interaction as a social determinant of health: Descriptive findings from the health and retirement study. *BMC Public Health, 18*(1), 1–7. https://doi.org/10.1186/s12889-018-5188-0

Murray, P., Rosenthal, K., & Pfaller, M. (2005). *Medical microbiology*. Maryland: Mosby.

National Academies of Sciences Engineering Medicine. (2018). Urbanization and slums: Infectious diseases in the Built environment. In *Proceedings of a Workshop*. Washington, DC: National Academies of Sciences Engineering Medicine.

Nolte, C. G., Dolwick, P. D., Fann, N., Horowitz, L. W., Naik, V., Pinder, R. W., ... Ziska, L. H. (2018). Air quality. In D. R. Reidmiller, C. W. Avery, D. R. Easterling, K. E. Kunkel, K. L. M. Lewis, T. K. Maycock, & B. C. Stewart (Eds.), *Impacts, risks, and adaptation in the United States: Fourth national climate assessment* (Vol. II, pp. 512–538). Washington, DC, USA: U.S. Global Change Research Program.

Oliveira, G. M., Vidal, D. G., & Ferraz, M. P. (2020). Urban lifestyles and consumption patterns. In W. L. Filho, A. M. Azul, L. Brandli, P. G. Özuyar, & T. Wall (Eds.), *Sustainable cities and communities. Encyclopedia of the UN sustainable development goals* (pp. 851–860). https://doi.org/10.1007/978-3-319-71061-7_54-1

Oliveira, G. M., Vidal, D. G., Ferraz, M. P., Cabeda, J. M., Pontes, M., Maia, R. L., ... Barreira, E. (2019). Measuring health vulnerability: An interdisciplinary indicator applied to Mainland Portugal. *International Journal of Environmental Research and Public Health, 16*(21), 1–18. https://doi.org/10.3390/ijerph16214121

One Health European Joint Programme. (2019). *The One Health European Joint Programme: Strategic Research Agenda*. Retrieved from https://www.rivm.nl/sites/default/files/2019-09/8106. RIVM Clickable PDF One Health EJP A4 TG.pdf.

Palumbi, S. R. (2001). Humans as the World's greatest evolutionary force. *Science, 293*(5536), 1786 LP–1790. https://doi.org/10.1126/science.293.5536.1786

Pearce-Duvet, J. M. C. (2006). The origin of human pathogens: Evaluating the role of agriculture and domestic animals in the evolution of human disease. *Biological Reviews of the Cambridge Philosophical Society, 81*(3), 369–382. https://doi.org/10.1017/S1464793106007020

Phelan, J. C., Link, B. G., & Tehranifar, P. (2010). Social conditions as fundamental causes of health inequalities: Theory, evidence, and policy implications. *Journal of Health and Social Behavior, 51*, 28–40. https://doi.org/10.1177/0022146510383498

Prussin, A. J., 2nd, Schwake, D. O., & Marr, L. C. (2017). Ten questions concerning the aerosolization and transmission of Legionella in the built environment. *Building and Environment, 123*, 684–695. https://doi.org/10.1016/j.buildenv.2017.06.024

Raviglione, M., & Krech, R. (June 2011). Tuberculosis: Still a social disease. *The International Journal of Tuberculosis and Lung Disease*, 6–8. https://doi.org/10.5588/ijtld.11.0158

Ridgley, H., Hands, A., Lovell, R., Petrokofsky, C., Stimpson, A., Feeley, A., … Brannan, M. (2020). *Improving access to greenspace: A new review for 2020.* Retrieved from https://assets.publishing.service.gov.uk/government/uploads/system/uploads/attachment_data/file/904439/Improving_access_to_greenspace_2020_review.pdf.

Rutstein, D. D., Berenberg, W., Chalmers, T. C., Child, C. G., Fishman, A. P., Perrin, E. B., … Evans, C. C. (1976). Measuring the quality of medical care. *New England Journal of Medicine, 294*(11), 582–588. https://doi.org/10.1056/NEJM197603112941104

Schlosberg, D. (2007). *Defining environmental justice: Theories, movements, and nature*. Oxford: Oxford University Press.

SDG 7 Technical Advisory Group, & H. P. F. on S. D. (HLPF). (2018). *Accelerating SDG7 achievement: Policy briefs in support of the first SDG7. Review at the UN high-level political Forum 2018.* Retrieved from https://sustainabledevelopment.un.org/content/documents/262818041SDG7_Policy_Brief.pdf.

Silva, I., Maia, R., Teixeira, M., Santos, J., & Goncalves, A. (2009). Gastric and colorectal cancers: Do patients change their lifestyle after the diagnosis?. In, *23rd annual conference of the European health psychology society: Vol. 24. Abingdon.* Taylor & Francis.

Simons, R. L., Lei, M.-K., Klopack, E., Beach, S. R. H., Gibbons, F. X., & Philibert, R. A. (2020). The effects of social adversity, discrimination, and health risk behaviors on the accelerated aging of African Americans: Further support for the weathering hypothesis. *Social Science & Medicine*, 113169. https://doi.org/10.1016/j.socscimed.2020.113169

Song, C., Ikei, H., Igarashi, M., Miwa, M., Takagaki, M., & Miyazaki, Y. (2014). Physiological and psychological responses of young males during spring-time walks in urban parks. *Journal of Physiological Anthropology, 33*(8). https://doi.org/10.1186/1880-6805-33-8

Song, C., Ikei, H., Igarashi, M., Takagaki, M., & Miyazaki, Y. (2015). Physiological and psychological effects of a walk in urban parks in fall. *International Journal of Environmental Research and Public Health, 12*(11), 14216–14228. https://doi.org/10.3390/ijerph121114216

Song, C., Joung, D., Ikei, H., Igarashi, M., Aga, M., Park, B.-J., … Miyazaki, Y. (2013). Physiological and psychological effects of walking on young males in urban parks in winter. *Journal of Physiological Anthropology, 32*(1), 18. https://doi.org/10.1186/1880-6805-32-18

Squillacioti, G., Bellisario, V., Levra, S., Piccioni, P., & Bono, R. (2019). Greenness availability and respiratory health in a population of urbanised children in North-Western Italy. *International Journal of Environmental Research and Public Health, 17*(1), 108. https://doi.org/10.3390/ijerph17010108

Stenvinkel, P. (2020). The One Health concept–the health of humans is intimately linked with the health of animals and a sustainable environment. *Journal of Internal Medicine, 287*(3), 223–225. https://doi.org/10.1111/joim.13015

Takano, T., Nakamura, K., & Watanabe, M. (2002). Urban residential environments and senior citizens' longevity in megacity areas: The importance of walkable green spaces. *Journal of Epidemiology and Community Health, 56*(12), 913 LP–918. https://doi.org/10.1136/jech.56.12.913

The White House. (2015). *National action plan for combating antibiotic-resistant bacteria.* Washington DC: The White House.

Tyrväinen, L., Ojala, A., Korpela, K., Lanki, T., Tsunetsugu, Y., & Kagawa, T. (2014). The influence of urban green environments on stress relief measures: A field experiment. *Journal of Environmental Psychology, 38*, 1–9. https://doi.org/10.1016/j.jenvp.2013.12.005

UNEP, & WHO. (2016). *Healthy environment, healthy people*. Retrieved from Nairobi https://www.unenvironment.org/news-and-stories/story/healthy-environment-healthy-people.

United Nations. (2015). *Transforming our world: The 2030 Agenda for sustainable development. Resolution adopted by the general assembly on 25 September 2015, A/RES/70/1*. Retrieved from http://www.un.org/en/development/desa/population/migration/generalassembly/docs/globalcompact/A_RES_70_1_E.pdf.

United Nations Development Programme. (2019). *Human development index (HDI)*. Retrieved July 22, 2019, from Human Development Reports website http://hdr.undp.org/en/content/human-development-index-hdi.

University of Leeds. (2015). *A brief guide to the benefits of urban green spaces*. Retrieved from http://leaf.leeds.ac.uk/wp-content/uploads/2015/10/LEAF_benefits_of_urban_green_space_2015_upd.pdf.

U.S. Global Change Research Program (USGCRP). (2018). *Impacts, risks, and adaptation in the United States: Fourth National climate assessment*. Retrieved from Washington, DC, USA https://nca2018.globalchange.gov/chapter/front-matter-about/.

Veflen, N., Scholderer, J., & Langsrud, S. (2020). Situated food safety risk and the influence of social norms. *Risk Analysis, 40*(5), 1092–1110. https://doi.org/10.1111/risa.13449

Veras, A. S. S., Vidal, D. G., Barros, N., & Dinis, M. A. P. (2020). Landscape sustainability: Contribution of Mucajaí-RR (Brazil) region. In W. Leal Filho, A. M. Azul, L. Brandli, P. G. Özuyar, & T. Wall (Eds.), *Responsible consumption and production, Encyclopedia of the UN sustainable development goals* (pp. 1–7). https://doi.org/10.1007/978-3-319-71062-4_82-1

Vidal, D. G., Barros, N., & Maia, R. L. (2020). Public and green spaces in the context of sustainable development. In W. Leal Filho, A. M. Azul, L. Brandli, P. G. Özuyar, & T. Wall (Eds.), *Sustainable cities and communities, Encyclopedia of the UN sustainable development goals* (pp. 1–9). https://doi.org/10.1007/978-3-319-71061-7_79-1

Vidal, D. G., Fernandes, C. O., Viterbo, L. M. F., Barros, N., & Maia, R. L. (2020). Healthy cities to healthy people: A grid application to assess the potential of ecosystems services of public urban green spaces in Porto, Portugal. *The European Journal of Public Health, 20*(Suppl. 2), 17. https://doi.org/10.1093/eurpub/ckaa040.050

Vidal, D. G., Fernandes, C. O., Viterbo, L. M. F., Barros, N., & Maia, R. L. (2021). Combining an evaluation grid application to assess ecosystem services of urban green spaces and a socio-economic spatial analysis. *International Journal of Sustainable Development & World Ecology, 28*(4), 291–302. https://doi.org/10.1080/13504509.2020.1808108

Vidal, D. G., Maia, R. L., Vilaça, H., Barros, N., & Oliveira, G. M. (2018). Green spaces and human rights: To an environmental justice in urban space. In *Book of abstracts of 2nd International conference in Ethics, Politics and culture: Human rights* (pp. 26–27). Porto: Instituto de Filosofia da Universidade do Porto.

Vidal, D. G., Pontes, M., Barreira, E., Oliveira, G. M., & Maia, R. L. (2018). Differential mortality and inequalities in health services access in Mainland Portugal. *Finisterra - Revista Portuguesa de Geografia, 53*(109), 53–70. https://doi.org/10.18055/Finis14118

Villeneuve, P. J., Jerrett, M., Su, J. G., Burnett, R. T., Chen, H., Wheeler, A. J., & Goldberg, M. S. (2012). A cohort study relating urban green space with mortality in Ontario, Canada. *Environmental Research, 115*, 51–58. https://doi.org/10.1016/j.envres.2012.03.003

Watts, N., Adger, W. N., Ayeb-Karlsson, S., Bai, Y., Byass, P., Campbell-Lendrum, D., ... Costello, A. (2017). The *Lancet* countdown: Tracking progress on health and climate change. *The Lancet, 389*(10074), 1151–1164. https://doi.org/10.1016/S0140-6736(16)32124-9

Whitehead, M., & Dahlgren, G. (2007). Part A: Concepts. In *Concepts and principles for tackling social inequities in health: Levelling up Part 1* (pp. 1–12). Retrieved from http://www.euro.who.int/__data/assets/pdf_file/0010/74737/E89383.pdf.

WHO Department of Public Health Environmental and Social Determinants of Health (PHE). (2016). *Ambient air pollution: A global assessment of exposure and burden of disease*. Retrieved from Geneva, Switzerland https://www.who.int/phe/publications/air-pollution-global-assessment/en/.

WHO Regional Office for Europe & WHO European Centre for Environment and Health. (2010). *WHO guidelines for indoor air quality: Selected pollutants*. Retrieved from http://www.euro.who.int/__data/assets/pdf_file/0009/128169/e94535.pdf.

Wood, S. L. R., & DeClerck, F. (2015). Ecosystems and human well-being in the sustainable development goals. *Frontiers in Ecology and the Environment, 13*(3), 123. https://doi.org/10.1890/1540-9295-13.3.123

World Cities Culture Forum. (2018). *% of public green space (parks and gardens)*. Retrieved May 28, 2019, from http://www.worldcitiescultureforum.com/data/of-public-green-space-parks-and-gardens.

World Health Organization. (1978). Declaration of Alma-Ata. In *International conference on primary health care*. Retrieved from https://www.who.int/publications/almaata_declaration_en.pdf?ua.

World Health Organization. (1986). *The Ottawa charter for health promotion*. Retrieved June 3, 2020, from Health promotion website https://www.who.int/healthpromotion/conferences/previous/ottawa/en/.

World Health Organization. (2006). *The control of neglected zoonotic diseases: A route to poverty alleviation*. Retrieved from https://www.who.int/zoonoses/Report_Sept06.pdf?ua=1.

World Health Organization. (2011). *NCDs and development. In Global status report on non-communicable diseases 2010. Description of the Global Burden of NCDs, Their Risk Factors and Determinants* (pp. 33–40). Retrieved from https://www.who.int/nmh/publications/ncd_report_chapter2.pdf?ua=1.

World Health Organization. (2012). Health indicators of sustainable cities in the context of the Rio+20 UN conference on sustainable development. *Initial Findings from a WHO Expert Consultation*, 17–18. WHO/HSE/PHE/7.6.2012f https://doi.org/10.1787/9789264122246-en.

World Health Organization. (2015). *WHO estimates of the global burden of foodborne diseases. Foodborne diseases burden epidemiology reference group 2007-2015*. Geneva: World Health Organization.

World Health Organization. (2016). *Ambient air pollution: A global assessment of exposure and burden of disease*. Geneva: World Health Organization.

World Health Organization. (2017a). *Urban green space interventions and health: A review of evidence*. Retrieved from http://www.euro.who.int/__data/assets/pdf_file/0010/337690/FULL-REPORT-for-LLP.pdf?ua=1.

World Health Organization. (2017b). Urban green spaces: A brief for action. In *Regional Office for Europe*. https://doi.org/10.1590/S1516-89132004000200018

World Health Organization. (2018). Air quality and health effects: WHO resources and support. In *First WHO global conference on air pollution and health: Improving air quality, combatting climate change – saving lives. Geneva.*

World Health Organization. (2020). *Managing public health risks at the human-Animal-environment interface*. Retrieved July 16, 2020, from Zoonoses website https://www.who.int/zoonoses/en/.

World Health Organization (WHO). (2017). *Annex 1. Compendium of Possible Actions to Advance the Implementation of the Ostrava Declaration*. Retrieved from http://www.euro.who.int/en/media-centre/events/events/2017/06/sixth-ministerial-conference-on-environment-and-health/documentation/declaration-of-the-sixth-ministerial-conference-on-environment-and-health/annex-1.-compendium-of-possible-actions-to-advance.

World Health Organization (WHO), FAO and World Organisation for Animal Health (OIE). (2019). *Taking a multisectoral, one health approach: A tripartite guide to addressing zoonotic diseases in countries*. Retrieved from https://www.oie.int/fileadmin/Home/eng/Media_Center/docs/EN_TripartiteZoonosesGuide_webversion.pdf.

Wüstemann, H., Kalisch, D., Kolbe, J., Wüstemanna, H., Kalischa, D., & Kolbeb, J. (2017). Access to urban green space and environmental inequalities in Germany. *Landscape and Urban Planning, 164*, 124–131. https://doi.org/10.1016/j.landurbplan.2017.04.002

Zinsstag, J., Schelling, E., Waltner-Toews, D., & Tanner, M. (2011). From "one medicine" to "one health" and systemic approaches to health and well-being. *Preventive Veterinary Medicine, 101*(3–4), 148–156. https://doi.org/10.1016/j.prevetmed.2010.07.003

Chapter 7

Environmental contaminants and antibiotic resistance as a One Health threat

Najla Haddaji[a, b]
[a]*Department of Biology, Faculty of Sciences, University of Ha'il, Ha'il, Kingdom of Arabia Saudi;*
[b]*Laboratory of Analysis, Treatment and Valorization of Pollutants of the Environment and Products, Faculty of Pharmacy, Monastir, Tunisia*

1. Introduction

The One Health approach aims to attain optimal health for people, animals and the environment (King et al., 2008). In fact, the environment is the most dynamic and consequently the most confounding sector of the One Health triad as evident from the examples of environmental contamination, antibiotic resistance and climate change. This environment is subject to variable weather conditions such as fluctuations in temperature, humidity and precipitation which affect bacterial ecosystems (Essack, 2018). As a result, The One Health concept gives the privilege to an interdependent relationship and collaboration between veterinarians, doctors, scientists and other professions in order to promote human, animal and ecosystem health. To better understand this concept, we have inspired the "One Health diagram" from the symbolic "umbrella" developed in 2014 (Gibbs, 2014) by One Health Sweden in cooperation with the One Health Initiative's autonomous pro bono team, which attempts to encompass all relevant aspects of the One Health movement (Fig. 7.1). This concept has traditionally focused on zoonoses and their relation to the detection and prevention of human diseases (Buttke, 2011). Zoonoses are diseases or infections that are transmitted from vertebrate animals to humans, and vice versa. The pathogens involved can be bacteria, viruses or parasites. The transmission of these diseases occurs either directly, during contact between an animal and a human being, or indirectly through food or through a vector (insect, arachnids, etc.) (Chomel, 2009; Damborg et al., 2016). An estimated 60% of known infectious diseases and up to 75% of new or emerging infectious diseases are zoonotic in origin (Daszak, 2008; Gowtage-Sequeria & Woolhouse, 2005). Globally, infectious diseases account

One Health. https://doi.org/10.1016/B978-0-12-822794-7.00010-1

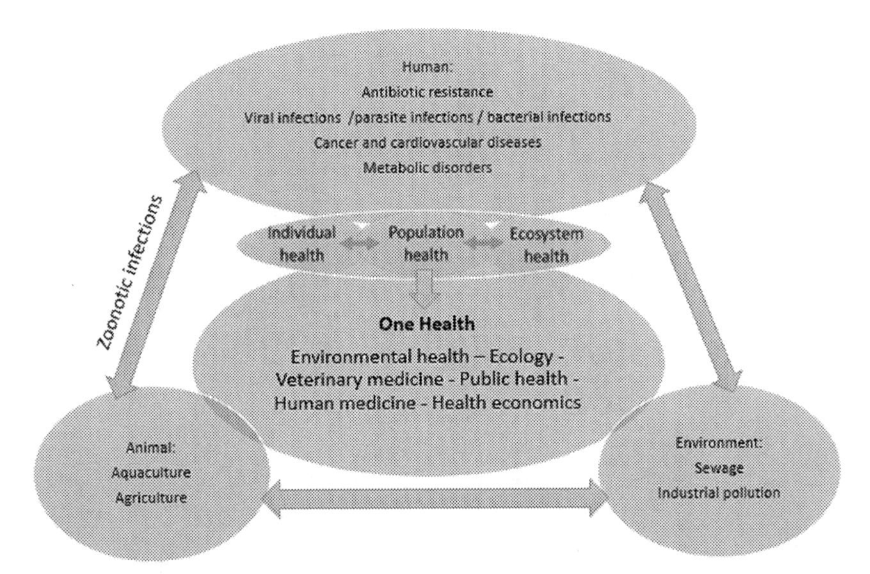

FIG. 7.1 The 'One Health diagram' was inspired by the 'One Health Umbrella' developed by the networks 'One Health Sweden' and 'One Health Initiative' to illustrate the scope of the 'One Health concept'. *Reproduced from Gibbs, E. P. J. (2014). The evolution of one health: A decade of progress and challenges for the future.* The Veterinary Record, 174*(4), 85–91.*

for 15.8% of all deaths and 43.7% of deaths in low-resource countries (GDaH, 2016; Wang, Naghavi, et al., 2016). Other studies focused on toxic exposures in animals and their relationship with chemical-associated human illness (Buttke, 2011; Council, 1991). Indeed, an environment shared between animals and humans allows potential exposure to the same toxic agents, which makes it essential to study diseases in animals in addition to humans in order to control them (Buttke, 2011). Human activity is the first responsible for several types of pollution such as contamination by pesticides, heavy metals, pharmaceuticals, personal care products and microorganisms associated with human waste flows and agriculture. This diversity of pollutants, which vary in concentration spatially and temporally, is a major issue for monitoring (Gillings et al., 2015). For example, chemical contaminants, especially synthetic organic compounds, will remain a serious problem, especially in heavily industrialized areas (Kennish, 2002).

Among the environmental contaminations, antimicrobial resistance remains one of the leading public health issues globally and an added risk of zoonotic bacterial pathogens. Unfortunately, this resistance has been described as the "quintessential One Health problem." However, transmission and persistence of antibiotic resistance is poorly understood (Robinson, Bu, et al., 2016). In this respect, an increase in production in aquaculture and agriculture is remarkable and it is due to the intensification of existing farming systems (and increased farm densities), resulting in higher risks of epidemics. This has led to an increased use of antibiotics, which are now commonly used, sometimes excessively and/or inefficiently, in a wide range of aquaculture systems and countries (Lulijwa, Rupia, & Alfaro, 2020). Additionally, use in subtherapeutic doses for prolonged periods creates optimal conditions for bacteria to implant antibiotic resistance genes. Then, those genes will be passed on to human pathogens or commensals via humans, contaminated food or the environment (Essack, 2018). Since antibiotics used in humans and animals are often analogues of each other, the transmission of resistance between humans and animals in the same environment becomes possible. This has been proven by a study showing a strong association between antibiotic consumption in cattle and clinical antibiotic resistance (Robinson, Wertheim, et al., 2016). Since environmental bacteria are quantitatively the most prevalent bacteria, they therefore serve as reservoirs of resistance genes that can be incorporated into human and animal pathogens over time. These reservoirs of resistance genes are increased by the influx of resistance genes from livestock and human waste into the environment. They are further increased by the entry of antibiotic residues from pharmaceutical industries, intensive animal husbandry and hospitals, which disrupt the microflora of the soil and water in addition to exerting selection pressure for the development of resistance (Robinson, Bu, et al., 2016). Generally, improper use is generally associated with production systems characterized by factors such as high densities, poor hygiene and insufficient animal health control and/or access to appropriate technologies (Henriksson et al., 2018). Antibiotic resistant bacteria accumulate in water, sediment and wildlife in and around farms, posing a risk for the effective treatment of infections in animals and humans (Ducrot et al., 2019).

The purpose of this chapter is to elucidate the different types of environmental contaminations as well as antibiotic resistance and other anthropogenic impacts as One Health threatening.

2. Environmental contaminants

The environment is shared between human and animals, being an important pathway for exposure to polluted air, noise, and hazardous contaminants. In their report on preventing disease through healthy environments, the WHO estimates environmental stressors are responsible for 12%–18% of all deaths in the 53 countries of the WHO Europe Region. Improving the quality of the environment in key areas such as air, water and noise can prevent disease and improve human health (Prüss-Üstün, Wolf, Corvalán, Bos, & Neira, 2016). The environment shared between animals and humans allows for potential exposure to the same toxic agents (Buttke, 2011). Two groups of communicable diseases have been identified: those transmitted from vertebrate animals to humans, which are, strictly speaking, zoonoses; and those common to man and animals. In the first group, animals play an essential role in maintaining infection in nature. In the second group, animals and humans are usually exposed to the same sources, such as soil, water, invertebrate animals, and plants (Acha & Szyfres, 2003). Therefore, environmental contamination is an issue which should be addressed under a One Health approach.

Air pollution is the single largest environmental health risk in Europe, and is associated with heart disease, stroke, and lung disease and lung cancer. Air pollution is determined as the presence of pollutants in the air in large quantities for long periods. Air pollutants are dispersed particles, hydrocarbons, carbon monoxide, carbon dioxide, nitrogen monoxide, nitrogen dioxide, sulfur trioxide, and others. (Manisalidis, Stavropoulou, Stavropoulos, & Bezirtzoglou, 2020). Exposure to air pollution is estimated to result in over 400 000 premature deaths in the EU each year (Schwela, 2021) and 4.2 million deaths worldwide (Organization, 2021). Airborne particulate matter is a mixture of solids and aerosols defined by their diameters, namely those <10 μm (PM10) which considered inhalable particulate matter, and fine particulate matter <2.5 μm (PM2.5). These particles are responsible for respiratory and cardiovascular diseases, dysfunctions of the central and reproductive nervous system, and cancer, by entering the respiratory system by inhalation (Manisalidis et al., 2020); they have been linked to lung cancer deaths in China (Kan, Chen, & Tong, 2012). Likewise, ozone can be harmful in high concentrations at ground level, also affecting the respiratory and cardiovascular system. However, it is interesting to note that cardiovascular disease has mostly been observed in developed and high-income countries rather than in low-income developing countries with high exposure to air pollution (Pena & Rollins, 2017). India, especially New Delhi, recorded extreme air pollution, when air quality reaches dangerous levels. Due to rapid industrialization, urbanization, increased use of motorcycle transport and combustion, pollution occurs in both urban and rural areas of the country (Kankaria, Nongkynrih, & Gupta, 2014). Indeed, many problems arise from the lack of planning of human spaces and activities. For instance, residential areas neighboring gas stations can be contaminated with benzene, toluene, ethylbenzene and xylene (Barros et al., 2019) while those near motorways are exposed to nitrogen dioxide and benzene (Barros, Fontes, Silva, & Manso, 2013). While PM2.5 is also released by traffic, exhaust emissions (i.e., from combustion) are decreasing due to increased motor efficiency and electric vehicles, highlighting the importance of controlling non-exhaust emissions (e.g., from abrasion) (GROUP, 2012). In addition, nitrogen oxide, sulfur dioxide, volatile organic compounds, dioxins and polycyclic aromatic hydrocarbons are all considered to be air pollutants harmful to humans (Manisalidis et al., 2020). It seems that mortality was closely related to the levels of fine, inhalable, and sulfate particles more than with the levels of total particulate pollution, aerosol acidity, sulfur dioxide, or nitrogen dioxide (Dockery, Pope, & Xu, 1993). Several serious diseases are at the origin of the exposure to these substances like the respiratory problems such as obstructive pulmonary disease, asthma, bronchiolitis as well as the lung cancer, the dysfunctions of the central nervous system and some skin diseases (Manisalidis et al., 2020). Beyond human health, exposure to high concentration of indoor fine particulate matter (PM2.5) was associated with respiratory disease in domestic cats (Lin, Lo, Wu, Chang, & Wang, 2018). Presence of pets in the household is also thought to attenuate asthma in children, making them more resilient to air pollution (Zeng et al., 2021). For metals like lead, when absorbed by the human body, can lead to direct poisoning or chronic poisoning, depending on exposure. Childhood lead poisoning (e.g., from paint or leaded gasoline) could be one of the factors in the origin of neurobehavioral disturbances leading to higher crime rates in adults, having a direct consequence on society beyond adverse health outcomes (Nevin, 2007). Decrease in air quality can also be associated with intensive rearing of animals for food production. For instance, dairy operations have a negative impact on air quality of neighboring houses up to 4.8 km away (D'Ann et al., 2011). A One Health approach would contribute to a better understanding of sources of environmental contaminants, both anthropogenic or biological, as well as shared adverse health outcomes in animals and humans.

The geographic distribution of many infectious diseases and contaminants is also influenced by climate change. Air pollution and climate change are closely linked, often originating from the same anthropogenic activities (D'amato et al., 2016). The incidence of mosquito-borne diseases, including malaria, dengue, and viral encephalitis, are among those diseases most sensitive to climate as their vectors may be able to expand to new habitats (Bezirtzoglou, Dekas, & Charvalos, 2011). Likewise, the warming of water due to climate change leads to a high incidence of water-borne infections. Recently in Europe, eradicated diseases seem to be emerging due to population migration, for example cholera, polio, tick-borne encephalitis and malaria (Castelli & Sulis, 2017). By changing meteorological parameters (e.g., temperature, wind, precipitation, solar radiation), climate change can impact water quality. These forcing factors, together with overloading of nutrient concentrations (e.g., nitrogen, phosphorus) can facilitate eutrophication and harmful algal blooms, compromising light penetration and reducing available oxygen, impacting the whole aquatic ecosystem (Nazari-Sharabian, Ahmad, & Karakouzian, 2018). Climate change is expected to increase the frequency and distribution of harmful toxin-producing algal blooms (Zaias, 2010). This increase is likely to affect human health directly through exposure to toxic cyanobacteria and indirectly through the consumption of mollusks that ingest these toxins. Clinical signs of such exposure range from mild skin or respiratory irritation (resulting from inhalation or skin contact) to severe gastrointestinal illness. Exposure often results in the death of animals that have ingested contaminated water. In fact, these animal deaths are often the first sign that the toxins have reached a level of public health concern and action (Buttke, 2011). The Florida Red Tide program piloted an early warning system in which lifeguards monitor beaches for dead fish, dead animals, and skin and respiratory irritation in bathers. Rescuers report all cases to a central system that issues public health warnings based on this information (Kirkpatrick et al., 2008). Early signs of environmental contamination are often observed in animals. This was the case of Minamata disease caused by methylmercury poisoning, occurring in 1956 in Minamata, Japan (Eto, 2000; Harada, 1995). Changes in a manufacturing process in a nearby factory led to the discharge of methylmercury over 20 years. Adverse human health outcomes, such as severe neurological symptoms and death of over a thousand individuals as a result of ingestion of contaminated seafood, were preceded by neurological disturbances in fish, shellfish, birds, and most notoriously cats. Cats played an important role in determining the etiology of the disease, as newly imported animals rapidly developed the disease. While the disease resulted from environmental contamination, animals played the role of a sentinel, contributing to a better understanding of the human disease, following a One Health approach.

A similar case was reported in Bangladesh involving sick children, where the disease was characterized by a sudden onset of respiratory distress and alteration or loss of consciousness in children from two neighboring villages. These cases have tested negative for various infectious causes of respiratory and neurological diseases (Buttke, 2011). What has been found is that this outbreak was preceded by the sudden death of calves and puppies in affected villages and surrounding farm fields, suggesting an environmental etiology. Clinical signs in animals and humans suggested toxicity of the cholinesterase inhibitor pesticides. Laboratory analyzes of human samples ultimately revealed that the likely agent was carbofuran, a carbamate-type pesticide. Carbamate pesticides have the ability of reversibly inhibiting the enzyme acetylcholinesterase, which removes the neurotransmitter acetylcholine from the synaptic cleft, necessary for allowing intermittent signaling of messages. Carbamates are used as pesticides following the same mechanism of action, which under high concentrations produce neurological signs on vertebrates, including domestic animals and humans as seen in Bangladesh. In Cameroon, the presence of toxic substances in food of animal origin consumed has been critical suggesting significant contamination by toxic metals, mycotoxins, veterinary drug residues, and pesticides (Pouokam et al., 2017). Improper handling of food can lead to exposure to various toxic contaminants. One frequent contaminant common to human and livestock is mycotoxins. Mycotoxins, which are natural toxins produced by fungal metabolism, remain a serious problem associated with food and farming systems for humans and animals worldwide (Viegas et al., 2020). As a result of contamination of animal feed, dairy farming is affected which poses a risk to productivity and performance. However, the ability of these contaminants to reach human food through animal by-products, such as meat and milk, is becoming a public danger (Changwa et al., 2018). Although the rumen is supposed to be a barrier against mycotoxin contamination, reports have demonstrated that the carry-over of mycotoxins into the milk is possible in some cases (Becker-Algeri et al., 2016; Flores-Flores, Lizarraga, de Cerain, & González-Peñas, 2015). Thirteen dairy farms were surveyed and 40 dairy feeds of varying nature collected and analyzed for mycotoxins in South Africa (Viegas et al., 2020) revealing the presence of persistent co-occurrence of multiple mycotoxins across samples, however, may elicit synergistic and/or additive effects in hosts, hence raising concerns about their impacts and how such interactions may affect the dairy livestock sector (Viegas et al., 2020). Since mycotoxins are not volatile, airborne dust and fungal spores and fragments act as carriers of mycotoxins to the lungs of farm works (Viegas, Viegas, & Oppliger, 2018). Concerning dermal contact, because workers normally handle the feedstuffs with their naked hands, it is important to consider that several mycotoxins can be absorbed by the skin (Viegas et al., 2020). Therefore, mycotoxins represent a danger to livestock, farm workers, and consumers of animal products, originating from contaminated animal feed which is also dependent on environmental conditions. Widespread use of antimicrobials in human health, livestock, and farming, and the resultant environmental contamination is also in the origin of antimicrobial resistant strains, described in the next section. These cases clearly highlight the benefits of addressing environmental issues by combining approaches of animal and human health.

Contamination is not limited to the effects of chemical or particulate contaminants. Noise exposure from transport sources and industry can lead to annoyance, sleep disturbance and related increases in the risk of hypertension and cardiovascular disease in humans (Schwela, 2021). Similarly, underwater noise pollution, caused by human activities (e.g., shipping, boating, drilling) can induce auditory injury, compromise communication, and impair predatory activities in aquatic animals (Faulkner, Farcas, & Merchant, 2018). Artificial light at night also has consequences in the physiology and behavior of organisms resulting in altered reproduction, compromising survival, and ultimately changing the ecosystem (Giavi, Blösch, Schuster, & Knop, 2020). Contaminants can also change the physical properties of habitats. This is, for instance, the case of microplastics (plastic pieces $\leq$5 mm), which can change water permeability of soils (de Souza Machado, Kloas, Zarfl, Hempel, & Rillig, 2018), thermal conductivity of sediments (Carson, Colbert, Kaylor, & McDermid, 2011), and provide novel and persistent substates to be colonized by unique communities of organisms (Oberbeckmann, Löder, & Labrenz, 2015). In the same way, anthropogenic disturbances affect several types of environments, contributing to habitat alteration and changes in the structure and dynamics of biotic communities as well as the biodiversity of an ecosystem, such as estuaries (Zinsstag, Crump, & Winkler, 2017). The environmental problems encountered in these systems are invariably the result of overpopulation and uncontrolled human activities (Zinsstag et al., 2017). As human activity extends into unexplored territories containing natural foci of infection, new zoonotic diseases are continually being recognized (Acha &

Szyfres, 2003). Similarly, there is no longer a pristine environment on Earth untouched by anthropogenic activity, as contaminants spread around the globe beyond human borders. Therefore, impacts of contamination are often global, involving simultaneously domestic animals, humans, and ecosystems.

Surveillance plays an important role in disease prevention and control. The One Health concept which united several disciplines working at local, national and global levels to achieve optimal health for people, animals and the environment (Association, 2008). This concept promotes collaboration between veterinarians, physicians, scientists and other professions to promote human, animal and ecosystem health (Association, 2008). The adoption of this concept is growing in developed countries during the last decade, mainly aiming to control zoonoses, but it is still not adopted in developing countries where these diseases have the greatest impact. However, the lack of basic health infrastructure in these countries means that the environment, human and animal health are all affected (Bidaisee & Macpherson, 2014). Environmental contamination could benefit from a One Health approach, especially when considering that humans and animals share the environment, food and water sources (Buttke, 2011). As previously exemplified, animals are often considered sentinels by providing an early warning system, mostly resulting from higher susceptibility, biologically compressed lifespans, and lack of confounding effects (Neo & Tan, 2017). Therefore, animal outbreaks of environmental diseases often precede human cases, which could benefit from a shared surveillance system. The One Health concept is also present in the concept "from Farm to Fork," a strategy adopted by the European Union to motivate sustainable food systems and traceability, highlighting the potential exposure of humans through the food web (Commission, 2020). Environmental contaminants can be monitored in human or animal populations by measuring specific hazards (i.e., environmental concentration), exposure (i.e., concentrations in the body), or outcomes (i.e., disease or measurable changes resulting from the exposure) (Thacker, Stroup, Parrish, & Anderson, 1996). The creation of a shared surveillance system under the One Health approach will lead to earlier detection and interventions on harmful environmental contaminants. While some contaminants are already being monitored (e.g., antimicrobial residues in meat in the European Union), others should be scrutinized creating a list of priorities with different monitoring strategies depending on the severity of outcomes produced (e.g., metals, acetylcholinesterase inhibitors). Finally, more information is needed to fully comprehend the impacts of environmental contamination in the ecosystems. One of the few topics addressed in more detail, resulting from healthcare, food production practices, and environmental contamination, is antimicrobial resistance, as detailed in the next section.

3. Antibiotic contamination and antimicrobial resistance: a threat to One Health

Over the recent years, the emergence of antibiotic-resistant bacteria has continuously increased due to the misuse and overuse of antibiotics for human therapy and livestock production. Thereby, this led to increased antimicrobial resistance in diverse environments (Berendonk et al., 2015; Garbisu et al., 2018). The transmission of antibiotic-resistant bacteria in the environment can increase the prevalence of resistance determinants in the human microbiome (Leonard, Zhang, Balfour, Garside, & Gaze, 2015). Indeed, the environmental antibiotic resistance is identified as the top of six emerging issues of concern (Ferri, Ranucci, Romagnoli, & Giaccone, 2017). In addition to climate change, water stress, and environmental degradation, widespread antibiotic resistance should be regarded as one of the global challenges that humans face in this century (Tiedje et al., 2019).

Several studies have identified the spread of antibacterial resistant genes in different environments, including wastewater and sludge (Manaia et al., 2018), livestock farms and soil (Zhu et al., 2013), river water and sediment (Muziasari et al., 2017), drinking water (Ma et al., 2017), glacier environments (Segawa et al., 2013), and even the Antarctic (Wang, Stedtfeld, et al., 2016).

The increasing of antibiotic resistance transform affected environments into large reservoirs of antibacterial resistant genes (Larsson et al., 2018; Manaia et al., 2018). Furthermore, the prevalence of mobile genetic elements, e.g., transposons, integrons, and plasmids, can promote the horizontal transfer of antibacterial resistant genes to other bacteria including human pathogens, exacerbating the antibacterial resistance issue (Gillings et al., 2015; Ma et al., 2017).

Previous studies on antibiotic resistance mainly focused on clinical microorganisms to address the direct threat of emerging antibiotic-resistant bacteria on public health. Furthermore, the emergence of resistant bacteria has continuously increased leading to the development of multidrug resistant bacteria (Ventola, 2015). With the spread of multidrug-resistant human pathogens in medical treatment, both medical and public concern has increased (Tiedje et al., 2019). Therefore, the control of antibiotic-resistant bacteria became a high priority in hospitals and other clinical settings (Ventola, 2016). Different multidrug resistant bacteria such as *Staphylococcus aureus* has become a leading cause of nosocomial and community-acquired infections (Van Duin & Paterson, 2016). However, the development of novel antibiotics has been decreased due to the low profitability in the pharmaceutical industry. Understanding the exact resistance mechanisms associated with the genomic and proteomic analyses is essential to design an effective therapeutic method against multidrug resistant bacteria (Ventola, 2016). This has led to the recognition of the interdependence of human medicine, agriculture, aquaculture and veterinary medicine, engaging scientists and practitioners across these disciplines in the study of the problem and its remedies in a cross-disciplinary manner. The United Nations through its agencies, the United Nations Environment Program (UNEP), the World Health Organization (WHO), the Food and Agriculture Organization (FAO), and the World Organization for Animal Health (OIE), work jointly to promote collective action to minimize the emergence and spread of antimicrobial resistance. Many studies illustrate that soil is the main reservoir of antibiotic remaining bacteria. It is a central component in One Health since it not only harbors a large natural resistome, but also receives antibiotic resistant bacteria and antibiotic resistant genes from both human and animal wastes, which can be returned to humans through vegetable and animal products, through surface, ground, and reclaimed water and via aerosols (Tiedje et al., 2019). The next section describes the development of antibiotics resistance in aquaculture.

3.1 Antimicrobial resistance in aquaculture

Aquaculture production has become an important means of food source for the human consumption throughout the world (Watts, Schreier, Lanska, & Hale, 2017). The use of antimicrobials in aquaculture should be seen within a general framework for risk analysis of antimicrobial resistance in relation to the use of antimicrobials in animals.

The use of antibiotics in aquaculture is strictly regulated and limited in some regions where only a limited number of drugs are approved. As an example, only five different antibiotics are authorized for use in aquaculture in the United Kingdom, compared to 13 different ones in China (Liu, Steele, & Meng, 2017). Several challenges that scientific and technological involved in understanding the variation of the use of antibiotics in aquaculture. Indeed, the lack of vaccination, high fish density, and sub-optimal fishing practices, that include underdeveloped feeding with unknown components that may contain antibiotics or other agents causing selection pressure for antibiotic resistance, may be in the origin of the spread of antimicrobial resistant agents. In addition, aquaculture systems integrating animal production wastes are efficient for nutrient cycling, but may have potential problems with the dissemination of antibiotic resistance (Cabello, Godfrey, Buschmann, & Dölz, 2016).

The application of antimicrobials in aquaculture depends on a variety of factors such as the type of pathogen present (and its sensitivities to antimicrobials), timing of treatment, host disease state and system parameters (salinity, temperature, photoperiod, etc.). However, the laws and regulations of the government organization for each country mainly govern this use (Smith, 2008). The use of antibiotics in aquaculture depends on local regulations, which vary considerably from one country to another. In some countries (especially Europe, North America, and Japan), regulations on the use of antibiotics are strict and only a few antibiotics are allowed. In Europe, for example, the European Directive on Veterinary Medicinal Products, as amended and codified in Directive 2001/82/EC banned the practice of non-therapeutic prophylactic use of antibiotics in 2001 (Koschorreck, Koch, & Rönnefahrt, 2002).

Antibiotics are the first line treatment for bacterial infections, and therefore play an essential role in modern medicine (Davies, Grant, & Catchpole, 2013). Antibiotic resistance is an ancient process and predates any clinical antibiotic usage (D'Costa et al., 2011). However, three types of drug resistant are identified: the extensive drug resistant (XDR), the multidrug resistant (MDR) and the pan-drug resistant (PDR). XDR was defined as acquired non-susceptibility to at least one agent in all but two or fewer antimicrobial categories (i.e., bacterial isolates remain susceptible to only one or two categories). MDR was defined as non-susceptibility to at least one agent in three or more antimicrobial categories) and PDR was defined as non-susceptibility to all agents in all antimicrobial categories (Magiorakos et al., 2012). Indeed, all these types of resistance are a cause of immense concern (Watts et al., 2017).

Antimicrobial resistance resulting from the use of antimicrobials in aquaculture poses a risk to public health by spreading. Indeed, the spread of this resistance to humans can be direct or indirect. Direct spread is due to the development of acquired resistance in bacteria in aquatic environments, that can infect humans, and indirect spread caused by horizontal gene transfer is due to the development of acquired resistance in bacteria in aquatic environments, whereby these resistant bacteria can act as a reservoir of resistance genes from which genes can be further disseminated and eventually end up in human pathogens (Organization, 2006).

3.2 Transmission of antimicrobial resistance in the environment

A recent study (Vk et al., 2019) has shown that the resistance of Gram-negative bacteria to several classes of drugs such as beta-lactams, fluoroquinolones, aminoglycosides and colistin is implicated in serious health complications in human. They also concluded that the reversible nature of the silent antibiotic resistance gene could cause the emergence of antibiotic resistance revertants in the gut upon antibiotic challenge. Which may prove that bacterial drug resistance is a growing threat to the human community (Vk et al., 2019). In addition, a study by Young et al. (2019), discussed the wide spreading of novel antimicrobial resistance genes. The study showed that complex transmission pathways are at the origin of the transmission of resistance between animals, humans and the environment (Young et al., 2019). Therefore, the results obtained characterized the sharing of antimicrobial resistance between humans and between human, animal and environmental niches at a well-identified site in Nepal (Young et al., 2019). Although the majority of deaths related to antibiotic resistance occur in hospitals and nursing home settings, antibiotic-resistant infections can happen anywhere, and are most common in the general community (Control & Prevention, 2019).

Current evidence suggests that widespread dependency on antibiotics and complex interactions between human health, animal husbandry and veterinary medicine, have contributed to the propagation and spread of resistant organisms (Levy, 2002; Organization, 2014; Rizzo et al., 2013).

To understand the contribution of environmental factors in the spread of antimicrobial resistance, some studies suggested the presence of an interdependence between antibiotics and the complex interactions between health, animal husbandry and veterinary medicine has contributed to the spread of resistant organisms (Fletcher, 2015). In addition, other studies examine the role of the environment, specifically water, sanitation and hygiene factors that contribute to the development of antimicrobial resistant bacteria. Indeed, the release of fecal and other pollutants, including antimicrobial compounds, into the environment can promote resistance by creating conditions favorable for the transfer or emergence of new resistance genes. Up to 80% of the administered dose of antimicrobial may be excreted as the active compound or metabolites depending on the class of antimicrobial and how it is used, and wastewater treatment is often insufficient or impossible. Likewise, antimicrobials in the water downstream of some antimicrobial manufacturing sites have been found at higher concentrations than in the blood of patients taking drugs (Larsson, 2013).

Understanding these elements is necessary to identify any modifiable interactions to reduce or interrupt the spread of resistance from the environment into clinical settings. Unfortunately, in previous years, limited attention has been paid to the role of environmental factors in the spread of resistance. Therefore, limited monitoring further complicated the issue (Organization, 2014).

The development of antimicrobial resistance is the cause of the ineffectiveness of routine treatment for a pathogen, resulting in difficulties in controlling infections, an increased risk of the infection spreading to others, and in some cases, an increased risk of death (Control & Prevention, 2019; Organization, 2014). The development of antibiotic resistance can occur naturally (intrinsically) because of spontaneous gene mutation in the absence of selective antibiotic pressure (Alanis, 2005). Bacteria develop resistance to the antibiotic (acquired) when at least one bacterium within a heterogeneous colony of bacteria carries the genetic determinant capable of expressing resistance to the antibiotic. Genetic determinants classify the type and intensity of resistance that is ultimately expressed by the bacterial cell. Regardless of how a gene is transferred to a bacterium, resistance develops because of efficient expression of the gene, which can spread and spread to other bacteria, allowing them to produce a tangible biological effect that inhibits the activity of the antibiotic, providing higher chance of survival (Alanis, 2005). Naturally resistant was found in the detection of a large amount of antibiotic resistance genes (e.g., the vancomycin resistance element VanA) in 30,000 year old Beringian permafrost sediments (D'Costa et al.,

2011). This allows us to conclude that antibiotic resistance genes reside in environmental bacteria that produce and release antibacterial to influence the microbial populations with which they compete for nutrients (Djordjevic, Stokes, & Roy Chowdhury, 2013). Indeed, this natural development of antibiotic resistance is a complex phenomenon, which has many influencing factors, of which the agricultural and aquaculture industry is considered an important contributor. Indeed, the food chain is considered the most responsible route for transfer to humans. Therefore, the natural environment eventually became a vector of this transfer, *Enteric bacteria* are introduced into the environment with human and animal feces, and people may be exposed to these bacteria through e.g. recreation in contaminated surface water, consumption of contaminated drinking water, fresh produce or (shell)fish, and inhalation of bioaerosols (Huijbers et al., 2015). A recent study in South Africa assessed the factors influencing the emergence and transfer of this resistance. The same conclusion as the old studies have been confirmed, such as insufficient surveillance, documentation and control of antibiotics in industry is at the origin of this development (Van den Honert, Gouws, & Hoffman, 2018). However, it is believed that this natural resistance may also play a role in the development of acquired resistance; and that both types of resistance can be transmitted horizontally or vertically (Alanis, 2005; D'Costa et al., 2011). At the molecular level, the same gene can be identified in highly independent pathogens in humans, in clinical settings, as well as bacteria that thrive in an agricultural setting, meat animals and pets (Djordjevic et al., 2013; Van den Honert et al., 2018). Resistance mechanisms carried by bacteria can be transferred to humans can lead to silent carriers which can later lead to seemingly unrelated infections and increased disease severity, leading to poorer outcomes for patients (Smith & Coast, 2013). In this context, resistance to antibiotics such as *Streptococcus* resistant to pyogenic sulfonamides in the 1930s (Levy, 1982) and penicillin resistant *S. aureus* in the 1940s (Barber & Rozwadowska-Dowzenko, 1948) were observed for the first time in hospitals where most drugs were prescribed. *Mycobacterium tuberculosis* developed resistance to streptomycin soon after the discovery of this antibiotic (Crofton & Mitchison, 1948). Multiple drug resistance emerged a few years later in the 1950s to early 1960s among several gastrointestinal/foodborne pathogens, including *Escherichia coli*, *Shigella* and *Salmonella* (FBI) (Watanabe, 1963). Multiple drugs resistance has continued to spread among common pathogens, with *Haemophilus influenzae* developing resistance to ampicillin (Leaves et al., 2000), chloramphenicol and tetracycline (Chopra & Roberts, 2001; Levy, Buu-Hoi, & Marshall, 1984) and an increased resistance to ampicillin observed, with ampicillin-resistant *Neisseria gonorrhea* emerging mainly in developed countries in the 1970s (Elwell, Roberts, Mayer, & Falkow, 1977; Levy & Marshall, 2004). Currently, all WHO regions report very high rates of resistance among common bacteria (for e.g. *E. coli*, *Klebsiella pneumoniae* and *S. aureus*) that are responsible for common infections like urinary tract infections, wound infections, bloodstream infections and pneumonia, which can be acquired in both healthcare and community settings (Organization, 2014). Resistance to antibiotics can also be passed from pets to humans. Some examples exist, such as vancomycin-resistant *Enterococcus faecium* (VRE) infections diagnosed in critically ill patients in European hospitals in the 1980s (Willems & Van Schaik, 2009). Some studies have been done to elucidate whether *Enterococci* in these domestic animals can serve as a reservoir for antibiotic resistance genes (Bates, Jordens, & Griffiths, 1994; Devriese et al., 1996). In cats, dogs, calves and pigs, about 30%−50% of all enterococci sampled were identified as *E. faecalis*, while *E. faecium* is less abundant at about 5%−15%. For poultry, the situation is radically different. Here, *E. faecium* is by far the most common enterococcus present in 60% of all isolates identified. *E. faecalis* occurs at much lower levels (<5% of the total Enterococcus population) (Gilmore et al., 2002). Resistant *Campylobacter* spp. can also be transmitted from puppies to Humans (Montgomery et al., 2018). Additionally, humans with multidrug resistant *S. aureus* (MRSA) infections should consider having their pets examined as potential MRSA reservoirs (Ferreira et al., 2011). Even though MRSA is not adapted to dogs, cats, and other animals, these animals can become colonized with MRSA and then serve as vectors for spread to humans in the same household (Sing, Tuschak, & Hörmansdorfer, 2008).

By considering the above factors and more, one can see why it is difficult to delineate clear pathways for the spread of antimicrobial resistance among food and other domestic animals, human populations and the environment. Antimicrobial resistance has been reported as a general problem in the hope of unraveling useful information and advice for the food animal production industry (Graham et al., 2019). Despite the studies carried out and published, the understanding of the mechanism of the spread of antibiotic resistance among human populations, the environment (soils, wastewater and surface water) and the production of animals for food still remains blurry. One of the challenges in clarifying this problem is to determine how to quantify the specific antibiotic resistant bacteria of concern within a complex chemical matrix and mixed microbial community, in particular identifying the presence of specific resistant bacteria from a defined source. Recent methods of tracking microbiome and resistance gene sources have been used to stochastically distinguish resistant bacteria of community and hospital origin (Dechesne et al., 2018). A recent study (Garcês et al., 2019) focused on characterizing the diversity of extended spectrum beta-lactamases in *E. coli* isolates from European free-tailed bats (*Tadarida teniotis*) in Portugal. This study suggested that wildlife is not exposed to antimicrobial agents used clinically, but can acquire antimicrobial resistant bacteria through contact with humans, pets and the environment, where water polluted with feces seems to be the most important vector of contamination (Garcês et al., 2019). Thus, it must be understood whether the resistant bacteria circulated from person to person or from animals and environment to person, or vice versa. The epidemiology of antibiotic resistant microorganisms at the human-animal-environment interface involves complex and largely unpredictable systems that include transmission routes of resistant bacteria, as well as resistance genes and the impact of selective pressures of antibiotics in various reservoirs (animals, humans and environment) (Purohit et al., 2017). Although the presence and patterns of commensal indicator bacteria resistant to antibiotics *E. coli* isolates from humans, animals and water should be studied together, i.e. - say using the One Health approach (Organization, 2015). However, a comparison of the antibiotic resistance profile among commensal coliforms and *E. coli* in humans, animals and water from the same community is already ongoing in a Indian rural community in order to determine the antibiotic resistance pattern (Purohit et al., 2017).

4. Conclusion

This work has discussed the risks to One Health of environmental contaminants and antibiotic resistance. Nonetheless, knowledge about the links between antimicrobial resistance and the environment remains a knowledge gap. In fact, environmental contaminants can be synthetic or natural substances present in amounts harmful to human health or the ecosystem. Chemical, physical and biological of environmental contaminants are varied, with risk depending on its characteristics. However, there is sufficient evidence indicating several environmental risks to human and animal health and the need for coordinated efforts to tackle these problems. The lack of quantitative data remains the key factor hampering antimicrobial resistance surveillance efforts. Risk assessment studies are needed to provide accurate estimates of the level of resistant bacteria in wastewater effluents that would not pose risks for human and environmental health. Wastewater treatment technologies that are capable of producing effluents with an acceptable level of resistant bacteria are also needed. All the literature proves that the inappropriate use and misuse of antibiotics results in the contamination of various aspects of the environment and this can lead to the introduction of resistance genes and resistant bacteria into the human food chain and clinical environments. Therefore, the provision and application of adequate standards and guidelines, basic food safety, secondary water and sewage/waste treatment, basic sanitation and hygiene practices are measures necessary to tackle this problem. Similarly, early detection of environmental contamination could benefit from shared surveillance between human and veterinary medicine. Therefore, a careful assessment of the situation is needed at community and national levels to determine the needs of each country and a multidisciplinary and intersectional approach between the sectors of health, veterinary medicine and water and wastewater is essential.

References

Acha, P. N., & Szyfres, B. (2003). *Zoonoses and communicable diseases common to man and animals* (Vol. 580). Pan American Health Org.

Alanis, A. J. (2005). Resistance to antibiotics: Are we in the post-antibiotic era? *Archives of Medical Research, 36*(6), 697–705.

Association, A. V. M. (2008). *One health: A new professional imperative. One Health initiative task force: Final report*. Schaumburg: American Veterinary Medical Association.

Barber, M., & Rozwadowska-Dowzenko, M. (1948). Infection by penicillin-resistant *Staphylococci. Lancet*, 641–644.

Barros, N., Carvalho, M., Silva, C., Fontes, T., Prata, J. C., Sousa, A., & Manso, M. C. (2019). Environmental and biological monitoring of benzene, toluene, ethylbenzene and xylene (BTEX) exposure in residents living near gas stations. *Journal of Toxicology and Environmental Health, Part A, 82*(9), 550–563.

Barros, N., Fontes, T., Silva, M., & Manso, M. C. (2013). How wide should be the adjacent area to an urban motorway to prevent potential health impacts from traffic emissions? *Transportation Research Part A: Policy and Practice, 50*, 113–128.

Bates, J., Jordens, J. Z., & Griffiths, D. T. (1994). Farm animals as a putative reservoir for vancomycin-resistant enterococcal infection in man. *Journal of Antimicrobial Chemotherapy, 34*(4), 507–514.

Becker-Algeri, T. A., Castagnaro, D., de Bortoli, K., de Souza, C., Drunkler, D. A., & Badiale-Furlong, E. (2016). Mycotoxins in bovine milk and dairy products: A review. *Journal of Food Science, 81*(3), R544–R552.

Berendonk, T. U., Manaia, C. M., Merlin, C., Fatta-Kassinos, D., Cytryn, E., Walsh, F., ... Pons, M.-N. (2015). Tackling antibiotic resistance: The environmental framework. *Nature Reviews Microbiology, 13*(5), 310–317.

Bezirtzoglou, C., Dekas, K., & Charvalos, E. (2011). Climate changes, environment and infection: Facts, scenarios and growing awareness from the public health community within Europe. *Anaerobe, 17*(6), 337–340.

Bidaisee, S., & Macpherson, C. N. (2014). Zoonoses and one health: A review of the literature. *Journal of Parasitology Research, 8*. https://doi.org/10.1155/2014/874345

Buttke, D. E. (2011). Toxicology, environmental health, and the "One Health" concept. *Journal of Medical Toxicology, 7*(4), 329–332.

Cabello, F. C., Godfrey, H. P., Buschmann, A. H., & Dölz, H. J. (2016). Aquaculture as yet another environmental gateway to the development and globalisation of antimicrobial resistance. *The Lancet Infectious Diseases, 16*(7), e127–e133.

Carson, H. S., Colbert, S. L., Kaylor, M. J., & McDermid, K. J. (2011). Small plastic debris changes water movement and heat transfer through beach sediments. *Marine Pollution Bulletin, 62*(8), 1708–1713.

Castelli, F., & Sulis, G. (2017). Migration and infectious diseases. *Clinical Microbiology and Infection, 23*(5), 283–289.

Changwa, R., Abia, W., Msagati, T., Nyoni, H., Ndleve, K., & Njobeh, P. (2018). Multi-mycotoxin occurrence in dairy cattle feeds from the gauteng province of South Africa: A pilot study using UHPLC-QTOF-MS/MS. *Toxins, 10*(7), 294.

Chomel, B. (2009). *Zoonoses* (p. 820). Encyclopedia of Microbiology.

Chopra, I., & Roberts, M. (2001). Tetracycline antibiotics: Mode of action, applications, molecular biology, and epidemiology of bacterial resistance. *Microbiology and Molecular Biology Reviews, 65*(2), 232–260.

Commission, E. (2020). Farm to fork strategy: For a fair, healthy and environmentally-friendly food system. DG SANTE/unit 'food information and composition. *Food Waste*.

C.F.D. Control & Prevention. (2019). *Antibiotic resistance threats in the United States, 2019*. US Department of Health and Human Services, Centres for Disease Control and.

Council, N. R. (1991). *Animals as sentinels of environmental health hazards*.

Crofton, J., & Mitchison, D. (1948). Streptomycin resistance in pulmonary tuberculosis. *British Medical Journal, 2*(4588), 1009.

D'amato, G., Pawankar, R., Vitale, C., Lanza, M., Molino, A., Stanziola, A., ... D'amato, M. (2016). Climate change and air pollution: Effects on respiratory allergy. *Allergy, asthma & Immunology Research, 8*(5), 391.

D'Ann, L. W., Breysse, P. N., McCormack, M. C., Diette, G. B., McKenzie, S., & Geyh, A. S. (2011). Airborne cow allergen, ammonia and particulate matter at homes vary with distance to industrial scale dairy operations: An exposure assessment. *Environmental Health, 10*(1), 1–9.

Damborg, P., Broens, E. M., Chomel, B. B., Guenther, S., Pasmans, F., Wagenaar, J. A., ... Vanrompay, D. (2016). Bacterial zoonoses transmitted by household pets: State-of-the-art and future perspectives for targeted research and policy actions. *Journal of Comparative Pathology, 155*(1), S27–S40.

Daszak, P. (2008). Global trends in emerging infectious diseases. *Nature, 451*(7181), 990–993.

Davies, D. S., Grant, J., & Catchpole, M. (2013). *The drugs don't work: A global threat*. Penguin UK.

D'Costa, V. M., King, C. E., Kalan, L., Morar, M., Sung, W. W., Schwarz, C., ... Debruyne, R. (2011). Antibiotic resistance is ancient. *Nature, 477*(7365), 457–461.

Dechesne, A., Li, L., He, Z., Nesme, J., Quintela-Baluja, M., Balboa, S., ... Smets, B. F. (2018). Patterns of permissiveness towards broad host range plasmids in microbial communities across the urban water cycle in Europe. In *Paper presented at the 17th International Symposium on microbial ecology*.

Devriese, L. A., Ieven, M., Goossens, H., Vandamme, P., Pot, B., Hommez, J., & Haesebrouck, F. (1996). Presence of vancomycin-resistant enterococci in farm and pet animals. *Antimicrobial Agents and Chemotherapy, 40*(10), 2285–2287.

Djordjevic, S. P., Stokes, H. W., & Roy Chowdhury, P. (2013). Mobile elements, zoonotic pathogens and commensal bacteria: Conduits for the delivery of resistance genes into humans, production animals and soil microbiota. *Frontiers in Microbiology, 4*, 86.

Dockery, D. W., Pope, C., 3rd, & Xu, X. (1993). An association between air pollution and mortality in six US cities. *New England Journal of Medicine, 329*(241), 1753–1759.

Ducrot, C., Adam, C., Beaugrand, F., Belloc, C., Bluhm, J., Chauvin, C., ... Fortané, N. (2019). Apport de la sociologie à l'étude de la réduction d'usage des antibiotiques. *INRA Productions Animales, 31*(4), 307–324.

Elwell, L. P., Roberts, M., Mayer, L. W., & Falkow, S. (1977). Plasmid-mediated beta-lactamase production in Neisseria gonorrhoeae. *Antimicrobial Agents and Chemotherapy, 11*(3), 528–533.

Essack, S. Y. (2018). Environment: The neglected component of the one health triad. *The Lancet Planetary Health, 2*(6), e238–e239.

Eto, K. (2000). Minamata disease. *Neuropathology, 20*, 14–19.

Faulkner, R. C., Farcas, A., & Merchant, N. D. (2018). Guiding principles for assessing the impact of underwater noise. *Journal of Applied Ecology, 55*(6), 2531–2536.

Ferreira, J. P., Anderson, K. L., Correa, M. T., Lyman, R., Ruffin, F., Reller, L. B., & Fowler, V. G., Jr. (2011). Transmission of MRSA between companion animals and infected human patients presenting to outpatient medical care facilities. *PLoS One, 6*(11), e26978.

Ferri, M., Ranucci, E., Romagnoli, P., & Giaccone, V. (2017). Antimicrobial resistance: A global emerging threat to public health systems. *Critical Reviews in Food Science and Nutrition, 57*(13), 2857–2876.

Fletcher, S. (2015). Understanding the contribution of environmental factors in the spread of antimicrobial resistance. *Environmental Health and Preventive Medicine, 20*(4), 243–252.

Flores-Flores, M. E., Lizarraga, E., de Cerain, A. L., & González-Peñas, E. (2015). Presence of mycotoxins in animal milk: A review. *Food Control, 53*, 163–176.

Garbisu, C., Garaiyurrebaso, O., Lanzén, A., Álvarez-Rodríguez, I., Arana, L., Blanco, F., … Alkorta, I. (2018). Mobile genetic elements and antibiotic resistance in mine soil amended with organic wastes. *The Science of the Total Environment, 621*, 725–733.

Garcês, A., Correia, S., Amorim, F., Pereira, J. E., Igrejas, G., & Poeta, P. (2019). First report on extended-spectrum beta-lactamase (ESBL) producing *Escherichia coli* from European free-tailed bats (*Tadarida teniotis*) in Portugal: A one-health approach of a hidden contamination problem. *Journal of Hazardous Materials, 370*, 219–224.

GDaH, C. (2016). *Global burden of disease study 2015 (GBD 2015) results*. Seattle, United States: Institute for Health Metrics and Evaluation (IHME).

Giavi, S., Blösch, S., Schuster, G., & Knop, E. (2020). Artificial light at night can modify ecosystem functioning beyond the lit area. *Scientific Reports, 10*(1), 1–11.

Gibbs, E. P. J. (2014). The evolution of one health: A decade of progress and challenges for the future. *The Veterinary Record, 174*(4), 85–91.

Gillings, M. R., Gaze, W. H., Pruden, A., Smalla, K., Tiedje, J. M., & Zhu, Y.-G. (2015). Using the class 1 integron-integrase gene as a proxy for anthropogenic pollution. *The ISME Journal, 9*(6), 1269–1279.

Gilmore, M. S., Clewell, D. B., Courvalin, P., Dunny, G. M., Murray, B. E., & Rice, L. B. (2002). *The enterococci: Pathogenesis, molecular biology, and antibiotic resistance* (Vol. 10). Washington, DC: ASM press.

Gowtage-Sequeria, S., & Woolhouse, M. (2005). Host range and emerging and reemerging pathogens. *Emerging Infectious Diseases, 11*(12), 1842–1847.

Graham, D. W., Bergeron, G., Bourassa, M. W., Dickson, J., Gomes, F., Howe, A., … Simjee, S. (2019). Complexities in understanding antimicrobial resistance across domesticated animal, human, and environmental systems. *Annals of the New York Academy of Sciences, 1441*(1), 17.

GROUP, A. (2012). *Fine particulate matter (PM 2.5) in the United Kingdom*. London: Department for Environment, Food and Rural Affairs.

Harada, M. (1995). Minamata disease: Methylmercury poisoning in Japan caused by environmental pollution. *Critical Reviews in Toxicology, 25*(1), 1–24.

Henriksson, P. J., Rico, A., Troell, M., Klinger, D. H., Buschmann, A. H., Saksida, S., … Zhang, W. (2018). Unpacking factors influencing antimicrobial use in global aquaculture and their implication for management: A review from a systems perspective. *Sustainability Science, 13*(4), 1105–1120.

Huijbers, P. M., Blaak, H., de Jong, M. C., Graat, E. A., Vandenbroucke-Grauls, C. M., & de Roda Husman, A. M. (2015). Role of the environment in the transmission of antimicrobial resistance to humans: A review. *Environmental Science & Technology, 49*(20), 11993–12004.

Kan, H., Chen, R., & Tong, S. (2012). Ambient air pollution, climate change, and population health in China. *Environment International, 42*, 10–19.

Kankaria, A., Nongkynrih, B., & Gupta, S. K. (2014). Indoor air pollution in India: Implications on health and its control. *Indian Journal of Community Medicine: Official Publication of Indian Association of Preventive & Social Medicine, 39*(4), 203.

Kennish, M. J. (2002). Environmental threats and environmental future of estuaries. *Environmental Conservation*, 78–107.

King, L. J., Anderson, L. R., Blackmore, C. G., Blackwell, M. J., Lautner, E. A., Marcus, L. C., … Ohle, J. (2008). Executive summary of the AVMA one health initiative task force report. *Journal of the American Veterinary Medical Association, 233*(2), 259–261.

Kirkpatrick, B., Currier, R., Nierenberg, K., Reich, A., Backer, L. C., Stumpf, R., … Kirkpatrick, G. (2008). Florida red tide and human health: A pilot beach conditions reporting system to minimize human exposure. *The Science of the Total Environment, 402*(1), 1–8.

Koschorreck, J., Koch, C., & Rönnefahrt, I. (2002). Environmental risk assessment of veterinary medicinal products in the EU—a regulatory perspective. *Toxicology Letters, 131*(1–2), 117–124.

Larsson, D. (2013). Pollution from drug manufacturing review and perspectives, perspectives. *Philosophical Transactions of the Royal Society B: Biological Sciences*. https://doi.org/10.1098/rstb

Larsson, D. J., Andremont, A., Bengtsson-Palme, J., Brandt, K. K., de Roda Husman, A. M., Fagerstedt, P., … Kuroda, M. (2018). Critical knowledge gaps and research needs related to the environmental dimensions of antibiotic resistance. *Environment International, 117*, 132–138.

Leaves, N., Dimopoulou, I., Hayes, I., Kerridge, S., Falla, T., Secka, O., … Crook, D. (2000). Epidemiological studies of large resistance plasmids in Haemophilus. *Journal of Antimicrobial Chemotherapy, 45*(5), 599–604.

Leonard, A. F., Zhang, L., Balfour, A. J., Garside, R., & Gaze, W. H. (2015). Human recreational exposure to antibiotic resistant bacteria in coastal bathing waters. *Environment International, 82*, 92–100.

Levy, S. (1982). Microbial resistance to antibiotics. An evolving and persistent problem. *Lancet, 2*, 83–88.

Levy, S. B. (2002). Factors impacting on the problem of antibiotic resistance. *Journal of Antimicrobial Chemotherapy, 49*(1), 25–30.

Levy, S. B., Buu-Hoi, A., & Marshall, B. (1984). Transposon Tn10-like tetracycline resistance determinants in Haemophilus parainfluenzae. *Journal of Bacteriology, 160*(1), 87–94.

Levy, S. B., & Marshall, B. (2004). Antibacterial resistance worldwide: Causes, challenges and responses. *Nature Medicine, 10*(12), S122–S129.

Lin, C. H., Lo, P. Y., Wu, H. D., Chang, C., & Wang, L. C. (2018). Association between indoor air pollution and respiratory disease in companion dogs and cats. *Journal of Veterinary Internal Medicine, 32*(3), 1259–1267.

Liu, X., Steele, J. C., & Meng, X.-Z. (2017). Usage, residue, and human health risk of antibiotics in Chinese aquaculture: A review. *Environmental Pollution, 223*, 161–169.

Lulijwa, R., Rupia, E. J., & Alfaro, A. C. (2020). Antibiotic use in aquaculture, policies and regulation, health and environmental risks: A review of the top 15 major producers. *Reviews in Aquaculture, 12*(2), 640–663.

Magiorakos, A.-P., Srinivasan, A., Carey, R. T., Carmeli, Y., Falagas, M. T., Giske, C. T., … Olsson-Liljequist, B. (2012). Multidrug-resistant, extensively drug-resistant and pandrug-resistant bacteria: An international expert proposal for interim standard definitions for acquired resistance. *Clinical Microbiology and Infection, 18*(3), 268–281.

Ma, L., Li, B., Jiang, X.-T., Wang, Y.-L., Xia, Y., Li, A.-D., & Zhang, T. (2017). Catalogue of antibiotic resistome and host-tracking in drinking water deciphered by a large scale survey. *Microbiome, 5*(1), 1–12.

Manaia, C. M., Rocha, J., Scaccia, N., Marano, R., Radu, E., Biancullo, F., … Zammit, I. (2018). Antibiotic resistance in wastewater treatment plants: Tackling the black box. *Environment International, 115*, 312–324.

Manisalidis, I., Stavropoulou, E., Stavropoulos, A., & Bezirtzoglou, E. (2020). Environmental and health impacts of air pollution: A review. *Frontiers in Public Health, 8*.

Montgomery, M. P., Robertson, S., Koski, L., Salehi, E., Stevenson, L. M., Silver, R., … Weisner, M. B. (2018). Multidrug-resistant *Campylobacter jejuni* outbreak linked to puppy exposure—United States, 2016–2018. *Morbidity and Mortality Weekly Report, 67*(37), 1032.

Muziasari, W. I., Pitkänen, L. K., Sørum, H., Stedtfeld, R. D., Tiedje, J. M., & Virta, M. (2017). The resistome of farmed fish feces contributes to the enrichment of antibiotic resistance genes in sediments below Baltic Sea fish farms. *Frontiers in Microbiology, 7*, 2137.

Nazari-Sharabian, M., Ahmad, S., & Karakouzian, M. (2018). Climate change and eutrophication: A short review. *Engineering, Technology & Applied Science Research, 8*(6), 3668.

Neo, J. P. S., & Tan, B. H. (2017). The use of animals as a surveillance tool for monitoring environmental health hazards, human health hazards and bioterrorism. *Veterinary Microbiology, 203*, 40–48.

Nevin, R. (2007). Understanding international crime trends: The legacy of preschool lead exposure. *Environmental Research, 104*(3), 315–336.

Oberbeckmann, S., Löder, M. G., & Labrenz, M. (2015). Marine microplastic-associated biofilms–a review. *Environmental Chemistry, 12*(5), 551–562.

Organization, W. H. (2006). *The world health report 2006: Working together for health.* World Health Organization.

Organization, W. H. (2014). *Antimicrobial resistance global report on surveillance: 2014 summary.* World Health Organization.

Organization, W. H. (2015). *Antimicrobial resistance draft global action plan on antimicrobial resistance*. Geneva: WHO.

Organization, W. H. (2021). *Air pollution.* https://www.who.int/health-topics/air-pollution#tab=tab_1.

Pena, M. S. B., & Rollins, A. (2017). Environmental exposures and cardiovascular disease: A challenge for health and development in low-and middle-income countries. *Cardiology Clinics, 35*(1), 71–86.

Pouokam, G. B., Foudjo, B., Samuel, C., Yamgai, P. F., Silapeux, A. K., Sando, J. T., … Frazzoli, C. (2017). Contaminants in foods of animal origin in Cameroon: A one health vision for risk management "from farm to fork". *Frontiers in Public Health, 5*, 197.

Prüss-Üstün, A., Wolf, J., Corvalán, C., Bos, R., & Neira, M. (2016). *Preventing disease through healthy environments: A global assessment of the burden of disease from environmental risks.* World Health Organization.

Purohit, M. R., Chandran, S., Shah, H., Diwan, V., Tamhankar, A. J., & Stålsby Lundborg, C. (2017). Antibiotic resistance in an Indian rural community: A 'One-Health' observational study on commensal coliform from humans, animals, and water. *International Journal of Environmental Research and Public Health, 14*(4), 386.

Rizzo, L., Manaia, C., Merlin, C., Schwartz, T., Dagot, C., Ploy, M., … Fatta-Kassinos, D. (2013). Urban wastewater treatment plants as hotspots for antibiotic resistant bacteria and genes spread into the environment: A review. *The Science of the Total Environment, 447*, 345–360.

Robinson, T. P., Bu, D., Carrique-Mas, J., Fèvre, E. M., Gilbert, M., Grace, D., … Kariuki, S. (2016). Antibiotic resistance is the quintessential One Health issue. *Transactions of the Royal Society of Tropical Medicine and Hygiene, 110*(7), 377–380.

Robinson, T. P., Wertheim, H. F., Kakkar, M., Kariuki, S., Bu, D., & Price, L. B. (2016). Animal production and antimicrobial resistance in the clinic. *The Lancet, 387*(10014), e1–e3.

Schwela, D. (2021). Review of environmental noise policies and economics in 2014-2016. *South Florida Journal of Health, 2*(1), 46–61.

Segawa, T., Takeuchi, N., Rivera, A., Yamada, A., Yoshimura, Y., Barcaza, G., … Ushida, K. (2013). Distribution of antibiotic resistance genes in glacier environments. *Environmental Microbiology Reports, 5*(1), 127–134.

Sing, A., Tuschak, C., & Hörmansdorfer, S. (2008). Methicillin-resistant *Staphylococcus aureus* in a family and its pet cat. *New England Journal of Medicine, 358*(11), 1200–1201.

Smith, P. (2008). Antimicrobial resistance in aquaculture. *Revue scientifique et technique (International Office of Epizootics), 27*(1), 243–264.

Smith, R., & Coast, J. (2013). The true cost of antimicrobial resistance: Richard Smith and Joanna Coast argue that current estimates of the cost of antibiotic resistance are misleading and may result in inadequate investment in tackling the problem. *BMJ, 346*.

de Souza Machado, A. A., Kloas, W., Zarfl, C., Hempel, S., & Rillig, M. C. (2018). Microplastics as an emerging threat to terrestrial ecosystems. *Global Change Biology, 24*(4), 1405–1416.

Thacker, S. B., Stroup, D. F., Parrish, R. G., & Anderson, H. A. (1996). Surveillance in environmental public health: Issues, systems, and sources. *American Journal of Public Health, 86*(5), 633–638.

Tiedje, J. M., Fang, W., Manaia, C. M., Virta, M., Sheng, H., Liping, M., … Edward, T. (2019). Antibiotic resistance genes in the human-impacted environment: A One Health perspective. *Pedosphere, 29*(3), 273–282.

Van Duin, D., & Paterson, D. L. (2016). Multidrug-resistant bacteria in the community: Trends and lessons learned. *Infectious Disease Clinics, 30*(2), 377–390.

Van den Honert, M., Gouws, P., & Hoffman, L. (2018). Importance and implications of antibiotic resistance development in livestock and wildlife farming in South Africa: A review. *South African Journal of Animal Science, 48*(3), 401–412.

Ventola, C. L. (2015). The antibiotic resistance crisis: Part 1: Causes and threats. *Pharmacy and Therapeutics, 40*(4), 277.

Ventola, C. L. (2016). Immunization in the United States: Recommendations, barriers, and measures to improve compliance: Part 1: Childhood vaccinations. *Pharmacy and Therapeutics, 41*(7), 426.

Viegas, S., Assunção, R., Twarużek, M., Kosicki, R., Grajewski, J., & Viegas, C. (2020). Mycotoxins feed contamination in a dairy farm—potential implications for milk contamination and workers' exposure in a One Health approach. *Journal of the Science of Food and Agriculture, 100*(3), 1118–1123.

Viegas, S., Viegas, C., & Oppliger, A. (2018). Occupational exposure to mycotoxins: Current knowledge and prospects. *Annals of Work Exposures and Health, 62*(8), 923–941.

Vk, D., Srikumar, S., Shetty, S., van Nguyen, S., Karunasagar, I., & Fanning, S. (2019). Silent antibiotic resistance genes: A threat to antimicrobial therapy. *International Journal of Infectious Diseases, 79*, 20.

Wang, H., Naghavi, M., Allen, C., Barber, R., Carter, A., Casey, D., … Coggeshall, M. (2016). Global, regional, and national life expectancy, all-cause mortality, and cause-specific mortality for 249 causes of death, 1980â 2015: A systematic analysis for the global burden of disease study 2015. *The Lancet, 388*(10053), 1459–1544.

Wang, F., Stedtfeld, R. D., Kim, O.-S., Chai, B., Yang, L., Stedtfeld, T. M., … Hashsham, S. A. (2016). Influence of soil characteristics and proximity to Antarctic research stations on abundance of antibiotic resistance genes in soils. *Environmental Science & Technology, 50*(23), 12621–12629.

Watanabe, T. (1963). Infective heredity of multiple drug resistance in bacteria. *Bacteriological Reviews, 27*(1), 87.

Watts, J. E., Schreier, H. J., Lanska, L., & Hale, M. S. (2017). The rising tide of antimicrobial resistance in aquaculture: Sources, sinks and solutions. *Marine Drugs, 15*(6), 158.

Willems, R. J., & Van Schaik, W. (2009). Transition of Enterococcus faecium from commensal organism to nosocomial pathogen. *Future Microbiology, 4*(9), 1125–1135.

Young, C., Kharmacharya, D., Bista, M., Sharma, A., Goldstein, T., Anthony, S., ... Johnson, C. (2019). Sharing of antimicrobial resistance genes among animals, humans, and the environment in Nepal: A one health case study. *International Journal of Infectious Diseases, 79*, 20.

Zaias, J., & Fleming, L. E. (2010). Toxic exposures: Harmful algal blooms. In P. Rabinowitz, & L. Conti (Eds.), *Human animal medicine: Clinical approaches to zoonoses, toxicants and other shared health risks* (p. 50). Maryland Heights: Saunders.

Zeng, X. W., Lodge, C. J., Lowe, A. J., Guo, Y., Abramson, M. J., Bowatte, G., ... Dharmage, S. C. (2021). Current pet ownership modifies the adverse association between long-term ambient air pollution exposure and childhood asthma. *Clinical and Translational Allergy, 11*(1), e12005.

Zhu, Y.-G., Johnson, T. A., Su, J.-Q., Qiao, M., Guo, G.-X., Stedtfeld, R. D., ... Tiedje, J. M. (2013). Diverse and abundant antibiotic resistance genes in Chinese swine farms. *Proceedings of the National Academy of Sciences, 110*(9), 3435–3440.

Zinsstag, J., Crump, L., & Winkler, M. (2017). Biological threats from a 'One Health' perspective. *OIE Revue Scientifique et Technique, 36*(2), 671–680.

Chapter 8

Climate change and its impacts on health, environment and economy

Jorge Rocha[a], Sandra Oliveira[a], Cláudia M. Viana[a] and Ana Isabel Ribeiro[b]
[a]*Centro de Estudos Geográficos, Instituto de Geografia e Ordenamento do Território, Universidade de Lisboa, Rua Branca Edmée Marques, Cidade Universitária, Lisboa, Portugal;* [b]*Epidemiologist and Health Geographer, Public Health Institute of University of Porto, Portugal*

1. Introduction

The welfare and stability of health systems depend on how they cope with climatic changes (Anderson & Bows, 2011). Climate change denotes a long-term change (normally for decades or longer) of the climate state that can be acknowledged through statistical methods (e.g. variability and/or the average of its properties) and have come to outline local, regional and global climates. Climate change can be triggered by natural processes, or by continuous anthropogenic modifications in land use/cover or in the atmosphere composition. Changes witnessed in the last century are mainly motivated by human actions (e.g. burning fossil fuel) that increased greenhouse gas concentration in the atmosphere and lead to the increase of average surface temperature (Masson-Delmotte et al., 2021; Overview: Weather, 2020).

Globally, an increase of 1°C in the average temperature since the beginning of the 20th century, has resulted in additional risks (Haustein et al., 2017; IPCC, 2018), such as emerging infectious diseases in areas unaffected before (Legendre et al., 2014; Watts et al., 2015), persistent drought and heatwaves, severe storms and floods, and threats to food security.

Average temperatures may increase between 1.4 and 5.8°C by the end of the 21st century, as suggested by predictive climate models (Huss, 2010; Wilkinson, 2006, p. 232). The Intergovernmental Panel on Climate Change (IPCC, 2018) highlights that, to bound global warming to 1.5°C, two conditions are required: a substantial cut in global annual emissions (by 2030), and a

One Health. https://doi.org/10.1016/B978-0-12-822794-7.00009-5

complete counterbalance of all the human related Greenhouse Gases (GHG) emissions by their removal from the atmosphere (e.g. carbon uptake), i.e. a net-zero range in 2050. At the same time, it acknowledges that no extent of climate change is harmless (IPCC, 2018). Human health and climate change are interconnected in complex and numerous ways (Watts et al., 2015) and their consequences are not equally scattered across the world. All populations are affected by climate change, although some are more vulnerable than others. Coastal, mountainous regions and large cities are the most susceptible areas (Harlan & Ruddell, 2011; Nicholls et al., 2007, pp. 315–356) and population groups such as children, the elderly, and people with prior conditions regarding cardiovascular, respiratory, oncological diseases and immune disorders, are more vulnerable (Gamble et al., 2016). In addition, places with less developed health infrastructure will have a lower capacity to prepare and respond to the challenges brought by climate change (WHO, 2015, p. 54). Extreme climatic events are expected to become more frequent and there will be a rise in sea level (Kirch, Menne, & Bertollini, 2005, p. 306). These changes can have direct and indirect effects in health conditions, being one of the most critical environmental threats in this 21st century (Wilkinson, 2006, p. 232).

Health can be affected directly through extreme temperatures and shifts in precipitation patterns, causing heatwaves or cold spells, storms, droughts, wildfires and floods, and a change in the patterns of disease spread by vectors and rodents. Indirectly, health can be affected by changes in food sources, disruption of methods of food production, and decreased economic productivity. Climate change can influence the health status of the future generations and weaken the advancements headed for the United Nations Sustainable Development Goals (SDGs) (Pecl et al., 2017). Ongoing climate change is a challenge on a global scale, with repercussions at the level of communities, but also to health services and their professionals (WHO Geneva, 2013). Such challenges demand the development of community-based programs, in which all individuals can participate, in a full exercise of citizenship.

2. Climate change and environmental conditions

Climate change modifies temperature, rainfall, moisture, and wind patterns. These changes affect the distribution of plants and animals around the globe, the availability and quality of natural resources, the occurrence of environmental hazards and the levels of atmospheric pollutants, all of which influence human health.

The number of natural disasters related to climate change has been increasing and it is expected that this will continue (Fang, Lau, Lu, Wu, & Zhu, 2019). These disasters result in an estimated annual value of more than 60 000 casualties worldwide (Ritchie & Roser, 2014). These occurrences are sources of deaths, but also of injuries, disabilities, illnesses and emotional distress.

2.1 Environmental hazards

Increases in the overall temperature of the atmosphere and oceans cause changes in wind, moisture, and heat circulation patterns, contributing to shifts in the occurrence of extreme weather events.

The impact of temperature on population morbidity and mortality can be seen from both a daily and a seasonal perspective. Seasonally, peaks of winter mortality caused by pneumonia and flu, heart and cerebrovascular diseases and diabetes are identified (Weinberger et al., 2017); stroke and ischemic heart disease have a pattern of mortality that can be related with daily and monthly variations in temperature (Li et al., 2018).

In the last years, there has been growing interest in investigating the relationship between climate and human mortality, focusing mainly on the climatic paroxysms of the summer and winter periods (heat and cold waves). Given their high frequency in the recent past, heatwaves are noteworthy; the prolonged heat in the summer of 2003 in Europe caused an increase in mortality rates by at least 4 times the levels expected for that time of the year - there were more than 70.000 excess deaths (Robine et al., 2008). In fact, an increase of 1°C above the climatological normal (representing a 30 years average of climatic variables) (Fig. 8.1), can increase mortality by 2%–5%, the impacts depending on the level of exposure (severity, frequency, duration

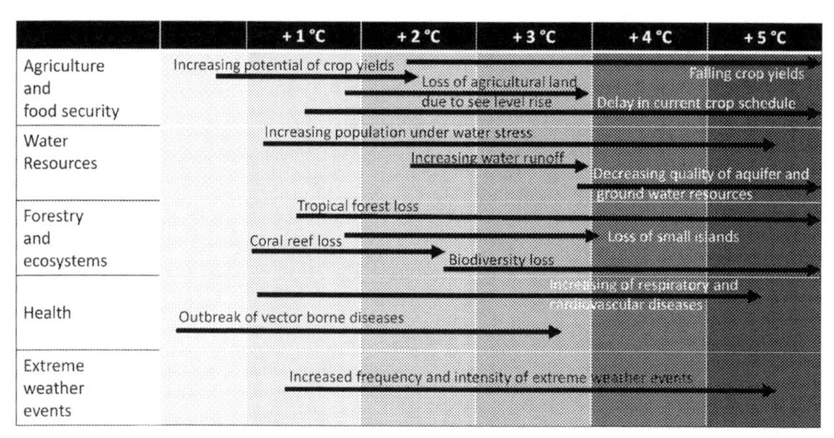

FIG. 8.1 Effects of temperature rise. *Adapted from Allen, M. R., Dube, O. P., Solecki, W., Aragón-Durand, F., Cramer, W. Humphreys, S., ... Kala, J. Framing and context. In: Masson-Delmotte, V., Zhai, P., Pörtner, H.-O., Roberts, D., Skea, J., Shukla, P. R., ... TW, editor (2018).* Global Warming of 15°C an IPCC Special Report on the impacts of global warming of 15°C above pre-industrial levels and related global greenhouse gas emission pathways, in the context of strengthening the global response to the threat of climate change, *IPCC. Available from: https://www.ipcc.ch/sr15/chapter/chapter-1/; Stern, N. (2007).* The economics of climate change: The stern review. *Cambridge: Cambridge University Press. Available from: https://www.cambridge.org/core/books/economics-of-climate-change/A1E0BBF2F0ED8E2E4142A9C878052204.*

of the events) and the vulnerability of the affected population (Lay et al., 2021). Thermal stress is more severe in cities, due to the "urban heat island" effect, a phenomenon that occurs when temperatures rise in cities in relation to their surroundings, due to changes in radiative and thermal properties caused by urban infrastructures and buildings (Oke, 1995). The urban heat island is usually more intense in weather conditions typical of summertime, with clear skies and calm winds, which can aggravate the effects of high temperatures and heatwaves, and substantially decrease the thermal comfort and health conditions of urban dwellers (Alcoforado & Andrade, 2008). Despite variations across cities and between seasons, temperatures can rise by more than 5°C in urban areas in relation to their surroundings, exacerbating as well the effects of ozone and suspended particles (Menne & Ebi, 2006, p. 449).

Droughts are periods of prolonged dry weather, with negative repercussions on food availability and increasing the risk of transmissible diseases associated to water availability and quality. The frequency of wildfires is expected to increase as drought conditions become more frequent, which can greatly reduce air quality. People living in drought conditions may also be more exposed to dust storms or flash floods, for example.

Over the last decades, we have also seen an increase in the number of heavy precipitation events (Papalexiou & Montanari, 2019). Floods and cyclones can cause widespread devastation, with human losses and material damages, including in health infrastructure (Doocy et al., 2013a,b). They affect the supply of water and food, as well as the distribution of vectors that can spread infectious diseases (Kirch et al., 2005, p. 306). Regarding sea level rise, in some areas of the globe like Bangladesh it has already led to the displacement of people from coastal shores who must move to other places, becoming climate refugees (Bose, 2013; Khan, Huq, Risha, & Alam, 2021).

2.2 Atmospheric pollutants

The accumulation of greenhouse gases in the atmosphere increases temperatures. The use of fossil fuels also favors the release of methane, carbon monoxide and dioxide, and particulate matter (PM). Although considering that air pollution is not all related to climate change, the World Health Organization (WHO) estimated that, in 2012, more than 6 million premature deaths were caused by human exposure to pollutants (WHO, 2013, 2015).

The accumulation of industrial chemicals in the atmosphere, in particular Chlorofluorocarbons (CFCs), has depleted the ozone layer in the stratosphere - a natural barrier to the ultraviolet (UV) rays that reach the Earth - resulting in higher levels of ultraviolet radiation (Merrill, 2008) with harmful consequences for human health and for the environment, for example by decreasing agricultural yields. Climate change also modifies the concentration and distribution of pollens and other allergens, by altering the range of allergenic species, the duration of the pollen season, the pollen production, and the release and dispersion of pollens in the atmosphere (Orru, Ebi, & Forsberg, 2017). Air quality is highly connected with weather conditions; pollutant emission, transport, dispersion, chemical transformation and deposition can be influenced by meteorological variables such as temperature, humidity, and air circulation (Orru et al., 2017). Reduced air quality will directly affect human health and ecosystems and impact climate in a feedback loop (Orru et al., 2017). For instance, $PM_{2.5}$ concentration is expected to increase about $\pm$ 1 $\mu g/m^3$ in the USA and Europe by 2030 (Jacob & Winner, 2009). In turn, these particles can absorb heat, thus increasing local temperatures. Another example is the near-surface ozone, whose rate of formation increases with temperature and insolation. Projections suggest that climate change will increase summertime surface ozone in polluted regions by 1–10 ppb (parts per billion) over the coming decades (Jacob & Winner, 2009). Ozone is a dangerous air pollutant leading to various health problems, such as aggravated asthma, reduced lung capacity and increased risk of respiratory infections, therefore the projected rise in ozone levels may increase the frequency of these health problems (Merrill, 2008).

Climate change is expected to increase the frequency and intensity of wildfires (Bowman et al., 2020), which can significantly reduce air quality and affect people's health, by releasing hazardous pollutants such as contaminant particles ($PM_{2.5}$ and PM_{10}), carbon monoxide and nitrogen oxides (Xu et al., 2020).

3. Climate change and human health

Climate change will affect human health in numerous ways and influence the distribution of an extensive variety of diseases (Fig. 8.2). These health effects result from complex relations among direct and indirect consequences of climate change, and are also linked with economic development, accessibility to health services and demographics (Watts et al., 2015).

The increase in temperature will have direct effects through the occurrence of extreme heat events, which are dangerous to health – even fatal. These events result in increased hospital admissions for heat-related illness, as well as cardiovascular and respiratory disorders. Extreme heat events can trigger a variety of heat stress conditions, such as heat stroke, the most serious heat-related disorder. During a heatstroke, hyperthermia is accompanied by the loss of body temperature regulation leading to its rapid increase, systemic inflammation, disseminated intravascular coagulation, and organ failure, which can result in dead. Small children, the elderly, and other vulnerable groups including people with chronic diseases, low-income populations, and outdoor workers, have a higher risk of suffering from heat-related illness.

As previously mentioned, a decrease in air quality is expected due to raising global temperatures (The Lancet Respiratory Med, 2018). One reason is because higher temperatures contribute to the build-up of harmful air

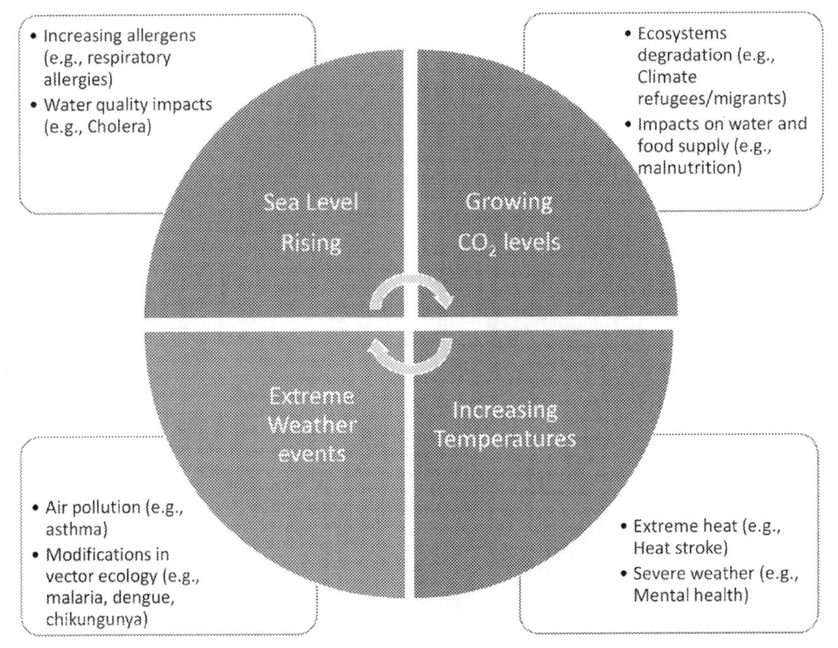

FIG. 8.2 Impacts of climate change in Human Health. *Adapted from CDC (2016).* Climate effects on health. *Available from https://www.cdc.gov/climateandhealth/effects/default.htm.*

pollutants (De Sario, Katsouyanni, & Michelozzi, 2013) and to increase allergens (Pascal, Wagner, Alari, Corso, & Le Tertre, 2021). For instance, longer warm seasons can mean longer pollen seasons – which can subsequently increase allergic and asthma episodes (D'Amato et al., 2015) and diminish productive work and school days.

The prevalence of allergies is related to the presence of allergens and pollen in the air, and their seasonal variations cause an increase in symptoms and the consumption of antihistamines. The impact of this situation is enormous, given that 300 million people worldwide suffer from asthma and 80 million Europeans suffer from various respiratory allergies (Nunes, Pereira, & Morais-Almeida, 2017). People with existing pollen allergies may have increased risk for acute respiratory effects, leading to illnesses and death, health changes resulting from atmospheric pollution, namely cardiovascular and respiratory, and even effects on mental health. When sensitive individuals are simultaneously exposed to allergens and air pollutants, allergic reactions often become more severe. The increase in air pollutants makes the effects of increased allergens associated with climate change even worse (Reinmuth-Selzle et al., 2017).

Climate change can also lead to heavier downpours and floods, with flood waters often containing a variety of contaminants. In some cases, floods can overwhelm a region's drainage or wastewater treatment systems, increasing the risk of exposure to bacteria, parasites, and other contaminants. The projected shifts in rainfall patterns are expected to cause heavy downpours that may increase exposure to diseases disseminated through drinking and recreational water. Floodwaters can contain disease-causing bacteria, as well as parasites and viruses, which can cause outbreaks of cholera and dysentery, among other water-borne diseases (Watson, Gayer, & Connolly, 2007). In addition, leachates resulting from downpours can be contaminated with other harmful pollutants including agricultural waste, chemicals, and raw sewage (Blum & Hotez, 2018). Flooded materials in homes, schools, and businesses can cause mould to grow and be inhaled, contributing to respiratory problems (Curriero, Patz, Rose, & Lele, 2001). Moreover, if initially, flooding after a heavy rainfall event may impact present mosquito breeding locations in a destructive way, after the flooding recedes the number of stagnant water ponds rises, promptly intensifying the mosquito's breeding capability. Approximately two weeks after the event, new adult mosquitoes arise and their abundance in the affected areas increases (Chowell, Mizumoto, Banda, Poccia, & Perrings, 2019).

Ground-level ozone (a key component of smog, a mix of smoke and fog that reduces visibility and increases air pollution) will tend to increase and is associated with many health problems, including diminished lung function, more frequent hospital admissions and emergency department visits for asthma, and increases in premature deaths (Kurt, Zhang, & Pinkerton, 2016).

Radiant heat and smoke inhalation due to wildfires can cause serious injury, illness or death to people directly exposed. Therefore, climate change and the consequent increase in wildfires can lead to an increase in the probability for acute (or sudden onset) breathing problems, respiratory and cardiovascular hospitalizations, and medical visits for lung illnesses, exacerbated for people who suffer from asthma, bronchitis, and other respiratory problems (Johnston, 2009).

The effects of climate change in human health are also linked to alterations in the distribution of plants and animals around the globe, since weather and environmental conditions influence the existence of suitable habitats for different species. As such, changes in weather parameters, namely temperature and rainfall, will modify the geographic extent of ecosystems and the life cycle of pathogens. This, in turn, influences the distribution of vectors, such as insects, snails or other cold-blooded animals (WHO, 2003), thereby altering the distribution of infectious diseases, which can reach regions and countries where they were absent before.

Climate change also impacts the quality and safety of both the availability and quality of natural resources, influencing the transmission of diseases related to water and food quality (Semenza et al., 2012). As the Earth's temperature rises, surface water temperatures in lakes and oceans also rise.

Warmer waters create a more hospitable environment for some harmful algae and other microbes to grow. Infections are occurring farther north, and warming waters may increase this particular risk (Cissé, 2019).

Certain marine bacteria, such as *Vibrio* species that make humans sick, are more likely to survive and grow as oceans get warmer. These can cause a range of human infections, including gastroenteritis, wound infections, septicaemia, and cholera. These bacteria are found in brackish marine waters and cases of infections are influenced by sea surface salinity, sea surface temperature, and chlorophyll concentrations (Deeb, Tufford, Scott, Moore, & Dow, 2018), conditions that are being modified by climate change.

4. Climate change and vector-borne diseases

Vector-borne diseases result from infections transmitted to humans and other animals by blood-feeding arthropods, such as ticks and mosquitoes. Ticks spend most of their lives in an external environment and thus are expected to be vulnerable to changes in climate. Many studies predict an increase in the risk of ticks and tick-borne diseases because, on the one hand, the prevalence, survival and diversity of ticks is expected to increase and, on the other hand, animal hosts and humans are expected to be in contact with ticks for a longer season, increasing the possibility of infection (Bouchard et al., 2019). For instance, in the USA in 2017, there were several cases of tick-borne diseases reported to the CDC (Center of Disease Control) and the number of reported cases of Lyme disease has tripled in the country since the late 1990s (Lyme and Other Tickb, 2019).

Climate change powers the spread and risk of numerous infectious diseases (Smith et al., 2014), mostly due to the geographic expansion of the environmental conditions that increase an area's suitability to vectors. Factors like temperature, humidity and precipitation define the suitability for mosquito-borne infectious diseases transmission (Rocklöv & Tozan, 2019), and these are becoming more favorable to the establishment of mosquitoes due to climate change, including in regions that used to be unsuitable, such as the British Isles and Central-Eastern Europe (Oliveira, Rocha, Sousa, & Capinha, 2021). Emerging infectious diseases are quickly increasing in frequency and/or geographic range (Mayer, Tesh, & Vasilakis, 2017). Epidemics of chikungunya, dengue, and Zika are sweeping through the world, and are part of a global public health crisis that places an estimated 3.9 billion people in 120 countries at risk (Brady et al., 2012). Chikungunya virus emerged in 2013, causing 1.8 million suspected cases from 44 countries and territories (Mordecai et al., 2017). Malaria and Dengue are endemic in several zones of the world and still contribute considerably to human morbidity and mortality, being children the most vulnerable.

Climate suitability for transmission is rising for most pathogens studied. For *Aedes aegypti* and *Aedes albopictus*, both vectors of dengue, the second highest daily vectorial capacity (number of potentially infectious bites that would arise from an infective event when all female mosquitoes biting a person/animal would become infected), was recorded in 2017, with the 2012−17 average being 7.2% and 9.8% above baseline (i.e., reference values for 1950−54) for *A. aegypti* and *A. albopictus*, respectively (Rocklöv & Tozan, 2019). This change emphasizes the continued upward trend of climate suitability for transmission of dengue, with 9 of the 10 most suitable years occurring since the year 2000 (Watts et al., 2019). Malaria and dengue stand up with regard to mortality, more than 400,000 deaths/year (World Malaria Report, 2020). In fact, it is the influence of climate change on the distribution and concentration of mosquitoes, such as *Anopheles*, *Aedes Albopictus* and *Aegypti*, among others, which greatly influences the distribution of infectious diseases (Environmental Risk, 2013).

Malaria suitability continues to increase in highland areas of Africa, with the 2012−17 average 29.9% above baseline. The percentage of coastal areas suitable for *Vibrio* infections from 2010 has increased at northern latitudes (40−70°N) by 3.8%, compared with the 1980s baseline, with 2018 the second most suitable year on record (5% above the baseline).

Dengue virus distribution and intensity has increased over the last three decades, infecting an estimated 390 million people (96 million clinical) per year (Bhatt et al., 2013) and the costs of treatment in endemic regions amounts to 1,9 mil billions EUR/year (Stahl et al., 2013). In the last years, Zika virus has also spread, causing 764,414 cases. The growing burden of these diseases (including links between Zika infection and both microcephaly and Guillain-Barreâ syndrome) (Rasmussen, Jamieson, & Honein, 2016), and their potential for spread into new areas, creates an urgent need for climate change models that can inform vulnerability and risk assessment, and guide interventions such as mosquito control, community outreach, and education.

Predicting the emerging of major vector-borne diseases, such as Chikungunya, Dengue and Zika, requires understanding the ecology of the vector species. These diseases are transmitted to humans by two mosquito species, *A. aegypti* and *A. albopictus*. *A. aegypti* is considered the most important vector, while *A. albopictus* is generally believed to be a less competent vector resulting in milder epidemics (Rezza, 2012). Each species displays a specific ecology, behavior and geographical distribution. *A. aegypti* prefers urban habitats, whereas *A. albopictus* is primarily a forest species that has become adapted to rural, suburban and urban human environments, being regarded as an important secondary vector (Messina et al., 2016). Historically, the *A. aegypti* has been considered the main dengue vector, because it has more vector competence than *A. albopictus,* and for being more common in urban areas, being closely affiliated with humans. *A. albopictus* has been progressively invading European countries since the 1990s, after an initial introduction in Albania (1979) and then in Italy (1990). *A. albopictus* is classified as one of the most dangerous species in the Global Invasive Species Program.

Presently, *A. albopictus* has been detected in over 20 European countries (ECDC, 2020), some of which with densely populated areas, and it has actually played a part in the transmission of the Chikungunya virus outbreaks in Italy (Rezza et al., 2007) and France (Grandadam et al., 2011). Recent studies based on climate change scenarios indicate that, by 2050, the suitability to *A. albopictus* will tend to increase and expand throughout the European territory (Oliveira et al., 2021), as shown in Fig. 8.3.

5. Climate change, forests and wildfires

Forests cover around 30% of the Earth's surface and harbor 80% of the world's terrestrial biodiversity (Bonan, 2008; FAO & UNEP, 2020), resources that humans use for food, medicine, shelter and clothing. Forests are vital for the stability of the Earth's atmosphere and climate; they are essential sinks of greenhouse gases, contributing to mitigate the effects of climate change. On the other hand, forest loss and deforestation due to logging, conversion to other land uses or wildfires, cause the release of the carbon dioxide that had been stored in the trees and the soil, which can aggravate climate change. Latest estimates indicate that activities related to forestry, agriculture and other land uses have a net contribution of about 23% of all anthropogenic emissions of greenhouse gases (Shukla et al., 2019). Between 2001 and 2015, about 27% of the existing forest was converted to other land use in a long-term or permanent way, such as for large-scale agriculture and palm oil plantations, for energy infrastructures or mining (Curtis, Slay, Harris, Tyukavina, & Hansen, 2018). In this case, forests become a direct contributor of carbon dioxide to the

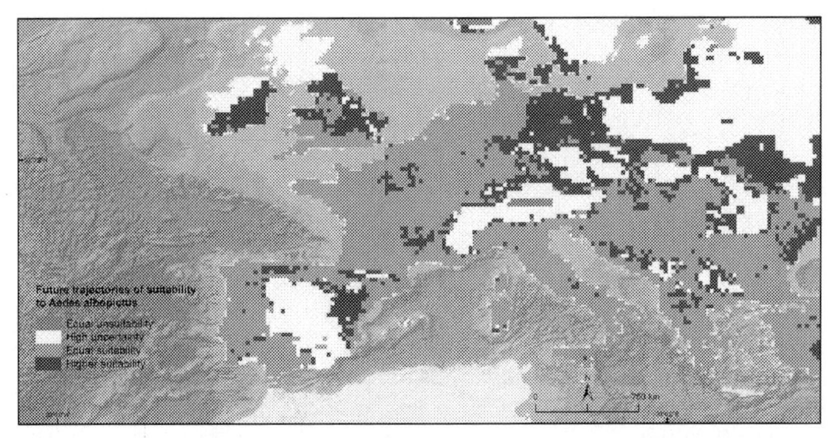

FIG. 8.3 Trajectories of suitability to *A. Albopictus* in Europe considering future climate scenarios. *Adapted from Oliveira, S., Rocha, J., Sousa, C.A., Capinha, C. (2021). Wide and increasing suitability for Aedes albopictus in Europe is congruent across distribution models.* Science Reports, 11*(1), 9916. Available from https://doi.org/10.1038/s41598-021-89096-5.*

atmosphere and to the intensification of global warming. In other cases, forest loss was only temporary, and forests have grown back after the occurrence of wildfires, or due to small-scale agriculture that is later abandoned, helping to capture carbon from the atmosphere.

The abandonment of farming activities, driven by sociodemographic changes in the last decades, are one of the causes for the substantial accumulation of vegetation (fuel loads) that feed large fires (Moreira et al., 2020). When high fuel loads are combined with extreme weather conditions, characterized by increasing temperatures, extensive drought conditions and strong winds, wildfires become more destructive and difficult to suppress, causing massive social, economic and environmental impacts (Bowman et al., 2017; Ruffault, Curt, Martin-Stpaul, Moron, & Trigo, 2018). Indeed, climate changes could influence fire activity in different ways regarding the vegetation productivity (rate of biomass generation) of the affected areas (Fig. 8.4). Fire systems in high-productivity regions are drought-driven, i.e., vulnerable to warming, whereas in low-productivity regions they are fuel-driven, i.e., respond mostly to fuel or land cover changes (Pausas & Ribeiro, 2013). The expected rise in temperatures and in droughts intensity and duration due to climate change, will increase the fire-proneness of vegetated areas around the world. These areas will become easier to ignite and burn, changing the rate of spread and intensity of the fires, which can surpass the existing firefighting abilities. Climate change also modifies the spatial distribution of vegetation types, and territories that are not fire-prone today can become so. The fire season (the time of the year when wildfires occur more often), will likely be longer, as warm and/or dry weather conditions extend beyond the usual dry or

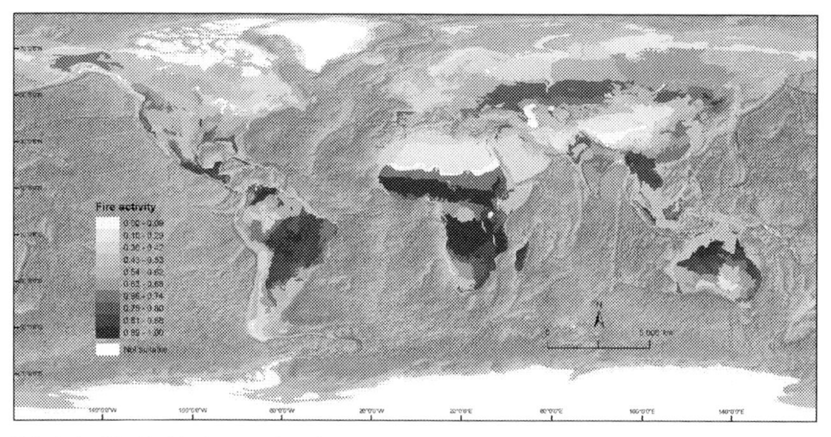

FIG. 8.4 Worldwide fire map based on ecological features. *Adapted from Pausas, J.G., Ribeiro, E. (June 1, 2013). The global fire–productivity relationship.* Global Ecology and Biogeography; *22(6), 728–736. Available from https://doi.org/10.1111/geb.12043.*

summer season (Moreira et al., 2020; Prichard, Stevens-Rumann, & Hessburg, 2017; Turco et al., 2014, 2018). Climate change will increase the number of days with weather conditions favorable to the ignition and propagation of a wildfire, and more extreme conditions (beyond average values) are expected. If these weather conditions are felt in fire-prone areas, where burnable vegetation predominates and topography is irregular, wildfires will become larger and more difficult to control.

Regarding human health, and besides the immediate loss of human lives and the negative effects on biodiversity, wildfires also release smoke and different pollutants that can spread over vast geographical areas and substantially decrease air quality. Particulate matter (fine and coarse), carbon monoxide and ozone, the latter resulting from a photochemical reaction between nitrogen oxides and volatile organic compounds under sunlight, are the most hazardous contaminants spread by wildfire smoke that affect human health (Xu et al., 2020). These particles can trigger or aggravate respiratory illnesses, such as asthma, chronic obstructive pulmonary disease (COPD) and reduced lung function, and recent estimates indicate an average of 339,000 resulting deaths per year (Johnston et al., 2012; Reisen, Duran, Flannigan, Elliott, & Rideout, 2015). Despite the uncertainties in calculating the effects of wildfire smoke in human health, due to the different particles considered, the diverse conditions of the affected territories and the rather unpredictable nature of wildfires, it is recognized that wildfires cause a serious health burden (Reid et al., 2016), exacerbated in extreme fire seasons as the one occurred in southeast Australia in 2019/2020 (Johnston et al., 2021). It was estimated that the wildfire smoke released during this severe season in Australia, was responsible for 417 excess deaths and more than 3000 hospitalizations for cardiovascular and respiratory problems (BorchersArriagada et al., 2020).

In a context of climatic changes and the ensuing increase in weather conditions favorable to wildfire occurrence, the negative impacts of wildfire emissions in human health are expected to worsen (Burke et al., 2021; Xu et al., 2020). Bearing in mind that wildfires release greenhouse gases to the atmosphere, which in turn aggravate climate change, this harmful cycle will be difficult to halt. Wildfire risk mitigation will require the implementation of strategies and interventions based on spatial planning and forest management, that enable quick and efficient firefighting actions and safety measures, such as fuel management, the creation of mosaics of different vegetation and discontinuities in forested areas, compatible with their economic and environmental uses, and the construction of shelters in human settlements to protect people.

In the last decades, the incidence of large and uncontrolled fires has risen. Recent disasters in Portugal in 2017, Greece in 2018 and Australia in 2019 (Gómez-González, Ojeda, & Fernandes, 2017; Nolan et al., 2020; Tedim et al., 2020; Turco et al., 2019), are examples of unprecedented events that have caused massive damages and hundreds of fatalities, giving insights on the negative impacts brought by climate change, even in countries where wildfires are frequent events. In 2018 in Sweden, where wildfires are less common, unusual low rainfall occurred in spring and summer, which is pointed out as a main factor for the occurrence of large fires in the country, that required international assistance (San-Miguel-Ayanz et al., 2019). Regarding wildfires and human health, other threats can also emerge. The large fires in Russia in 2010, the most extreme since 1972 and fueled by high temperatures and severe drought conditions (Schmuck et al., 2010, p. 2011), have raised concerns with regard to the possible resuspension of radionuclides in the atmosphere. These radionuclides have been stored in the soil and plants over extensive areas since the Chernobyl accident in 1986 and would be released if the wildfires reached these contaminated areas (Ager et al., 2019). This, and all other challenges, will have to be included in the forest, wildfire and human health strategies that need to be adopted to tackle climate change impacts throughout the world.

6. Climate change and food security

The "State of food security and nutrition in the world" FAO (Food and Agriculture Organization) report, revealed that in 2018 about 9.2% of the world population may have suffered from hunger and about 17.2% experienced food insecurity at moderate levels (FAO, 2020). Both figures reveal that about 26% of the world population have been exposed to food insecurity at some level, due to the reduction in the amount of food consumed and the lack of regular access to sufficient quantity and nutritious food (FAO, 2020). There are, therefore, different scales of food insecurity worldwide, from poor quality food consumption to large-scale levels of hunger (FAO, 2021; GroupEnding Poverty, 2015). The situation may be chronic or transient and access may be limited within a specific period, due to lack of financial and other resources.

The concept of food security is based on four fundamental principles: availability, accessibility, use/utilization, and stability (Ericksen, 2008). Therefore, when it comes to food security, the question is not exclusively how much will be produced, but who will have access, how it will be used and with what stability that same production will be achieved. Those concerns reach another dimension, with the seemingly consensual acknowledgment that agri-food sector became the economic activity most affected by climate change (Abd-Elmabod et al., 2020; Porter et al., 2014; Schmidhuber & Tubiello, 2007).

Indeed, it is recognized that climate change, including rising temperatures, precipitation patterns changes and extreme events (particularly floods and droughts), increase the risk of food insecurity via different routes and at various stages in the agri-food chain (e.g., from primary production to consumption) (Ahmed, Wang, You, & Yu, 2016; Basso, Hyndman, Kendall, Grace, & Robertson, 2015; Kukal & Irmak, 2018; Leng & Hall, 2019). Furthermore, it is estimated that the most relevant impacts of climate change will occur in low latitude countries, with tropical climates. The most affected would be the populations in the poorest areas of sub-Saharan Africa, and South and Southeast Asia, especially those whose subsistence depends on agriculture and where the largest number of chronically malnourished people is currently occurring (FAO, 2020; Nations World Urbaniza, 2019).

Effectively, hunger, poverty and climate change are internally linked issues that go hand in hand, being a fundamental priority preserving food security and eradicating hunger, together with reducing the vulnerability inherent to food production methods and the effects of climate change (FAO, 2020).

According to the "State of Agricultural Commodity Market" FAO report, climate change is evidently linked with food security and agriculture, as not only is responsible for major changes in food production systems but also in their geographies, bringing new challenges to food security (FAO, 2020, p. 164). For instance, rising temperature influences the phenology (the timing of life cycle events and their seasonal variations), which reduces the cycle length of croplands and, therefore, may result in reduced agricultural productivity, affecting the local economies and compromising food security in certain regions of the globe (FAO, 2020, p. 164). For example, dairy cow related-production will likely require the adaptation of the grassland species that feed them and the introduction of measures to safeguard the animals wellbeing in face of challenging climatic conditions such as extreme heat (Gauly et al., 2013). In addition, expected changes in temperature and precipitation will have an effect on CO_2 levels, hindering the photosynthetic capacity of croplands and, therefore, influencing agricultural productivity (CCSP, 2008; USGCR, 2014). For example, the United Nations Environment Program (2006 Annual Report, 2006) climate projections indicate that the effect of climate change on cereal production by 2080 can be considerably high in some regions (e.g., in some African and South America countries the total losses could reach 50%) (Fig. 8.5).

However, we cannot fail to mention the duality linked to the main reasons behind food insecurity and climate change. On the one hand, under current practices, the agri-food system is responsible for approximately a third of global greenhouse gas emissions (Crippa et al., 2021), being urgent to reduce the global emissions; on the other hand, the agri-food system will have to increase agricultural production by 70% to provide the necessary amount of food for a growing population, projected to reach about nine billion by 2050 (FAO, 2017), and in a context of finite resources. This is a double challenge faced by the agri-food sector, since a reduction in food system greenhouse gas emissions will only be possible with a change in the sector's paradigm and long-established practices (EEA, 2017; Foley et al., 2011; Nationsorld Urbaniza, 2019; van de Kamp et al., 2018).

Indeed, it will be necessary to invest in a sustainable strategy for the current agri-food production, with practical effects not only on the quality of the food consumed, but also on the food we intend to consume in the future

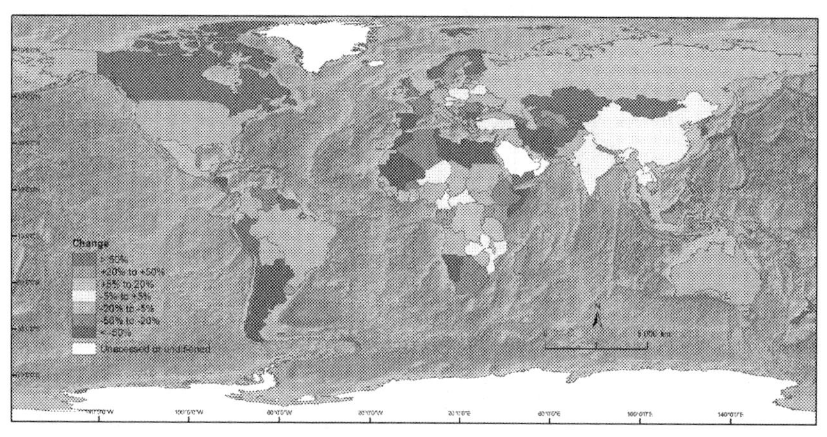

FIG. 8.5 Projected percentage gains and losses in rainfed cereal production potential by 2080. *Adapted from UNEP (2006) annual report. Available from: https://www.unep.org/resources/annual-report/unep-2006-annual-report*

(Lampridi, Sørensen, & Bochtis, 2019; Michel-Villarreal, Hingley, Canavari, & Bregoli, 2019). It is urgent to take measures to make agriculture more sustainable, productive and resilient (EEA, 2017); otherwise, the impacts of climate change will seriously jeopardize food production, particularly in countries and regions that already face high food insecurity (GroupEnding Poverty, 2015; Nationsransforming o, 2015). For instance, a strategy for adaptation and mitigation of climate change in agriculture regards the improvement of water management, by adopting large-scale use of irrigation which protects agriculture from fluctuations in rainfall (Dickie et al., 2014). In addition, adapted crops, which require less herbicides or fertilizers, are more resistant to extreme climatic events such as floods and droughts (Howden et al., 2007).

The consequences of climate change on food security go beyond the access to sufficient, safe and nutritious food, and include also food safety and the subsequent adverse effects on human health. Moreover, a structural change in the patterns of production and consumption, and a strong commitment to sustainable practices, will result in the global achievement of the second objective of the Sustainable Development Goals (SDGs), that seeks the end of hunger, to ensure food security and the improvement of nutrition, and the promotion of sustainable agriculture (United Nations, 2015). Thinking about our agricultural systems today requires more than ever a global and integrated vision, capable of understanding our food system from a broader perspective (Allen & Prosperi, 2016; Müller et al., 2020; Yu et al., 2012). The agricultural activity may represent a heavy influence on the (un)sustainability of the planet and restructuring the agri-food system will be a complex process, due to the wide range of stakeholders involved, the multiplicity of agricultural and food production systems, and the differences in global ecosystems (GroupEnding Poverty, 2015; Society, 2009). It is necessary to change the paradigm to a fully sustainable agricultural model that protects natural resources, that generates equitable socioeconomic development and that allows adaptation to climate change and the mitigation of its effects. Designing local, regional and national adaptation plans to climate change and food security are fundamental pillars for this process (Müller et al., 2020).

According to FAO, a change of paradigm is required for sustainable, inclusive and resilient food systems, based on productive techniques less dependent on the intensive use of chemicals and natural resources (FAO, 2017). The success of agricultural transformation depends largely on the possibility of smallholders in adapting to climate change. Some alternatives are described as economically viable ways to help farmers adapt and put in practice a more resilient agriculture, such as the adoption of smart practices (e.g., the use of nitrogen-efficient and tolerant crops) (FAO, 2017; Li et al., 2020). It is estimated that this practice would allow, for example, to reduce the number of people at risk of malnutrition by more than 100 million (FAO, 2017). Moreover, poor or low-quality diets that do not follow proper nutritional rules, cause micronutrient deficiencies and increase the risk of diseases such as coronary heart disease, stroke, and diabetes (Willett et al., 2019), causing high morbidity particularly in developed countries, where there are also high amounts of wasted food that need to be mitigated to ensure the food system's sustainability. Overall, to transform this paradigm, it is essential to promote local, diverse, and sustainable agriculture that respects the territory, and understand international trade as a complement to local production. The local and national systems need to be strengthened to adapt to the climate crisis and to prepare to provide diverse diets for the world's population. The diversity of diets can help farmers to diversify their risks, to open markets for different food crops, break their dependence on commercial crops and increase biodiversity and resilience (SocietyReaping the b, 2009). The food system can be improved by producing sustainable, nutritious food, reducing food loss and waste, innovating food distribution systems and markets and improving diet and nutrition (Foley et al., 2011; FAO, 2017; GroupEnding Poverty, 2015).

7. One Health and climate change

As described in the previous sections, climate change has an impact on multiple systems, which would be better addressed through the integrated and intersectoral approach of One Health. The impacts of climate change are wide, from modifications in vectors, reservoirs, and pathogens lifecycles, to impacts on animal and plant diseases, disruption of ecosystems, alteration and destruction of habitats, and trophic cascades (i.e., side-effects of reducing a trophic level from an ecosystem, leading to an increase in other species) (Patz & Hahn, 2013). Aspects related to food and water, complex and closely interlinked to climate, would benefit the most from a One Health approach. For instance, the dominating practice of intensive production of ruminant livestock fed on cereal food resources contributes to climate change through methane production (Zinsstag et al., 2018). Conversely, sustainable livestock farming in semi-arid and highland areas, currently practiced by communities as subsistence farming, is highly susceptible to the increasing intensities of droughts and heatwaves potentiated by climate change, leading to the loss of animals and food scarcity. Therefore, besides climate scientists, ecologists, and animal and human health professionals, climate change must be addressed in the socioeconomic sector, also involving sociologists and politicians. Engineers will also take part in solutions, especially when considering the sensitivity of water and sanitation systems to extreme events (e.g., hurricanes, flooding). For instance, drinking water safety may also be compromised after high precipitation events if these wash off human and animal feces. Finally, a better understanding of the disease emergence and geographical distribution in a changing climate would be achieved through improved public health surveillance, accounting for human and animal health and the dynamic nature of ecosystems, integrating rapid communications through technology. Thus, the necessary transdisciplinary approach to climate change at a global level is a One Health approach.

8. Conclusions

Climate change is recognized as one of the major problems, affecting society, the environment and the economy, today and in the future. Coping with the challenges brought by climate change will require the involvement of every sector and individual (Biermann & Boas, 2010).

Human action has caused great damage to the environment, through its predatory conduct and disregard for natural resources conditions, and the impacts are being perceived differently in the last years, when climate change effects have become stronger in the most diverse segments of society, although some suffering more intense impacts than others. It is undeniable that climate change is already having important consequences on human health, but the extent of these impacts is still uncertain and requires further research.

Several authors argue that the number of extreme events resulting from climatic changes will continue to increase (Goodess, 2013), likely transforming some places on Earth intolerable to humans (Belasen et al., 2013). Besides the increase in extreme phenomena, the negative impacts on human lives and assets have also become more serious. This is due not only to the increasing severity of natural disasters and their impacts, but also to the increased exposure of populations to such events (Oliver-Smith, 2012) and the failure of general planning and emergency systems to respond to these issues (Black, Arnell, Adger, Thomas, & Geddes, 2013).

It is also notorious that climate change will foster a higher incidence of some diseases, related to changes in global weather parameters, such as temperature, and in environmental conditions. At the same time, food production systems and the way man interacts with the environment through his work, will also need to be adapted to the new climatic conditions. Forest fires cause economic and social losses and have adverse effects on the environment, due to the large amounts of CO_2 they release and the negative consequences on biodiversity. Even grassland and savannah fires, mostly occurring for land clearing for agricultural purposes, and prescribed burning applied as a mechanism to reduce fuel accumulation and prevent larger and uncontrollable fires, may cause negative impacts regarding the release of GHG and pollutants harmful to human health.

The effects of climate change on human health and the possibilities for adaptation call for further studies because it is very difficult to accurately predict, today, both the precise effects of climatic changes in the society and the environment, and the human capacity to adapt to these changes. However, this context is not a justification for avoiding taking suitable measures and seek possible alternatives, because climate change impacts are already occurring, and it is crucial to minimize the effects on humanity, bearing in mind a precautionary mode for the future. There is no doubt that climate change is happening and if we fail to act, the consequences will be more devastating. Actions to mitigate climate change and adapt to new conditions are, in fact, beneficial in many ways and for different sectors, fostering the sustainable use of natural resources, the maintenance of ecosystems balance and the improvement of living conditions in human communities. And this belief may help us promoting global action to face the ongoing and urgent climatic crisis.

Funding

Ana Isabel Ribeiro was supported by National Funds through FCT, under the program of 'Stimulus of Scientific Employment – Individual Support' within the contract CEECIND/02386/2018. Ana Isabel Ribeiro was also funded by FEDER through the Operational Program Competitiveness and Internationalization and national funding from the Foundation for Science and Technology – FCT (Portuguese Ministry of Science, Technology and Higher Education) under the Unidade de Investigação em Epidemiologia - Instituto de Saúde Pública da Universidade do Porto (EPIUnit) (UIDB/04750/2020). Sandra Oliveira and Jorge Rocha were financed by national funds through FCT—Portuguese Foundation for Science and Technology, I.P., under the framework of the project "TRIAD - health Risk and social vulnerability to Arboviral Diseases in mainland Portugal" [PTDC/GES –OUT/30210/2017] and by the Research Unit UIDB/00295/2020 and UIDP/00295/2020. Cláudia M. Viana is financed by the Portuguese Foundation for Science and Technology (FCT) [grant number SFRH/BD/115497/2016 to Cláudia M. Viana] and by the Center for Geographical Studies—Universidade de Lisboa and FCT [grant number UID/GEO/00295/2019].

References

Abd-Elmabod, S. K., Muñoz-Rojas, M., Jordán, A., Anaya-Romero, M., Phillips, J. D., Laurence, J., … de la Rosa. (2020). Climate change impacts on agricultural suitability and yield reduction in a Mediterranean region. *Geoderma, 374*, 114453. https://doi.org/10.1016/j.geoderma.2020.114453

Ager, A. A., Lasko, R., Myroniuk, V., Zibtsev, S., Day, M. A., Usenia, U., … Evers. (2019). The wildfire problem in areas contaminated by the Chernobyl disaster. *The Science of the Total Environment, 696*(August), 133954. https://doi.org/10.1016/j.scitotenv.2019.133954

Ahmed, K. F., Wang, G., You, L., & Yu, M. (February 2016). Potential impact of climate and socioeconomic changes on future agricultural land use in West Africa. *Earth System Dynamics, 7*(1), 151–165. Available from https://esd.copernicus.org/articles/7/151/2016/.

Alcoforado, M. J., & Andrade, H. (2008). Global warming and the urban heat island. In *Urban ecology* (pp. 249–262). Springer.

Allen, M. R., Dube, O. P., Solecki, W., Aragón-Durand, F., Cramer, W., Humphreys, S., … Kala, J. (2018). Framing and context. In V. Masson-Delmotte, P. Zhai, H.-O. Pörtner, D. Roberts, J. Skea, P. R. Shukla, et al.T.W. (Eds.), *Global Warming of 15°C an IPCC Special Report on the impacts of global warming of 15°C above pre-industrial levels and related global greenhouse gas emission pathways, in the context of strengthening the global response to the threat of climate change*. IPCC. Available from https://www.ipcc.ch/sr15/chapter/chapter-1/.

Allen, T., & Prosperi, P. (May 1, 2016). Modeling sustainable food systems. *Environmental Management, 57*(5), 956–975.

Anderson, K., & Bows, A. (January 13, 2011). Beyond 'dangerous' climate change: Emission scenarios for a new world. *Philosophical Transactions of The Royal Society A Mathematical Physical and Engineering Sciences, 369*(1934), 20–44. https://doi.org/10.1098/rsta.2010.0290. Available from.

Basso, B., Hyndman, D. W., Kendall, A. D., Grace, P. R., & Robertson, G. P. (June 2015). Can impacts of climate change and agricultural adaptation strategies Be accurately quantified if crop models are annually Re-initialized?P. Spanoghe (Ed.). *PLoS One, 10*(6). e0127333–e0127333. Available from https://dx.plos.org/10.1371/journal.pone.0127333.

Belasen, A. R., & Polachek, S. W. (2013). Natural disasters and migration. In A. F. Constant, & K. F. Zimmermann (Eds.), *International handbook on the economics of migration. Cheltenham, UK.: Edward Elgar Publishing, Inc* (pp. 309–330).

Bhatt, S., Gething, P. W., Brady, O. J., Messina, J. P., Farlow, A. W., Moyes, C. L., … Hay, S. I. (2013). The global distribution and burden of dengue. *Nature, 496*(7446), 504–507. https://doi.org/10.1038/nature12060. Available from.

Biermann, F., & Boas, I. (February 1, 2010). Preparing for a warmer world: Towards a global governance system to protect climate refugees. *Global Environmental Politics, 10*(1), 60–88. https://doi.org/10.1162/glep.2010.10.1.60. Available from.

Black, R., Arnell, N. W., Adger, W. N., Thomas, D., & Geddes, A. (2013). Migration, immobility and displacement outcomes following extreme events. *Environmental Science & Policy, 27*, S32–S43. Available from https://www.sciencedirect.com/science/article/pii/S1462901112001475.

Blum, A. J., & Hotez, P. J. (July 2018). Global "worming": Climate change and its projected general impact on human helminth infections. *PLoS Neglected Tropical Diseases, 12*(7), e0006370.

Bonan, G. B. (2008). Forests and climate change: Forcings, feedbacks, and the climate benefits of forests. *Science, 320*(5882), 1444–1449.

Borchers Arriagada, N., Palmer, A. J., Bowman, D. M. J. S., Morgan, G. G., Jalaludin, B. B., & Johnston, F. H. (2020). Unprecedented smoke-related health burden associated with the 2019–20 bushfires in eastern Australia. *Medical Journal of Australia, 213*(6), 282–283.

Bose, S. (December 1, 2013). Sea-level rise and population displacement in Bangladesh: Impact on India. *Maritime Affairs: Journal of the National Maritime Foundation of India, 9*(2), 62–81. https://doi.org/10.1080/09733159.2013.848616. Available from.

Bouchard, C., Dibernardo, A., Koffi, J., Wood, H., Leighton, P. A., & Lindsay, L. R. N. (April 2019). Increased risk of tick-borne diseases with climate and environmental changes. *Canada Communicable Disease Report, 45*(4), 83–89.

Bowman, D. M. J. S., Kolden, C. A., Abatzoglou, J. T., Johnston, F. H., van der Werf, G. R., & Flannigan, M. (2020). Vegetation fires in the Anthropocene. *Nature Reviews Earth & Environment, 1*(10), 500–515. https://doi.org/10.1038/s43017-020-0085-3. Available from.

Bowman, D. M. J. S., Williamson, G. J., Abatzoglou, J. T., Kolden, C. A., Cochrane, M. A., & Smith, A. M. S. (2017). Human exposure and sensitivity to globally extreme wildfire events. *Nature Ecology & Evolution, 1*(3), 1–6.

Brady, O. J., Gething, P. W., Bhatt, S., Messina, J. P., Brownstein, J. S., Hoen, A. G., … Hay, S. I. (2012). Refining the global spatial limits of dengue virus transmission by evidence-based consensus. *PLOS Neglected Tropical Diseases, 6*(8), 1–15. https://doi.org/10.1371/journal.pntd.0001760. Available from.

Burke, M., Driscoll, A., Heft-Neal, S., Xue, J., Burney, J., & Wara, M. (2021). The changing risk and burden of wildfire in the United States. *Proceedings of the National Academy of Sciences of the United States of America, 118*(2), 1–6.

CCSP. (2008). *The effects of climate change on agriculture, land resources, water resources, and biodiversity in the United States. A report by the U.S. Climate change science program and the subcommittee on global change research. Washington, DC, USA*. Available from https://www.globalchange.gov/browse/reports/sap-43-effects-climate-change-agriculture-land-resources-water-resources-and.

CDC. Climate effects on health.(2016). Available from https://www.cdc.gov/climateandhealth/effects/default.htm.

CDC. Lyme and other tickborne diseases increasing. *CDC Newsroom*, (2019).

Chowell, G., Mizumoto, K., Banda, J. M., Poccia, S., & Perrings, C. (June 2019). Assessing the potential impact of vector-borne disease transmission following heavy rainfall events: A mathematical framework. *Philosophical Transactions of the Royal Society of London B Biological Sciences, 374*(1775), 20180272.

Cissé, G. (2019). Food-borne and water-borne diseases under climate change in low- and middle-income countries: Further efforts needed for reducing environmental health exposure risks. *Acta Tropica, 194*, 181–188. Available from https://www.sciencedirect.com/science/article/pii/S0001706X18309525.

Crippa, M., Solazzo, E., Guizzardi, D., Monforti-Ferrario, F., Tubiello, F. N., & Leip, A. (March 2021). Food systems are responsible for a third of global anthropogenic GHG emissions. *Nature Food, 2*(3), 198–209. Available from https://www.nature.com/articles/s43016-021-00225-9.

Curriero, F. C., Patz, J. A., Rose, J. B., & Lele, S. (August 2001). The association between extreme precipitation and waterborne disease outbreaks in the United States, 1948-1994. *American Journal of Public Health, 91*(8), 1194–1199. Available from https://pubmed.ncbi.nlm.nih.gov/11499103.

Curtis, P. G., Slay, C. M., Harris, N. L., Tyukavina, A., & Hansen, M. C. (2018). Classifying drivers of global forest loss. *Science, 361*(6407), 1108–1111.

D'Amato, G., Holgate, S. T., Pawankar, R., Ledford, D. K., Cecchi, L., Al-Ahmad, M., … Annesi-Maesano, I. (2015). Meteorological conditions, climate change, new emerging factors, and asthma and related allergic disorders. A statement of the World Allergy Organization. *World Allergy Organization Journal, 8*(1), 1–52. https://doi.org/10.1186/s40413-015-0073-0

De Sario, M., Katsouyanni, K., & Michelozzi, P. (September 1, 2013). Climate change, extreme weather events, air pollution and respiratory health in Europe. *European Respiratory Journal, 42*(3), 826 LP–843. Available from http://erj.ersjournals.com/content/42/3/826.abstract.

Deeb, R., Tufford, D., Scott, G. I., Moore, J. G., & Dow, K. (December 2018). Impact of climate change on Vibrio vulnificus abundance and exposure risk. *Estuaries and Coasts Journal of the Coastal and Estuarine Research Federation, 41*(8), 2289–2303. Available from https://pubmed.ncbi.nlm.nih.gov/31263385.

Dickie, A., Streck, C., Roe, S., Zurek, M., Haupt, F., & Dolginow, A. (2014). *Strategies for mitigating climate change in agriculture: Recommendations for philanthropy – executive summary.* Available from www.agriculturalmitigation.org.

Doocy, S., Daniels, A., Murray, S., & Kirsch, T. D. (April 16, 2013). The human impact of floods: A historical review of events 1980-2009 and systematic literature review. *PLoS Currents, 5.* ecurrents.dis.f4deb457904936b07c09daa98ee8171a. Available from https://pubmed.ncbi.nlm.nih.gov/23857425.

Doocy, S., Dick, A., Daniels, A., & Kirsch, T. D. (April 2013). The human impact of tropical cyclones: A historical review of events 1980-2009 and systematic literature review. *PLoS Currents*, 5.

ECDC. (2020). Aedes albopictus—current known distribution: September 2020. *Mosquito Maps.* [cited 2021 Aug 9]. Available from https://www.ecdc.europa.eu/en/disease-vectors/surveillance-and-disease-data/mosquito-maps.

ECDC. Environmental risk mapping: Aedes albopictus *in Europe Stockholm.*(2013). Available from http://www.ecdc.europa.eu/en/publications/Publications/climate-change-environmental-risk-mapping-aedes.pdf.

EEA. (2017). *Food in a green light - a systems approach to sustainable food.* Luxembourg: Publications Office of the European Union. Available from https://www.eea.europa.eu/publications/food-in-a-green-light.

Ericksen, P. J. (February 2008). Conceptualizing food systems for global environmental change research. *Global Environmental Change, 18*(1), 234–245.

Fang, J., Lau, C. K. M., Lu, Z., Wu, W., & Zhu, L. (2019). Natural disasters, climate change, and their impact on inclusive wealth in G20 countries. *Environmental Science and Pollution Research International, 26*(2), 1455–1463. https://doi.org/10.1007/s11356-018-3634-2. Available from.

FAO. (2017). *The future of food and agriculture - trends and challenges. Rome.* [cited 2020 May 27]. Available from www.fao.org/publications.

FAO. (2020). *The state of agricultural commodity markets 2020. The state of agricultural commodity markets 2020.* Rome, Italy: FAO.

FAO, IFAD, UNICEF, WFP, WHO. (2020). *The state of food security and nutrition in the world 2020. Transforming food systems for affordable healthy diets.* Rome: FAO. https://doi.org/10.4060/ca9692en. Available from.

FAO UNICEF, WFP and WHO I. The state of food security and nutrition in the world 2017. Building resilience for peace and food security. (2017). Rome: FAO. Available from http://www.fao.org/3/a-I7695e.pdf.

FAO. Crop Prospects and food situation No.1, 03. Rome, Italy.(2021). Available from http://www.fao.org/3/cb3672en/cb3672en.pdf.

FAO and UNEP. (2020). *The state of the world's forests 2020.* Rome: Forests, biodiversity and people.

Foley, J. A., Ramankutty, N., Brauman, K. A., Cassidy, E. S., Gerber, J. S., Johnston, M., … Zaks, D. P. M. (2011). Solutions for a cultivated planet. *Nature, 478*(7369), 337–342. https://doi.org/10.1038/nature10452

Gamble, J. L., Balbus, J., Berger, M., Bouye, K., Campbell, V., Chief, K., … AFW. (2016). Populations of concern. The impacts of climate change on human health. In A. Crimmins, J. Balbus, J. L. Gamble, C. B. Beard, J. E. Bell, D. Dodgen, … L. Ziska (Eds.), *The impacts of climate change on human health in the United States: A scientific assessment* (pp. 247–286). Washington, DC, USA: USGCRP. Available from https://health2016.globalchange.gov/low/ClimateHealth2016_09_Populations_small.pdf.

Gauly, M., Bollwein, H., Breves, G., Brügemann, K., Dänicke, S., Daş, G., … Wrenzycki, C. (2013). Future consequences and challenges for dairy cow production systems arising from climate change in Central Europe – a review. *Animal, 7*(5), 843–859. https://doi.org/10.1017/S1751731112002352

Gómez-González, S., Ojeda, F., & Fernandes, P. M. (July 2017). Portugal and Chile: Longing for sustainable forestry while rising from the ashes. *Environmental Science & Policy, 2018*(81), 104–107.

Goodess CM. (2013). How is the frequency, location and severity of extreme events likely to change up to 2060? *Environmental Science & Policy, 27*(Suppl. 1), S4–S14.

Grandadam, M., Caro, V., Plumet, S., Thiberge, J.-M., Souarès, Y., Failloux, A.-B., … Desprès, P. (2011). Chikungunya virus, southeastern France. *Emerging Infectious Diseases Journal, 17*(5), 910. https://doi.org/10.3201/eid1705.101873

Group WB. (2015). *Ending poverty and hunger by 2030 an agenda for the global food system.* Available from http://documents1.worldbank.org/curated/en/700061468334490682/pdf/95768-REVISED-WP-PUBLIC-Box391467B-Ending-Poverty-and-Hunger-by-2030-FINAL.pdf.

Harlan, S. L., & Ruddell, D. M. (2011). Climate change and health in cities: Impacts of heat and air pollution and potential co-benefits from mitigation and adaptation. *Current Opinion in Environmental Sustainability, 3*(3), 126–134. Available from https://www.sciencedirect.com/science/article/pii/S1877343511000029.

Haustein, K., Allen, M. R., Forster, P. M., Otto, F. E. L., Mitchell, D. M., Matthews, H. D., & Frame, D. J. (2017). A real-time global warming index. *Science Reports, 7*(1), 15417. https://doi.org/10.1038/s41598-017-14828-5

Howden, S. M., Soussana, J.-F., Tubiello, F. N., Chhetri, N., Dunlop, M., & Meinke, H. (December 11, 2007). Adapting agriculture to climate change. *Proceedings of the National Academy of Sciences, India, 104*(50), 19691 LP–19696. Available from http://www.pnas.org/content/104/50/19691.abstract.

Huss, A. (February 1, 2010). Environmental epidemiology—principles and methods. RM Merrill. *International Journal of Epidemiology, 39*(1), 319–320. https://doi.org/10.1093/ije/dyp005. Available from.

IPCC. (2018). *Global warming of 1.5 °C.* Geneva: Switzerland. Available from https://www.ipcc.ch/sr15/download/.

Jacob, D. J., & Winner, D. A. (2009). Effect of climate change on air quality. *Atmospheric Environment, 43*(1), 51–63. Available from https://www.sciencedirect.com/science/article/pii/S1352231008008571.

Johnston, F. H. (September 2009). Bushfires and human health in a changing environment. *Australian Family Physician, 38*(9), 720–724.

Johnston, F. H., Borchers-Arriagada, N., Morgan, G. G., Jalaludin, B., Palmer, A. J., Williamson, G. J., & Bowman, D. M. J. S. (2021). Unprecedented health costs of smoke-related PM 2.5 from the 2019–20 Australian megafires. *Nature Sustainability, 4*(1), 42–47. https://doi.org/10.1038/s41893-020-00610-5

Johnston, F. H., Henderson, S. B., Chen, Y., Randerson, J. T., Marlier, M., DeFries, R. S., ... Brauer, M. (2012). Estimated global mortality attributable to smoke from landscape fires. *Environmental Health Perspectives, 120*(5), 695–701. https://doi.org/10.1289/ehp.1104422

Khan, M. R., Huq, S., Risha, A. N., & Alam, S. S. (June 18, 2021). High-density population and displacement in Bangladesh. *Science, 372*(6548), 1290 LP–1293. Available from http://science.sciencemag.org/content/372/6548/1290.abstract.

Kirch, W., Menne, B., & Bertollini, R. (2005). *Extreme weather events and public health responses*. Berlin Heidelberg: Springer-Verlag.

Kukal, M. S., & Irmak, S. (December 2018). Climate-driven crop yield and yield variability and climate change impacts on the U.S. Great plains agricultural production. *Science Reports, 8*(1), 1–18. Available from http://www.nature.com/scientificreports/.

Kurt, O. K., Zhang, J., & Pinkerton, K. E. (March 2016). Pulmonary health effects of air pollution. *Current Opinion in Pulmonary Medicine, 22*(2), 138–143. Available from https://pubmed.ncbi.nlm.nih.gov/26761628.

Lampridi, M. G., Sørensen, C. G., & Bochtis, D. (2019). Agricultural sustainability: A review of concepts and methods. *Sustainable Times, 11*(18).

The Lancet respiratory medicine. Breathing on a hot planet. *The Lancet Respiratory Medicine, 6*(9), (2018), 647. Available from https://www.sciencedirect.com/science/article/pii/S2213260018303382.

Lay, C. R., Sarofim, M. C., Vodonos Zilberg, A., Mills, D. M., Jones, R. W., Schwartz, J., & Kinney, P. L. (June 1, 2021). City-level vulnerability to temperature-related mortality in the USA and future projections: A geographically clustered meta-regression. *The Lancet Planetary Health, 5*(6), e338–e346. https://doi.org/10.1016/S2542-5196(21)00058-9

Legendre, M., Bartoli, J., Shmakova, L., Jeudy, S., Labadie, K., Adrait, A., ... Claverie, J. M. (March 18, 2014). Thirty-thousand-year-old distant relative of giant icosahedral DNA viruses with a pandoravirus morphology. *Proceedings of the National Academy of Sciences, India, 111*(11), 4274 LP–4279. https://doi.org/10.1073/pnas.1320670111

Leng, G., & Hall, J. (March 2019). Crop yield sensitivity of global major agricultural countries to droughts and the projected changes in the future. *The Science of the Total Environment, 654*, 811–821.

Li, T., Horton, R. M., Bader, D. A., Liu, F., Sun, Q., & Kinney, P. L. (March 2018). Long-term projections of temperature-related mortality risks for ischemic stroke, hemorrhagic stroke, and acute ischemic heart disease under changing climate in Beijing, China. *Environment International, 112*, 1–9.

Li, M., Xu, J., Gao, Z., Tian, H., Gao, Y., & Kariman, K. (December 2020). Genetically modified crops are superior in their nitrogen use efficiency-A meta-analysis of three major cereals. *Science Reports, 10*(1), 1–9. https://doi.org/10.1038/s41598-020-65684-9. Available from.

IPCC. Annex VII: Glossary. In Masson-Delmotte, V., Zhai, P., Pirani, A., Connors, S. L., Péan, C., Berger, S., et al.Yelekçi RY and BZ, O. (Eds.), *Climate change 2021: The physical science basis contribution of working group I to the sixth assessment report of the Intergovernmental Panel on climate change*, (2021). Cambridge: Cambridge University Press.

Mayer, S. V., Tesh, R. B., & Vasilakis, N. (February 2017). The emergence of arthropod-borne viral diseases: A global prospective on dengue, chikungunya and zika fevers. *Acta Tropica, 166*, 155–163.

Menne, B., & Ebi, K. L. (2006). *Climate change and adaptation strategies for human health.* Darmstadt: Steinkopff Verlag.

Merrill, R. M. (2008). Environmental epidemiology: Principles and methods. *Jones & Bartlett Learning*.

Messina, J. P., Kraemer, M. U. G., Brady, O. J., Pigott, D. M., Shearer, F. M., Weiss, D. J., et al. (April 2016). Mapping global environmental suitability for Zika virusM. Jit (Ed.). *Elife, 5*, e15272. https://doi.org/10.7554/eLife.15272. Available from.

Michel-Villarreal, R., Hingley, M., Canavari, M., & Bregoli, I. (February 2019). Sustainability in alternative food networks: A systematic literature review. *Sustainability, 11*(3), 859. Available from http://www.mdpi.com/2071-1050/11/3/859.

Mordecai, E. A., Cohen, J. M., Evans, M. V., Gudapati, P., Johnson, L. R., Lippi, C. A., … Weikel, D. P. (April 27, 2017). Detecting the impact of temperature on transmission of Zika, dengue, and chikungunya using mechanistic models. *PLOS Neglected Tropical Diseases, 11*(4), e0005568. https://doi.org/10.1371/journal.pntd.0005568

Moreira, F., Ascoli, D., Safford, H., Adams, M. A., Moreno, J. M., Pereira, J. M. C., … Fernandes, P. M. (2020). Wildfire management in Mediterranean-type regions: Paradigm change needed. *Environmental Research Letters, 15*(1), 011001. https://doi.org/10.1088/1748-9326/ab541e

NASA. Overview: Weather, global warming and climate change.(2020). [cited 2021 Aug 1]. Available from https://climate.nasa.gov/resources/global-warming-vs-climate-change.

Nations U. World urbanization Prospects: The 2018 revision (ST/ESA/SER.A/420). (2019). New York: United Nations. Available from https://population.un.org/wup/Publications/Files/WUP2018-KeyFacts.pdf.

Müller, B., Hoffmann, F., Heckelei, T., Müller, C., Hertel, T. W., Polhill, J. G., … Webber, H. (July 1, 2020). Modelling food security: Bridging the gap between the micro and the macro scale. *Global Environmental Change, 63*, 102085. https://doi.org/10.1016/j.gloenvcha.2020.102085

Nations U. Transforming our world: The 2030 agenda for sustainable development. *A/RES/70/1*, (2015). Available from https://sustainabledevelopment.un.org/content/documents/21252030 Agenda for Sustainable Development web.pdf.

Nicholls, R. J., Wong, P. P., Burkett, V., Codignotto, J. O., Hay, J., McLean, R. F., … Woodroffe, C. D. (2007). *Coastal systems and low-lying areas*. In Cambridge University Press. Available from http://pubs.er.usgs.gov/publication/70204340.

Nolan, R. H., Boer, M. M., Collins, L., Resco de Dios, V., Clarke, H., Jenkins, M., … Bradstock, R. A. (2020). Causes and consequences of eastern Australia's 2019−20 season of mega-fires. *Global Change Biology, 26*(3), 1039−1041. https://doi.org/10.1111/gcb.14987

Nunes, C., Pereira, A. M., & Morais-Almeida, M. (January 6, 2017). Asthma costs and social impact. *Asthma Research and Practice, 3*, 1. Available from https://pubmed.ncbi.nlm.nih.gov/28078100.

Oke, T. R. (1995). The heat island of the urban boundary layer: Characteristics, causes and effects. In *Wind climate in cities* (pp. 81−107). Springer.

Oliveira, S., Rocha, J., Sousa, C. A., & Capinha, C. (2021). Wide and increasing suitability for *Aedes albopictus* in Europe is congruent across distribution models. *Science Reports, 11*(1), 9916. https://doi.org/10.1038/s41598-021-89096-5. Available from.

Oliver-Smith. (November 1, 2012). A debating environmental migration: SOCIETY, nature and population displacement in climate change. *Journal of International Development, 24*(8), 1058−1070. https://doi.org/10.1002/jid.2887. Available from.

Orru, H., Ebi, K. L., & Forsberg, B. (2017). The interplay of climate change and air pollution on health. *Current Environmental Health Reports, 4*(4), 504–513. https://doi.org/10.1007/s40572-017-0168-6. Available from.

Papalexiou, S. M., & Montanari, A. (June 1, 2019). Global and regional increase of precipitation extremes under global warming. *Water Resources Research, 55*(6), 4901–4914. https://doi.org/10.1029/2018WR024067. Available from.

Pascal, M., Wagner, V., Alari, A., Corso, M., & Le Tertre, A. (2021). Extreme heat and acute air pollution episodes: A need for joint public health warnings? *Atmospheric Environment, 249*, 118249. Available from https://www.sciencedirect.com/science/article/pii/S1352231021000674.

Patz, J. A., & Hahn, M. B. (2013). Climate change and human health: A one health approach. *Current Topics in Microbiology and Immunology, 366*, 141–171.

Pausas, J. G., & Ribeiro, E. (June 1, 2013). The global fire–productivity relationship. *Global Ecology and Biogeography, 22*(6), 728–736. https://doi.org/10.1111/geb.12043. Available from.

Pecl, G. T., Araújo, M. B., Bell, J. D., Blanchard, J., Bonebrake, T. C., Chen, I.-C., … Williams. (March 31, 2017). Biodiversity redistribution under climate change: Impacts on ecosystems and human well-being. *Science, 355*(6332), eaai9214. https://doi.org/10.1126/science.aai9214

IPCC. Food security and food production systems. In Porter, J. R., Xie, L., Challinor, A. J., Cochrane, K., Howden, S. M., Iqbal, M. M., … Travasso, M. I. (Eds.), *Climate change 2014: Impacts, adaptation, and vulnerability Part A: Global and sectoral aspects contribution of working group II to the fifth assessment report of the Intergovernmental Panel on climate change (Field, CB, Barros, VR, Dokken, DJ, Ma)*, (2014). IPCC.

Prichard, S. J., Stevens-Rumann, C. S., & Hessburg, P. F. (2017). Tamm review: Shifting global fire regimes: Lessons from reburns and research needs. *Ecological Management, 396*, 217–233.

Rasmussen, S. A., Jamieson, D. J., & Honein, M. A. (April 13, 2016). Petersen LR. Zika virus and birth defects — reviewing the evidence for causality. *The New England Journal of Medicine, 374*(20), 1981–1987. https://doi.org/10.1056/NEJMsr1604338. Available from.

Reid, C. E., Brauer, M., Johnston, F. H., Jerrett, M., Balmes, J. R., & Elliott, C. T. (2016). Critical review of health impacts of wildfire smoke exposure. *Environmental Health Perspectives, 124*(9), 1334–1343.

Reinmuth-Selzle, K., Kampf, C. J., Lucas, K., Lang-Yona, N., Fröhlich-Nowoisky, J., Shiraiwa, M., … Pöschl, U. (2017). Air pollution and climate change effects on allergies in the Anthropocene: Abundance, interaction, and modification of allergens and adjuvants. *Environmental Science & Technology, 51*(8), 4119–4141. https://doi.org/10.1021/acs.est.6b04908

Reisen, F., Duran, S. M., Flannigan, M., Elliott, C., & Rideout, K. (2015). Wildfire smoke and public health risk. *International Journal of Wildland Fire, 24*(8), 1029–1044.

Rezza, G. (January 2012). *Aedes albopictus* and the reemergence of Dengue [cited 2015 Jan 16] *BMC Public Health, 12*(1), 72. Available from http://www.biomedcentral.com/1471-2458/12/72.

Rezza, G., Nicoletti, L., Angelini, R., Romi, R., Finarelli, A. C., Panning, M., … Cassone. (2007). Infection with chikungunya virus in Italy: An outbreak in a temperate region. *Lancet, 370*(9602), 1840–1846. https://doi.org/10.1016/S0140-6736(07)61779-6

Ritchie, H., & Roser, M. (2014). Natural disasters. *Our World Data.*

Robine, J.-M., Cheung, S. L. K., Le Roy, S., Van Oyen, H., Griffiths, C., Michel, J.-P., & Herrmann, F. R. (2008). Death toll exceeded 70,000 in Europe during the summer of 2003. *Comptes Rendus Biologies, 331*(2), 171–178. https://doi.org/10.1016/j.crvi.2007.12.001

Rocklöv, J., & Tozan, Y. (May 10, 2019). Climate change and the rising infectiousness of dengue. *Emerging Topics in Life Sciences, 3*(2), 133–142. Available from https://pubmed.ncbi.nlm.nih.gov/33523146.

Ruffault, J., Curt, T., Martin-Stpaul, N. K., Moron, V., & Trigo, R. M. (2018). Extreme wildfire events are linked to global-change-type droughts in the northern Mediterranean. *Natural Hazards and Earth System Sciences, 18*, 847–856.

San-Miguel-Ayanz, J., Durrant, T., Boca, R., Libertà, G., Branco, A., de Rigo, D., … Leray, T. (2019). Forest fires in Europe, Middle East and north Africa 2018. *EUR 29856 EN*. https://doi.org/10.2760/1128

Schmidhuber, J., & Tubiello, F. N. (2007). Global food security under climate change. *Proceedings of the National Academy of Sciences of the United States of America* (Vol. 104, 19703–19708.

Schmuck, G., San-Miguel-Ayanz, J., Camia, A., Durrant, T., Santos de Oliveira, S., Boca, R., … Schulte, E. (2010). *Forest fires in Europe* (p. 2011). JRC. http://effis.jrc.ec.europa.eu/docs/fire-reports/forest-fires-in-europe-2010.pdf.

Semenza, J. C., Herbst, S., Rechenburg, A., Suk, J. E., Höser, C., Schreiber, C., & Kistemann, T. (2012). Climate change impact assessment of food- and waterborne diseases. *Critical Reviews in Environmental Science and Technology, 42*(8), 857–890. https://doi.org/10.1080/10643389.2010.534706

IPCC IP on CC. Climate change and land: An IPCC special report on climate change, desertification, land degradation, sustainable land management, food security, and greenhouse gas fluxes in terrestrial ecosystems. In Shukla, P. R., Skea, J., Buendia, E. C., Masson-Delmotte, V., Pörtner, H.-O., Roberts, D. C., et al. (Eds.), *Climate change and land*, (2019).

Smith, K., Woodward, A., Campbell-Lendrum, D., Chadee, D., Honda, Y., Liu, Q., … Aranda, C. (2014). Human health: Impacts, adaptation, and co-benefits. In *Climate change 2014: Impacts, adaptation, and vulnerability Part A: Global and sectoral aspects contribution of working group II to the fifth assessment report of the Intergovernmental Panel on climate change* (pp. 709–754). Cambridge University Press.

Society TR. (2009). Reaping the benefits. In *Science and the sustainable intensification of global agriculture*. Available from https://royalsociety.org/-/media/Royal_Society_Content/policy/publications/2009/4294967719.pdf.

Stahl, H.-C., Butenschoen, V. M., Tran, H. T., Gozzer, E., Skewes, R., Mahendradhata, Y., … Farlow, A. (2013). Cost of dengue outbreaks: Literature review and country case studies. *BMC Public Health, 13*, 1048. https://doi.org/10.1186/1471-2458-13-1048

Stern, N. (2007). *The economics of climate change: The stern review*. Cambridge: Cambridge University Press. Available from https://www.cambridge.org/core/books/economics-of-climate-change/A1E0BBF2F0ED8E2E4142A9C878052204.

Tedim, F., Leone, V., McCaffrey, S., McGee, T. K., Coughlan, M., Correia, F. J., & Magalhães, C. G. (2020). Safety enhancement in extreme wildfire events. In *Extreme wildfire events and disasters* (pp. 91–115). Elsevier.

Turco, M., Jerez, S., Augusto, S., Tarín-Carrasco, P., Ratola, N., Jiménez-Guerrero, P., … Trigo, R. M. (2019). Climate drivers of the 2017 devastating fires in Portugal. *Scientific Reports, 9*(1), 1–8. https://doi.org/10.1038/s41598-019-50281-2

Turco, M., Llasat, M. C., von Hardenberg, J., & Provenzale, A. (2014). Climate change impacts on wildfires in a Mediterranean environment. *Climatic Change, 125*(3–4), 369–380.

Turco, M., Rosa-Cánovas, J. J., Bedia, J., Jerez, S., Montávez, J. P., Llasat, M. C., & Provenzale, A. (2018). Exacerbated fires in Mediterranean Europe due to anthropogenic warming projected with non-stationary climate-fire models. *Nature Communications, 9*(1), 1–9. https://doi.org/10.1038/s41467-018-06358-z

Zinsstag, J., Crump, L., Schelling, E., Hattendorf, J., Maidane, Y. O., … Cissé, G. (2018). Climate change and one health. *FEMS Microbiology Letters, 365*(11). https://doi.org/10.1093/femsle/fny085. Available from.

UNEP. 2006 annual report.(2006). Available from https://www.unep.org/resources/annual-report/unep-2006-annual-report.

USGCRP. Agriculture. (2014). Climate change impacts in the United States: The third national climate assessment. In *National climate assessment. Washington, DC, USA: U.S. Global change research program* (pp. 150–174).

van de Kamp, M. E., van Dooren, C., Hollander, A., Geurts, M., Brink, E. J., van Rossum, C., … Temme, E. H. M. (February 2018). Healthy diets with reduced environmental impact? – the greenhouse gas emissions of various diets adhering to the Dutch food based dietary guidelines. *Food Research International, 104*, 14–24. https://doi.org/10.1016/j.foodres.2017.06.006

Watson, J. T., Gayer, M., & Connolly, M. A. (January 2007). Epidemics after natural disasters. *Emerging Infectious Diseases, 13*(1), 1–5.

Watts, N., Adger, W. N., Agnolucci, P., Blackstock, J., Byass, P., Cai, W., … Costello, A. (2015). Health and climate change: Policy responses to protect public health. *Lancet, 386*(10006), 1861–1914. https://doi.org/10.1016/S0140-6736(15)60854-6

Watts, N., Amann, M., Arnell, N., Ayeb-Karlsson, S., Belesova, K., & Montgomery, H. (2019). The 2019 report of the Lancet countdown on health and climate change: Ensuring that the health of a child born today is not defined by a changing climate. *Lancet, 394*(10211), 1836–1878. https://doi.org/10.1016/S0140-6736(19)32596-6

Weinberger, K. R., Haykin, L., Eliot, M. N., Schwartz, J. D., Gasparrini, A., & Wellenius, G. A. (2017). Projected temperature-related deaths in ten large U.S. metropolitan areas under different climate change scenarios. *Environment International, 107*, 196–204. Available from https://www.sciencedirect.com/science/article/pii/S016041201730750X.

WHO. (2003). *Climate change and human health: Risks and responses*. Geneva: Switzerland. Available from https://www.who.int/globalchange/publications/climchange.pdf.

WHO. (2013). In P. P. Geneva (Ed.), *Protecting health from climate change: Vulnerability and adaptation assessment. WHO - world Health Organization.* (p. 62). Geneva: World Health Organization. Available from https://apps.who.int/iris/handle/10665/104200.

WHO. (2015). *Operational framework for building climate resilient health systems*. World Health Organization. Available from https://www.who.int/publications/i/item/operational-framework-for-building-climate-resilient-health-systems.

WHO. World malaria report 2020: 20 years of global progress and challenges. In . *World malaria report 2020: 20 years of global progress and challenges*, (2020).

Wilkinson, P. (2006). *Environmental epidemiology (Understanding public health)* (1st ed.). Open University Press.

Willett, W., Rockström, J., Loken, B., Springmann, M., Lang, T., Vermeulen, S., … Murray, C. J. L. (2019). Food in the Anthropocene: The EAT–*Lancet* commission on healthy diets from sustainable food systems. *Lancet, 393*(10170), 447–492. https://doi.org/10.1016/S0140-6736(18)31788-4

Xu, R., Yu, P., Abramson, M. J., Johnston, F. H., Samet, J. M., Bell, M. L., … Guo, Y. (2020). Wildfires, global climate change, and human health. *New England Journal of Medicine, 383*(22), 2173–2181.

Yu, Q., Wu, W., Yang, P., Li, Z., Xiong, W., & Tang, H. (2012). Proposing an interdisciplinary and cross-scale framework for global change and food security researches. In , *Vol. 156. Agriculture, ecosystems and environment* (pp. 57–71). Elsevier.

Chapter 9

Degradation of ecosystems and loss of ecosystem services

Kahrić Adla[a], Kulijer Dejan[b], Dedić Neira[c] and Šnjegota Dragana[d]
[a]Department for Genetics and Biomedical Engineering, Center for Marine and Freshwater Biology Sharklab ADRIA, Sarajevo, Bosnia and Herzegovina; [b]Natural History Department, National Museum of Bosnia and Herzegovina, Sarajevo, Bosnia and Herzegovina; [c]Department of Botany and Zoology, Masaryk University, Brno, Czechia; [d]Faculty of Natural Sciences and Mathematics, University of Banja Luka, Banja Luka, Bosnia and Herzegovina

1. Introduction

Ecosystems provide a range of various services fundamental for human well-being, health, and livelihoods. They maintain biodiversity and the production of ecosystem goods such as food, water, industrial products, biomass fuels, natural fiber and many other goods, services, and cultural services important for human economy.

Millennium Ecosystem Assessment (MA) assessed the consequences of ecosystem change for human well-being and scientific basis for action needed to enhance the conservation that should be long-term sustainable and functional for human well-being. Thus, they based their assessment on peer-reviewed scientific information in a form that is relevant as a policy instrument which is available for the government for achieving further property management of ecosystem services. MA was called for by the United Nations Secretary – General Kofi Annan – in 2000 and initiated in 2001 with the overall aim to contribute in improving decision-making concerning ecosystem management and human well-being. Nowadays, the MA is a major UN-sponsored effort that analyzes the impact of human actions on ecosystems and human well-being, describing ecosystem services as "the benefits people derive from ecosystems" (MA, 2003).

Usually there is misunderstanding in the definition and terms that describe ecosystem services, such as "goods", "services" and "cultural services" which are often treated separately for better and simpler understanding, but MA use all these benefits together as „ecosystem services" because it is hard to identify

One Health. https://doi.org/10.1016/B978-0-12-822794-7.00008-3

provided benefits as a "goods" or "services". Both of the mentioned terms are often used together as "ecosystem goods and services" which, on the other hand, very often exclude the cultural services or similar benefits. In this chapter, all benefits will be considered as ecosystem services in order to better understand their features and essence. Therefore, the ecosystem services have been categorized by different factors and characteristics including functional grouping (such as regulation, habitat, production), organizational grouping (such as services associated with certain species that regulate some input), and descriptive grouping (renewable resources goods, social, and culture services). Regarded mentioned features MA identified four major categories of ecosystem service: provisioning, regulating, cultural, and supporting services.

Provisioning services are services provided by nature as products. This includes food products derived from plants or animals (such as cotton). It is any type of benefit that people could take, extract and use from nature, which also include drinking water as one of the most important benefits for human well-being. Oils are also included, where palm oil is one of the most produced and internationally traded edible oils. Various plants and mushrooms are useful in traditional medicine or for medicines and pharmaceuticals important for human health. Fuel, clothes (skins and shells as animal products), but also genetic resources used for animal and plant breeding and biotechnology, belong to this category. Provided services have direct benefits for livelihoods in rural regions, while in urban areas these services can be found in markets.

Regulating services are services that provide benefits from the regulation of ecosystem processes. Maintaining the air and soil quality are very useful but often invisible regulation services which losses could be difficult to restore. Trees and other vegetation have an important role in climate regulation, especially in air quality where they remove pollutants from the atmosphere. Green smart cities are very useful examples of creating sustainable effective cities which could provide a healthy environment. Moreover, vegetation prevents soil erosion and has a main role in land degradation or loss of soil fertility. This is regulated through natural biological processes such as nitrogen fixation. More detailed, these services include climate regulation, water regulation, erosion control, water purification and waste treatment, regulation of human disease, storm protection, biological control, as well as pollination which is mainly provided by insects, but also by some birds and bats which are essential for horticulture production.

Cultural services are services provided by nonmaterial benefits through spiritual and aesthetic experiences, cultural diversity, cultural heritage values, traditional knowledge systems, inspiration, and social relations. Through these services, humans create a connection with nature. However, this is not a modern movement, but it has been identified at the beginning of mankind in the oldest civilizations. Moreover, tourism and recreation are also considered as cultural services.

Supporting services are services that are necessary for the production of all other ecosystem services through which are provided living spaces for plants or animals or including the process which are necessary for their existence such as primary production, production of atmospheric oxygen, nutrient cycling, or water cycling. Because of the significant importance in all processes allowing nature's main basic processes, without supporting services, other services would not exist.

Considering the anthropogenic pressure on ecosystems, the highest impact is conducted by the local community, which has a significant role in the use of natural resources. Effective management of natural resources is rapidly needed because of the increased human impact on the natural environment. Globally, economies demand high consumption of products which are produced and extracted from nature or they demand using other valuable ecosystem services for human needs, where they tag the price of each nature's benefits. From this scenario, it is expectable that our planet needs help for proper and sustainable use of their natural resources. Connection between humans and ecosystems should be based on integrations between ecosystems and society through the balance where micro and macroeconomics should be functional for both ecosystems and society. However, a framework of ecosystem services use, needs to be established and implemented by scientists, decision-makers, and government. Also, legislation needs to be customized by the challenges and threats of the area, which should be obtained through assessment of ecosystem services. Moreover, making ecosystem services functional will require moving from proposed frameworks to practical realization through an environmental decision-making process in a way that is credible and sustainable. This would include work, willingness, and unity of multistakeholders including natural scientists, economists, social scientists to choose and assess the value of those services. The main obstacles of these processes include social and political challanges based on property establishment of services management. Realizing the importance of sustainable ecosystem services and natural resources, it is expected that their loss is primarily caused by human activities. Consequently, their loss is mainly caused by biodiversity and habitat loss, requiring prioritizing of protection and conservation measures. Different diseases, including zoonosis, are also increased by human activities. Considering the main anthropogenic pressures, in this chapter are described four main threats for human and animal health that are caused by ecosystem services loss: habitat loss, climate changes with global warming, invasive species, pollution and overexploitation and trade. It provides the overview of each threat caused by human activities that act as vectors for disease, including the scenario of increased zoonotic disease, which are also included (MA, 2003).

2. Habitat loss

The loss of habitat is one of the major threats to many species and their survival. Usually there is misunderstanding in two terms, habitat loss and habitat fragmentation or destruction. Some authors treat these terms together as a same process, while others treat them separately as two different processes. (e.g., Dodd & Smith, 2003). However, habitat loss could be caused by destruction, fragmentation, or degradation of habitats where ecosystems dramatically changed due high anthropogenic pressures.

According to Wilcove (1986) *habitat fragmentation* presents the division of habitat into smaller and more isolated fragments from each other with the matrix of habitats unlike the original. These fragments act like biogeographic islands (MacArthur & Wilson, 1967) where consistently negative effects result in long-term changes to their structure and function (Fahrig, 2003) on the way that smaller and more isolated habitats become more vulnerable to the intrusion of invasive species and infectious diseases, affecting native hosts of plants and animals (Allan, Keesing, & Ostfeld, 2003; Keesing, Holt, & Ostfeld, 2006; Suzán et al., 2008). Nowadays, many roads, lands, and water areas are fragmented. Therefore, the aquatic ecosystems could be presented as a good example of habitat fragmentation through dams and water diversions. Rivers in Mediterranean-climate are more heavily impounded where their hydrology is more affected than other different climate regions. Inducing the sediment load and flow reduction could have ecological impacts. In this context, no matter how native species of Mediterranean-climate rivers are adapted to survive, exotic species that were excluded by the highly variable flow regime may have higher probability of survival under the new stable flow regime (Kondolf, Podolak, & Grantham, 2012). Furthermore, these fragmented habitats may not be enough to support some species needs that require large territory for food and partners. Also, it is a challenge to survive for migratory species.

Habitat destruction is a process when natural habitat such as a forest or wetland is dramatically changed that could no longer support species. Humans, in different ways, destroy habitats in wetlands, rivers, and forests. This has a harmful impact on biodiversity loss, including valuable species, such as endangered red-cockaded woodpeckers, a bird that used to inhabit the forest of North Carolina. The Endangered Species Act (ESA) classified the red-cockaded woodpecker as protected species. As killing of this species is illegal, so it is illegal to destroy its habitat. Pine trees are common in North Carolina forest, but due to human activities pine trees were lost, and reduced the population of red-cockaded woodpeckers (Lueck and Michael, 2003). Therefore, it is expected that habitat destruction leads to biodiversity loss.

Habitat degradation is a set of processes where habitat quality is reduced and caused by natural processes (heat, cold) and human activities (forestry, agriculture, urbanization) (IPBES, 2019). Therefore, when habitats become degraded, they cannot support native species.

Species at higher trophic levels are lost more rapidly than species from lower trophic levels due to habitat loss. Moreover, species will disappear from the food chain from highest to lowest ranks because of the ecosystem services loss which are essential for species survive, especially for predators. Therefore, the ecosystem services could respond differently to loss of habitat relating to the biodiversity and their trophic levels. It can be explained through examples of primary production where losses of few species could result in drastic decrease in ecosystem services. This could be applied for mangroves or seagrass which have a main role in primary production. Due to high anthropogenic pressure the loss of biodiversity and ecosystem function in estuarine and coastline ecosystems is decreased, affecting the critical benefits or ecosystem services. They are one of the most used and threatened ecosystems globally which consequently caused negative impacts on the salt marshes, mangroves, coral reefs, and seagrass as an important habitats and nursery areas for numerous vertebrates and invertebrates' species, including species that support important nearshore fisheries. Through fisheries maintenance, these nursery habitats provide one of the most important services.

Mangroves act as storm barriers, protecting inland areas from flooding and erosion and have an important role as a carbon storage. Most of this carbon is stored in the soil beneath mangrove trees. In the beginning of the 21st century, global mangrove losses slowed globally due to mangrove deforestation. However, some countries such as Myanmar, Malaysia, Cambodia, Indonesia and Guatemala still have an increased rate of mangrove loss (Hamilton & Casey, 2016). Between 2000 and 2015 up to 30−122 Tg of carbon has been released due to mangrove forest loss. (Sanderman, 2018). Climate changes could also have an impact on mangroves through increased temperature and storminess, sea level rise, changes in ocean currents and in precipitation and increased CO_2. For example, mangroves are sensitive to sea level rise where flooding can lead to death or changes in species composition which further reduce productivity and ecosystem services (Castañeda-Moya et al., 2013).

Given the emphasis on mangroves as valuable species habitat, the biodiversity protection is partially based on the presume that habitat loss has an impact on biodiversity loss, which could be resulted in the loss of ecosystem function and many ecosystem services that provide to society (Costanza & Folke, 1997).

Duarte (2000) suggested that an increasing species richness should be on average linked to an increase in the ecosystem functions which are presented in society. Thus, this should lead to much effective achievement of land-use and resources that provide sustainability. This argument also could be extended to marine ecosystems.

According to the Loh et al. (2015), habitat modifications are responsible for nearly half of emerging zoonoses. Zoonoses are diseases or infections caused by various organisms, such as viruses, bacteria, and fungi, which are transmitted from animals[1] to humans. Among all emerging diseases 60% are zoonoses with wildlife origin that have the highest impact and threat to the health of the world population (Jones et al., 2008) causing around one billion cases of disease every year and millions of deaths (Morse et al., 2012). According to the World Health Organization (WHO) zoonoses are numerous and their research is of the greatest interest in human and veterinary medicine. Zoonoses have received increasing attention since the end of the 20th century. Although it seemed that the production and extensive use of antibiotics and vaccines helped humanity to outfight them, growing antibiotic resistance among bacterial pathogens and the increase in the outbreak of old and new zoonoses showed the opposite trend (Johnson et al., 2015).

It is still not quite certain how biodiversity changes affect zoonoses. It seems that this depends highly on the type of pathogen transmission. For directly transmitted diseases such as HIV, measles and human tuberculosis, a change in biodiversity may have no effect at all. For pathogens such as the West Nile virus, on the contrary, changes in biodiversity may have effects because this virus infects not only humans or primates, but also several bird species. This applies also to the hanta virus, which infects not only humans but also several mammals, and to leptospirosis, which is transmitted by rat urine and excrements (Van Langevelde & Mendoza, 2021).

The transmission of zoonoses is generally higher in areas with the high human population density, where infectiveness increases due to the intensification of human-animal contacts (Keusch, Pappaioanou, Gonzalez, Scott, & Tsai, 2009). These contacts emerge due to the constant growth of the human population[2] and, therefore, higher requirements for the land source. Human activities have significantly transformed three-quarters of the land and two-thirds of the ocean, changing the planet to such an extent as to determine the creation of a new era named as the Anthropocene (Almond, Grooten, & Peterson, 2020). Among these activities, deforestation, invasion of natural vegetated areas by human communities, and climate change are listed as those with the most significant influence on the landscape modifications that are leading to the natural habitat loss and fragmentations (Suzan et al., 2012). Land modifications bring wildlife, livestock, and humans to closer proximity, facilitating the diseases spreading among them (Fig. 9.1) (Jones et al., 2013).

1. The transmission of a pathogen from one host species to another is called spillover. This process can happen i) as the direct contact (as in the case of rabies), ii) through other organisms recognized as vectors that carry the disease agent (e.g. mosquitoes, ticks), or iii) through environmental carriers and food items. Some pathogens, like Ebola and the current coronavirus, allow the human-to-human transmission. These are the most dangerous for humans because the epidemics can spread very quickly turning to pandemics.
2. The exponential growth of the human population, from around 1 billion in 1900 to 6.5 billion in 2006 is influencing major ecological changes and drastic wildlife habitat reduction (United Nations, 2007).

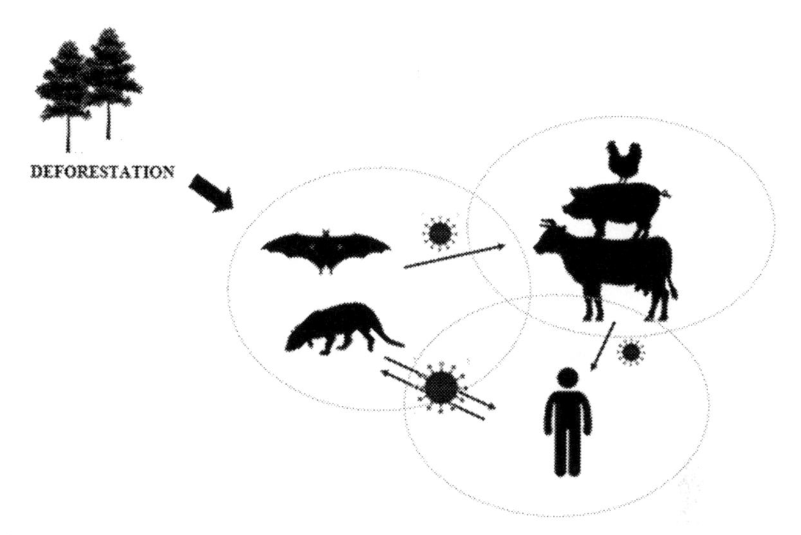

FIG. 9.1 Pathogen spillover from wild animals to humans and livestock, as the consequence of deforestation.

Pathogen transmission from wildlife to domestic animals and humans, and the opposite, trigger epidemics and pandemics worldwide. At the time of writing, humanity was witnessing the pandemic of the coronavirus disease, caused by SARS-CoV-2, a virus with a likely zoonotic origin (Rothan & Byrareddy, 2020). It is suspected, based on the results of the virus genome sequencing and evolutionary analysis, that bats are natural hosts of virus origin, often transmitting it via an unknown intermediate to humans (Guo et al., 2020). Bat species influenced by the human exploitation of habitats are the hosts of zoonotic viruses such as Hendra virus, Nipah virus, Ross River virus, Australian bat Lyssavirus, and Rabies virus. Habitat degradation involves human intrusion within these habitats, as well as movement of bats into new habitat, which increases the likelihood of human contact with an infected bat. Intrusion often takes the form of habitat destruction by deforestation, and conversion of the landscape for agriculture or human activities (Barker, 2013). Lu et al. (2020) found a high correspondence between the human SARS-CoV-2 genome and the coronavirus genome found in a bat in the Chinese province of Yunnan, suggesting that differences in the genetic sequence could have passed through an intermediate host before affecting humans. The area of the disease escalations seems to be the seafood and animal wholesale market in Wuhan, Hubei Province, China that sells exotic animals for consumption (Bevenuto et al., 2020; Bogoch et al., 2020; Lu et al. 2020). The current pandemic shows that the exploitation of the wildlife, together with loss of natural habitat, can increase the risk of spillover events, therefore representing not only conservation issues, but also the important drivers of zoonotic diseases transmission (Johnson et al., 2015) that have enormous influence on human health worldwide.

Human activities, in general, may have promoted direct or indirect contact between humans and an animal reservoir of viruses (Bausch & Schwarz, 2014), whereas the loss of forest can facilitate the spread of the disease to non-forest areas (Wallace et al., 2015; Rogalski et al., 2017). As the habitat modifications are responsible for facilitating emerging zoonosis, one of the best examples of the habitat modification effect to the zoonosis increase and outbreak is deforestation of tropical forests. Tropical forests are home for millions of various species; among them many are still unknown and others present undiscovered viruses, bacteria, fungi, and parasites. The destruction and degradation of these forests exposes humans to a closer contact with the pathogens hosted by wild species, making them more vulnerable to infection. Such an example can be seen in Ebola outbreaks in West Africa, happening due to the deforestation of the tropical African rainforests through the loss, regrowth, and fragmentation from human activities, resulted in creating of favorable conditions for the transmission of Ebola virus from its reservoir hosts into the humans. Bats are thought to be the likely reservoirs of the Ebola virus, or at least play a part in its transmission. Even moderate forest fragmentation increases the abundance and the diversity of certain bat species that successfully adapt to the fragmented landscapes and brings the Ebola virus in closer proximity to the humans (Bausch & Schwarz, 2014; Wallace et al., 2015).

Modifications of habitats eventually reduce the plant diversity of the specific area influencing the presence, abundance, and distribution of mosquitoes affecting the malaria transmission (Yasuoka & Levins, 2007). This can be seen through the increase of the new type of malaria in Malaysian Borneo, or, in the Peruvian Amazon, where deforested areas, compared to still intact forests, have a higher density of *Anopheles darling* - the most efficient local mosquitoes for transmitting malaria (Vittor et al., 2006). Although primarily controlled, the malaria rate in Brazil started increasing in parallel with rapid forest clearing and expansion of agriculture, resulting, at the turn of the century, in over 600,000 cases a year in the Amazon basin. According to the epidemiologists, clearing patches of the forest appears to create ideal habitat along forest edges for the mosquito *Anopheles darlingi* - the most important transmitter of malaria in the Amazon as well, to breed closer to humans exposing them to the *Plasmodium*, a pathogen causing malaria. The edges of the remaining natural areas are considered as the major launch pads for novel viruses that may spill over to humans (Dobson et al., 2020).

Untouched habitats often have a high diversity of both animals and pathogens (Han et al., 2016, Mollentze & Streicker, 2020). Deforestation and the shift of natural areas to human-dominated areas result in the large-scale loss or degradation and fragmentation of habitats and wildlife populations (Fahrig, 2003, Newbold et al., 2015). The resulting remnants of natural areas show increased risk for zoonotic diseases. Fragmented habitats also lead to an increase of host movement from the patches of nature into areas used for livestock and urban settlements (Suzán et al., 2008), resulting in the zoonoses outbreaks, as can be seen from the following examples.

The *Aedes aegypti,* mosquito that transmits Zika virus, represents a high risk for global transmission. This virus originates from Africa, in the Zika forest in Uganda, where it was discovered in 1947 in a *Rhesus* monkey. In May 2015, the first local cases of this virus were recorded in Brazil, exceeding 1.5 million cases in December of the same year. Urbanization in Brazil has led to the deforestation of large green areas, destroying the ecosystems in which the mosquitos and its predators used to reproduce. This process clearly influences *A. aegypti* (Degallier et al., 2012; de Rocha Taranto et al., 2015), that, due to the loss of the natural habitats, adapts and reproduces well within human settlements exposing humans to the Zika virus, but also to dengue, chikungunya and yellow fever. Yellow fever is an acute disease caused by Yellow fever virus that affects humans and non-human primates in South American and African countries (Monath, 2001; Vasconcelos, et al., 2003). Despite the existence of a safe vaccine, massive Yellow fever outbreaks occurred recently in Angola, Democratic Republic of Congo (Maguire & Heymann, 2016), and Brazil (MS-BR, 2017, 2018, 2019) as a result of the human invasions to the forests of South America and Africa and direct and indirect contact with disease reservoirs.

However, mosquitoes are not the only animals transmitting pathogens to people. Around 60% of various infectious diseases that occur in people, all of which originated in forest-dwelling animals - are transmitted by a range of other animals, the vast majority of them wildlife. Rodents are well-known reservoir hosts for various pathogens (e.g. Rabitsch et al., 2017). In West Africa, Liberia, deforestation for the establishment of palm oil plantations is directly affecting the rise of the Lassa fever. Mice transmitting the *Lassa* virus cause this disease. They are attracted to the palm fruits from the plantations and settlements and can transmit the virus to humans through contamination of food and other human-used objects, with the urine/feces. In Liberia, 36% of infected people died so far from this virus (Gibb et al., 2017).

Deforestation and urbanization are among the major factors influencing the worldwide epidemic of leishmaniasis - the tropical and subtropical disease triggered by an intracellular *Leishmania* parasite transmitted to humans by the bite of a sand fly, mainly *Phlebotomus* and *Lutzomyia* genera (Europe, Northern Africa, the Middle East, Asia, and part of South America) (Torres-Guerrero et al., 2017). The disease has a worldwide distribution, as has been found in about 90 countries, currently affecting about 6 million people (Kernif, Leulmi, Raoult, & Parola, 2016; Reithinger et al., 2007). The study of this disease in Brazil, which showed a very high frequency of *Lutzomyia antunesi* in fragmented forests and semi-urban areas, and *Lutzomyia aragoi* in completely urbanized regions, emphases their great adaptability to the suburban and urban areas. Such adaptability and consequently new epidemiological dynamics are consequences of the human-induced progressive destruction (e.g., deforestation and urbanization) of their natural habitat that results in closer contact of humans with parasites they transmit and along with that spreading of this disease.

The increase of the diseases transmitted via ticks is also directly connected with the humans' occupations of the wildlife habitats (Eisen et al., 2017). Lyme disease, for example, caused via *Borrelia burgdorferi*, is transmitted to humans and other animals via ticks. In the United States, in the late 20th century, Lyme disease was recognized as an important emerging infection (Steere, et al., 2001). The transmission of *B. burgdorferi* to the humans by *Ixodes scapularis* in the east and *Ixodes pacificus* in the west of the US was facilitated with changes in land use (e.g. reforestation and subsequent suburban development). These changes significantly increased the infectiveness of humans, living within the modified areas, with this bacterium, due to their higher interactions with ticks (Brownstein, Skelly, Holford, & Fish, 2005). An increase of the Lyme disease in humans with residences located in forested areas, compared to those inhabiting non-forested areas, were observed by Glass et al. (1995) in Baltimore County, Maryland, and Kitron and Kazmierczak (1997) in Wisconsin counties.

Intensification of agriculture and livestock farming are also very important drivers of the zoonoses transmission. With the expansion of agricultural areas, the wildlife areas decline, bringing humans, livestock, and wild animals - together with the pathogens they host - into closer proximity. This increases the opportunity for the spillover of pathogens into the humans or livestock (Jones et al., 2013), therefore, directly or indirectly affecting human health. According to the previous authors, the impact of the intensification of the livestock farming to the zoonoses transmission can be drastically reduced by using less workers per animal, however, various examples still show its effect on the zoonoses increase (Gilchrist et al., 2007; Graham et al., 2008). For example, the outbreak of the Nipah Virus disease that happened in Malaysia during 1998–99 (Epstein, Field, Luby, Pulliam, & Daszak, 2006), and has been directly linked with intensive livestock farming. Intensification of the livestock (pigs) farming directly affected Malaysian forests - subsequently cleared for the establishment of the livestock farms. This deforestation had a direct impact on the outbreak of the Nipah Virus when fruit bats – hosted the Nipah virus came in closer proximity to the pigs and humans from farms due to the loss of their natural habitat. Eventually, the bats infected pigs from the large farm in northern Malaysia with the Nipah virus. Further transport of these pigs to southern Malaysia, to smaller but higher density farms, had a direct effect in transmitting the virus and spreading the infection. While pigs that consumed the infected fruits suffered from respiratory disease, for humans that were in close contact with infected pigs it was fatal.

Urbanization is a significant factor reducing natural habitats of various species, such as coyotes, foxes, and wild boars. These species, consequently

TABLE 9.1 Human activities that might be potential drivers of the ecological changes that affect the emergence of the zoonoses within specific geographic regions.

Human activities	Disease	Geographic region
Deforestation and water projects	Malaria	Tropical[a]
Dam building and irrigation	Schistosomiasis	Tropical
Urbanization	Dengue fever	Tropical
Deforestation and mining	Rabies	Tropical
Habitat fragmentation	Lyme disease	North America and Europe
Deforestation	Ebola	Africa

[a]*America, Asia and Africa.*
Modified from the Millennium Ecosystem Assessment.

and intensively, colonize urban areas and spread zoonoses, and serve as a well-known reservoir hosts for various pathogens (Van Langevelde & Mandoza, 2021). Wild boars are hugely involved in the transmission of foodborne zoonotic diseases such as brucellosis, salmonellosis, tuberculosis, yersiniosis, toxoplasmosis, trichinellosis and hepatitis E into the urban areas (Fredriksson-Ahomaa, 2019). About 8% of the reported red foxes in Estonian largest urban areas exhibit symptoms of sarcoptic mange, a disease that also infects domestic animals, especially dogs, and which can be transmitted to humans (Plumer et al., 2014). Moreover, a substantial fraction of red foxes in Estonia was infected with the life-threatening tapeworm *Echinococcus multilocularis*, the causative agent of alveolar echinococcosis. Therefore, urban foxes may represent a source of serious infectious diseases for pets and humans. Analyses of the coyotes' (*Canis latrans*) feces in semi urban areas in Manitoba (Canada) showed the presence of the tapeworms *E. multilocularis* and *E. canadensis* in it. These tapeworms may affect humans and dogs (Tse et al., 2019; Van Langevelde & Mendoza, 2021).

The Millennium Ecosystem Assessment (Reid & Mooney, 2016) evaluated human activities that might be potential drivers of the ecological changes that affect the emergence of the zoonoses, and some of these can be found in Table 9.1.

3. Climate changes – with global warming

Climate change has several impacts on the environment and ecosystems. From sea rising, seawater intruding freshwater that can cause many species to relocate or die, and affecting the local food chain. Besides climate change, the environment is under stress caused by human activities. The factors that influence global changes in climate and environment are carbon dioxide production, the destruction of the ozone layer, the release of harmful chemical substances of various origins into nature (nitrates, phosphorus, heavy metals, etc.), non-biodegradable materials anarchically released in nature, and industrial petroleum waste from the pressure of exploitation of energy resources (oil, natural gas, coal, etc.).

A specific factor is also human development in coastal areas where it affects the species there and erosion caused by pressures on coastal environments. These factors act cumulatively. Modifications of ecosystems caused by humans also restrict the ecosystem's role as a buffer for extreme events like storms and wildfires. For example, forests act like wind blockers; islands act as storm buffers for costs, or wetlands that absorb water and prevent large wildfires. Forests regulate the world's climate by absorbing large amounts of carbon dioxide, preventing a greenhouse effect. Forests also provide habitat for many plant and animal species. Effects of climate change have a different effect on different kinds of forests. The most vulnerable are the subarctic and rain forests in the area of the Amazon, which can experience a high level of extinction events. Destruction of forests has a great impact on local communities and the release of stored carbon (e.g., in organic matter), contributing to a faster greenhouse effect. Not only forests, green areas generally play an important role in climate regulation, i.e., in mitigating and slowing down climate change. Thus, they mitigate the health consequences of climate change and are mitigated through the ability of plants to absorb carbon from greenhouse gases (Coutts & Hahn, 2015).

Global warming has been on the rise in the past decades and is expected to rise progressively in the next few years. This results from the greenhouse effect of carbon dioxide and deforestation (Jimenez-Clavero, 2012; Solomon et al., 2007; Watson et al., 2001). The changes in climate and temperatures also affect ecosystems, resulting in changes in disease epidemiology and exposure to causative agents (Ghazali et al., 2018). Climatic changes have a strong influence on zoonotic disease epidemiology. Climate change changes and creates niches for vectors and alters the temporal and spatial distribution of disease (Lafferty, 2009). It also influences the geographical distribution of vectors/intermediate host distribution, including invertebrate hosts (insects), rodents, and migratory birds. A good example is the zoonotic bacteria *Chlamydia,* which birds carry.

Global warming directly influences the abundance, survival, and distribution of pathogens and their vectors. According to the WHO, 20 core Neglected Tropical Diseases are candidates to reemerge in the developed countries due to global warming: the dengue fever and rabies (viral diseases), Buruli ulcer (*Mycobacterium ulcerans* infection), trachoma (*Chlamydia trachomatis*), yaws (spirochete), treponematoses and leprosy (bacterial diseases), Chagas disease, trypanosomiasis, leishmaniasis, dracunculiasis, cysticercosis, echinococcosis, foodborne trematodiases, lymphatic filariasis, onchocerciasis (river blindness), schistosomiasis and soil-transmitted helminthiases: ascariasis, hookworm, trichuriasis (Mackey et al., 2014). In addition, the soil itself is a "habitat" for a large number of microorganisms and viruses that are pathogenic. They can inhabit the soil in different ways, such as sewage sludge, municipal wastewater, waste on the land, etc. Climate change in itself actually accelerates and makes this process easier. (Brevik & Burgess, 2015).

The danger of climate change and rise in temperatures is the melting of the ice layer that has existed for thousands of years. The melting of glaciers also causes the destruction of freshwater ecosystems of the rivers and surrounding areas. Over a billion humans depend on the rivers made by these glaciers for drinking water, sanitation, agriculture as well for electric power. Viable bacteria can be isolated from ice samples as old as 25,000 years (Katayama et al., 2007). Scientists in Siberia detected *Variola* in a frozen mummy (Biagini et al., 2012), and NASA revived bacteria in Alaskan ice that has been frozen for the last 32,000 years. The bacteria were also isolated from ice samples from Antarctica older than 8 million years (Bidle, Lee, Marchant, & Falkowski, 2007). There are examples of people getting infected from an ancient bacteria, as in 2016 in Syberia, where one person died and 20 were hospitalized due to anthrax infection from spores *Bacillus anthracis* after the ice melted. These spores can be transported by floods or insects. It is possible that also smallpox is frozen in human bodies frozen under the Siberian ice (Antonenko et al., 2013; Walsh, Willem De Smalen, & Mor, 2018).

Furthermore, global warming will strongly increase the worldwide prevalence of enteric/diarrheal diseases and can also strongly influence the rise of fatal epidemics caused by the fatal thermophilic free-living ameba, the *Naegleria fowleri,* which causes fatal meningoencephalitis (Huizinga & McLaughlin, 1990; Sykora, Keleti, & Martinez, 1983). A change in climate or hydrological regime, including a change in precipitation or a change in the volume or seasonal presence of surface water, may cause a change in the population numbers of vectors and may alter the rates of human exposure to pathogens, as may changes to the trophic structure of a wetland as a result of over predation. Wetlands are loci for communicable disease; microorganisms (the pathogens), pollution, or toxicants are transmitted through water, people, animals as vectors, surfaces, foods, sediments, or air—any or all of which can be associated with wetlands (Horwitz & Finlayson, 2011).

Vector distribution along with disease risk should also rise for vector-borne zoonotic diseases. Transmission of diseases is most likely to occur when there are extreme changes of temperature (14−18°C at the lower end and 35−40°C at the upper end). The most significant vector densities should occur at 30−32°C. Especially mosquitos, ticks, and sandflies are impacted due to their life cycle that is dependent on temperature (Githeko, Lindsay, Confalonieri, & Patz, 2000) that would be discussed below.

Mosquitoes transmit parasitic diseases such as malaria and filariasis, which are spread due to global warming. Besides mosquitoes, there are other external parasites such as ticks which also expanded their habitat due to global warming. The ticks also survive the winter much better, and they spawn earlier. The risk for humans is the longer period of attack (Randolph, 2004). However, the tick-borne disease caused by *Anaplasma phagocytophilum* is a serious zoonotic disease that infects humans, cattle, equines, and canines worldwide. The disease can be transmitted by different tick species endemic in Europe (e.g., Ixodes ricinus), North America (e.g. *I. scapularis, I. pacificus,* and *Ixodes spinipalpis*), and Asia (e.g., *Ixodes persulcatus*). Climate and the local habitat change could interact and affect the dispersal and movement of the species involved in the transmission cycle of Lyme disease in a complex manner. For example, a warmer climate with milder winters and earlier spring snowmelt may shift the phenology of the white-footed mouse breeding activity and dispersal phenology, with higher activity and movement earlier in the season. Increased activity in turn alters the rate of the encounter between this host and its pathogens, affecting the dynamics of the transmission cycle of *B. burgdorferi* (Ogden & Lindsay, 2016). On the other hand, fragmentation of host habitat may reduce host population size, limit host dispersal, and alter host densities and diversity, reducing the encounter rate between hosts and pathogens and the transmission rate of the pathogen.

Many zoonotic diseases have increased across Europe. Genetic impoverishment of wildlife populations in degraded and fragmented habitats can imply low immunity and increase pathogens' competence. Recent reports listed the main tick-borne pathogens in Europe, such as *Rickettsia, A. phagocytophilum, B. burgdorferi, Babesia spp., Borrelia miyamotoi, Bartonella henselae, Candidatus N. mikurensis, Francisella tularensis*, and the viruses Crimean-Congo hemorrhagic fever virus (CCHFV) and the tick-borne encephalitis (TBE).

In wildlife, global warming benefits the change in rodent distribution maps (Moore, Shrestha, Tomlinson, & Vuong, 2011). Life-history variables (e.g., habitat type or diet) were correlated with changes in the size of the range, location, or habitat composition and reflected broad changes in the relative distribution of vegetation types. Granivores and herbivores used fewer new habitat associations after global climate changes than did omnivores or insectivores, as did terrestrial rodents compared with fossorial or arboreal rodents. The rodents are known to spread disease agents like leptospirosis, Sin Nombre virus, Hantaan virus, and *Yersinia pestis*, which over dried urine or indirectly. Higher temperatures and especially a decrease in snowfall forces the rodents to find shelter in human habitats. This, in particular, was the cause of the hantavirus outbreak in Scandinavia (Evander & Ahlm, 2009). Although to the wide range of symptoms, some infected people may have no symptoms at all.

A tiny flying blood-sucking dipteran from sandy areas, known as sandflies, *Phlebotomus papatasi*, are also much more active in the environment with higher temperatures and are responsible for the transmission of leishmaniasis. Rioux et al. (1984) found that the raising of temperature significantly increased the overall proportion of infected sandflies. Heat, humidity, and sufficient organic matter are the main effectors of the larval development of sand fly species, while the increasing carbon dioxide levels are usually unfavorably affecting the development of the insect larvae. Leishmaniasis is a vector-borne disease caused by intracellular protozoa of the genus *Leishmania*. It is transmitted by the bite of infected female sandflies. The parasites of this disease have faster development rates (Ready, 2010). The vectors spread into neighboring regions as shown by the potential spread of leishmaniasis in North America due to vector distribution and expansion (González et al., 2010). In tropical America, the transmission of leishmaniasis is believed to have traditionally been restricted to humid sylvatic habitats in which humans were exposed to the parasite during forest-related activities. However, human-induced habitat transformation has facilitated the rapid invasion of some vector and mammal species into non-sylvatic habitats thereby increasing both human exposure and risk of infection. The dynamics of the disease are correlated with population fluctuations in reservoirs and vectors and strongly correlated with environmental changes and climatic factors. Because climatic factors can lead to species' range shifts, analyses of vector and reservoir species distributional responses to climate change scenarios provide insight into how the spatial epidemiology of leishmaniasis may be affected by climate change (Dobson et al., 2020; Koch et al., 2017).

From above, it is obvious that some zoonoses have an outbreak in different areas worldwide. However, nowadays, COVID-19 is considered a pandemic of 2020, where disease etiology and spread features are still under investigation affecting the human population globally.

Tosepua et al. (2020) found that the average temperature was correlated with COVID-19, with the lowest average temperature of 26.1°C and the highest temperature of 28.6°C. This correlation is in line with previous research that shows the relationship between weather transmission and Syncytial Virus Respiration (RSV) (Vandini et al., 2013) and SARS (Tan et al., 2005). Temperature is also the environmental driver of the COVID-19 outbreak in China (Shi et al., 2020). The regression equation shows how temperature, relative humidity, and wind speed affected SARS transmission (Yuan et al., 2006). Many of the root causes of climate change also increase the risk of pandemics. Deforestation, which occurs mainly for agricultural purposes, is the largest cause of habitat loss worldwide. Loss of habitat forces animals to migrate and potentially contact other animals or people and share germs. Large livestock farms can also serve as a source for the spillover of infections from animals to people. Less demand for animal meat and more sustainable animal husbandry could decrease emerging infectious disease risks and lower greenhouse gas emissions.

Another effect of rising temperatures is on the freshwater cycle. Higher air temperatures can hold larger amounts of water, making rainfall more extreme and causing droughts and floods. The lakes and rivers provide fresh water for humans and animals and are already under pressure from drainage, building dams, pollution, extraction, and invasive species. Climate change will only accelerate the destruction of lakes and rivers. Rising temperatures that come with climate change also give rise to world ocean levels which not only causes the flooding but presents a risk for waterborne zoonoses. Wetlands are thought to be a source of waterborne infection diseases and the source of vectors. They are a sanitation challenge for the safe disposal of human excreta and make access to health services more difficult for those living in and around them. In contrast, the upstream factors are not perceived as such a threat and are considered less problematic.

Higher water temperature levels cause oceans to absorb more carbon dioxide making them acidic and toxic for many species. Temperature rise in oceans of above 2°C will cause coral reefs to disappear. This will have an impact on wildlife and fishing as a food source for humans. The higher temperature and acidity of the oceans will kill microorganisms that are expelled by the coral reefs. These microorganisms are essential for the health of the reefs. Not being able to protect their health will cause more coral reefs to die off. Higher carbon intake of oceans additionally changes the pH value of water (IPCC - Climate Change, 2007). *Vibrio* population increases by the symbiotic relationship between plankton and vibrio. Open wounds on swimmers or fish consumers can attract the infection. Zooplankton will bloom when the water temperature increases and is related to *Vibrio cholerae*. (Redshaw, Stahl-Timmins, Fleming, Davidson, & Depledge, 2013). Survival of *Vibrio cholerae* in vegetables was shown recently and is related to reduced microbial diversity caused by increased use of pesticides and fertilizers. However, diverse microbiomes probably cannot prevent an outbreak of emerging diseases (Bruggen et al., 2019).

Increased rainfall leads to more crops and food, which directly gives rise to the rodent population. Rodent urine is rich with leptospirosis bacteria and the occurrence of flooding after heavy rainfall causes a wider spread of this disease. Very likely environmental changes increased the vector (rodent) population, which facilitated transmission. Flooding gives rise to waterborne zoonoses through the expansion in the number and range of vector habitats. It is believed that heavy rainfall upstream to water treatment facilities is responsible for the cryptosporidium outbreak in Milwaukee, USA (Mac Kenzie et al., 1994). Epidemiological studies found that *Ebola haemorrhagic* fever and Marburg fever outbreaks are closely linked to climate changes and especially to rainfall/dryness patterns (Schmidt et al., 2017). This is one example that calls for ecosystem services with soil stewardship in this zoonotic, waterborne bacterial disease, Leptospirosis. Leptospirosis outbreaks have been linked with flooding, impeded soil hydrology, and erosion that mobilizes bacteria into waterways, while flood attenuation is a regulating ecosystem service. Therefore, land use and soil management that will protect soil functions will benefit the delivery of ecosystem services as well as animal and human health (Keitha et al., 2016).

The El Niño-Southern Oscillation (ENSO) cycle is a global climatic phenomenon consisting of hot and warm phases and contributing to increased extreme weather events resulting from biodiversity changes (Kovats, Bouma, Hajat, Worrall, & Houries, 2003). The ENSO has contributed to heavy rainfalls and Rift Valley fever outbreaks in East Africa (Gould et al., 2009). ENSO and global warming is also expected to influence fascioliasis in the Andes (Mas-Coma, 2009). However, the impact of global warming on the ENSO is as yet unknown (McPhaden, Zebiak, & Glantz, 2006).

4. Overexploitation

Environmental policy-making challenges are considered through society adaptation in sustainable methods and legislation of using natural resources. Excessive pressure on wildlife and plants is mostly correlated with overexploitation. Thus, the exploitation of ecosystems by humans has long-lasting consequences of nature resources and ecosystem services and requires sustainable use.

Food is fundamental for human well-being, but increased human population globally is considered as a problem for natural resources worldwide. This has a negative impact on plants and wildlife populations especially in poor countries with high biodiversity where their economy mostly depends on exploitation of natural resources. In this context, exploitation means the action of excessively using lands or seas to extract products, such as wild animals, plants and other products that comprise food and medicine (Reynolds & Peres, 2006). However, overexploitation, which refers to exploiting natural resources to an excessive degree, has become the second most significant threat of wild plants and animals globally. Human needs and desire for effective economic stability increased the exploitation of natural resources. This initial step consequently implements illegal activities through illegal hunting, overfishing as well as deforestation which have an impact on biodiversity.

Globally, overexploitation could lead to resource destruction, including species extinctions. These species are threatened by hunting, fishing, and other forms of exploitation. Most activities, such as fishing and hunting, do not affect only target species but also those which are accidently caught which could further affect the structure and functioning of terrestrial and aquatic ecosystems (Reynolds & Peres, 2006).

4.1 Global trade of wildlife

Global trade seems to be one of the major threats that can lead to overexploitation. Global trade has existed for centuries and probably for about 3000 years but very often is associated with overexploitation and illegal hunting. What has changed through these years are its intensity, which has negative impact on biodiversity, habitats, ecosystem services, and human society. Global trade could have both, direct and indirect impacts on ecosystem and ecosystem services, especially in exporting countries if their policy and management is not well regulated. For example, increased trade for forest products could lead to degradation of forest in countries with unregulated and poor systems, but on the other hand it increased profit. However, for long term processes, loss of ecosystem services have greater damage to economic profits enabled by natural resources. From this point of view, major impacts inflicted on ecosystems are caused by fisheries, shrimp and fish farming, agro-fuels, flower and vegetable, tropical timber, palm oil, and soybean cultivation.

Legal and illegal global trade of wildlife showed as very profitable yielding billions of dollars annually to the countries possessing the areas rich with the wildlife such as developing countries with extraordinary biodiversity (CRS, 2008). According to Congressional Research Service (CRS, 2008) the United States, the People's Republic of China, and the European Union are countries showing the greatest demands for specific animal parts for usage in e.g. zootherapeutics, human consumption, as symbols of wealth and as exotic pets. The illegal and uncontrolled trade of wild animals and direct contact with animal parts exposes humans to close contact with viruses and other pathogens hosted by them. Contact with various wild species such as bats, Asian palm civets, monkeys, pangolins and others can initiate and contribute further to the spread of various zoonoses (Johnson et al., 2015), as can be seen from the 2019 coronavirus disease pandemic.

Besides bats, Malayan pangolins *Manis javanica* also showed as the hosts of the coronavirus, similar to the SARS-CoV-2 (Zhang, 2004). Pangolins are over-hunted due to their usage in traditional Asian medicine, and the consumption of their meat is considered as a delicacy in some Asian and African communities. This results in the international over-trade of this species simultaneously affecting the spreading of the coronaviruses across the world. Although the species has been protected since 2016 the trade continued. Along with that, humans remain exposed to the virus hosted by this species. Rabies was introduced in the mid-Atlantic states in the 1970s when hunting areas were re-populated with raccoons trapped in rabies-endemic zones of the southern United States (Woodford & Rossiter, 1993). In Eastern Europe, raccoon dogs *Nyctereutes procyonoides* are becoming a new reservoir for rabies, next to established red fox reservoir, as raccoon dogs have spread into new habitats from accidental release of animals raised for fur trade (Gylys et al., 1998). Translocation of hares from central and Eastern Europe for sporting purposes affected the introduction of *Brucella suis* biovar 2 to western Europe, and subsequent encroachment of this brucellosis strain into the wild boar population of western Europe (Godfroid et al., 2005). During 1993–2003, *B. suis* biovar 2 infections were reported in more than 40 outdoor-rearing pig farms in France.

Live animal markets, wildlife hunting, intensive wildlife farming (farming of e.g. deer, rodents, civets, mongooses, fur mammals, ostriches), and domestic animals are the most common animal-human interfaces for the emerging zoonotic diseases; resulting in spillover to humans be transmitted through consumption, medicinal use, handling of the living animal or slaughtering and/or preparation of the meat for sale or consumption (Van Langevelde & Mendoza, 2021).

Bushmeat consumption, especially of primates, is increasing zoonoses risk, as can be seen from HIV and Ebola outbreaks (Chapman, Gillespie, & Goldberg, 2005; Daszak, 2006; Peeters et al. 2002). Ebola outbreak was explained earlier as the consequence of the tropical forests degradation; however, it is also linked with the hunting, butchering, and consumption of meat from the infected wild animals, particularly in Gabon and the Republic of Congo (Leroy et al. 2004; Gummov, 2010). Although humans historically hunted wild animals for meat, today, bushmeat consumption is drastically escalating throughout the world (Friant et al., 2020). In the rural and poor areas, bushmeat is used as a major source of nutrition. In urban areas, it is used as the sign of wealth, sold at higher prices despite often ignoring food safety standards, therefore directly increasing the spread of the pathogens among humans (Van Vliet et al., 2017). Bushmeat consumption, urbanization and road constructions in Central Africa were the primary drivers for the global human immunodeficiency virus-1 (HIV-1) escalation to a pandemic (Hahn, Shaw, De Cock, & Sharp, 2000). HIV-1 emerged from chimpanzees in Africa, spilling over to humans repeatedly before its global spread. The initial phase of emergence was driven by bushmeat hunting and was the primary driver of its emergence. The second phase of emergence was driven by increased urbanization and road expansion in Central Africa beginning in the 1950s, and dispersal of index cases harboring prototype HIV-1 infections that were transmissible from person to person. The virus then entered the rapidly expanding global air travel network and became pandemic, with its emergence in North America, Europe, and Asia, accelerated by changes in sexual behavior, drug use, trade-in blood derivatives, and population mobility.

Consumption of raw meat generally presents one of drivers for the zoonoses increase (Macpherson, 2005). This requires the existence of the so-called "wet-markets" that are selling live animals (Macpherson, 2005) and bringing humans to the more direct contact with pathogens hosted via these animals.

Intensified international wildlife trade is certainly one of the most prominent drivers for the occurrence of zoonotic diseases outside their native environments. Among the most prominent examples of a zoonotic disease vector being introduced to Europe is the Asian tiger mosquito, *Aedes albopictus*. This insect is an aggressive daytime biting mosquito that is emerging throughout the world as a public health threat due to its relevance in (among others)

West Nile virus, dengue virus and chikungunya virus outbreaks (Bonizzoni et al., 2013). The mosquito was introduced with the used tire and ornamental plant trade from its native range in East Asia and has established on every continent except Antarctica during the last decades, including Europe (1979), the continental USA (1985) and Central and South America (1980–1990s) (Medlock et al., 2012).

Severe acute respiratory syndrome (SARS) is a viral respiratory illness caused by a coronavirus, called SARS-associated coronavirus (SARS-CoV). SARS-CoV was first reported in Asia in February 2003 (CDC). The disease escalation is directly linked with the trade and consumption of the commercial bushmeat in Asia (Bell, Roberton, & Hunter, 2004; Donnelly et al., 2003; Peiris et al., 2003) from where it spread to more than two dozen countries in North America, South America, Europe, and further in Asia before the global outbreak in 2003. The virus developed in Chinese wet markets in both animals and workers (Guan et al., 2003) and further research showed that the origin of the SARS, in addition to markets, might be the wildlife farms and restaurants in China (Bell et al., 2004). Tu et al. (2004) that studied civets as a carrier of this virus found that those at the farms were free of the virus, whereas after their transport to the markets, were positive for the virus probably due to the contact with other infected animals. The SARS outbreak escalated from the markets in Guangdong, China, where the masked palm civets transmitted the virus from bats, the original reservoir of the virus, to the humans due to their frequent contact (Li et al., 2005).

The farming of wild animal species is also a very important driver of zoonoses transmission (Chomel, Belotto, & Meslin, 2007). A great example of this is the emergence of bovine tuberculosis in captive deer populations. In natural habitats, deer are less likely to be affected to any major extent by disease, due to the controlled population size. However, in captivity, during intensive farming of this animal, reemergence of the disease, that had been controlled from their domestic animal reservoirs, becomes the major concern (Wilson, 2002).

It is very important to understand the opposite process as well, where human activities increase the risk of wildlife infection. This can create new reservoirs of human pathogens, as can be seen from the recent outbreak of tuberculosis caused by *Mycobacterium tuberculosis* in mongooses *Mungos mungo* in Botswana and suricates *Suricata suricatta* in South Africa. Tuberculosis outbreak was caused in both cases by humans. Banded mongooses were observed feeding regularly at these garbage pits and would therefore be exposed to human excretions and any infectious material from tuberculosis-infected humans. Suricates were not observed feeding in garbage pits; however, they were seen foraging around roads and investigating human sputum. This outbreak was one of the first documented spillovers of a human disease within a wildlife population (Chomel et al., 2007).

4.2 Trade and overexploitation of aquatic wildlife

Frozen shrimps and prawn cover 25% of all EU imports of fish and fishery products in 2009 (53) where following five countries are considered as main for EU import: Ecuador (14%), India (13%), Argentina (12%), Bangladesh (9%), Thailand (7%). Increased shrimp farming caused ecological negative effects, loss of mangrove ecosystems, nutrients enrichment, eutrophication of coastal water, persistence of chemicals and toxicity, introduction of exotic species. Ponds for shrimp and fish destroy 20–50% of mangroves worldwide (Primavera, 1997) which are important for reduction of shoreline and river-bank erosion, stabilization sediments and absorption of pollutants, as well as providing timber and wild fish.

However, overexploitation of marine wildlife is frequent, especially in Asia where they exploit and trade endangered wildlife products. Moreover, shark fin and beche-de-mer (dried body wall of sea cucumbers) became one of the most traded products in China's markets for dried seafood.

Although it was expected that global sea cucumber and shark fin production would increase under the economic condition in China, so far it has not been the case. Actually, sea cucumber production is still available, while shark fin production has fallen because of resource constraints. Therefore, shark species are at the highest risk of extinction. However, shark fins are used in traditional cuisine as important luxury food, especially as a shark fin soup. Regarded the importance of shark fins for tradition and society in Asian countries, notably China, Clarke, Milner-Gulland, and Cemare (2007) consider that most shark populations would continue with overexploitation as long as fish activities provide economic value. Furthermore, as long as there is a high price of this luxury good, the fishing activities will be unregulated. Generally, overfishing could cause serious problems on ecosystem balance and species extinctions. Furthermore, consuming raw fish could have caused physiological problems, as well as diseases, including zoonotic.

Wildlife trade in aquatic ecosystems could be presented as a potential source of zoonotic pathogens. Fish-borne zoonoses caused by cestodes could infect humans which belong to order *Diphyllobothriidea*. It has been estimated 20 million people worldwide are infected by these zoonoses. Although it is recorded in colder areas such as subarctic and arctic areas, it has been considered that zoonoses could be introduced into the new geographic parts. Beside the facts that this zoonosis could be expanded through consumption as a raw food or human migration, it could be expanded through global trade.

Although there are a high number of fishbone-zoonoses, there are 15 diphyllobothriidean tapeworms species capable of infecting humans. *Diphyllobothrium latum* (Linnaeus, 1758) Lühe, 1910 is considered as zoonosis with an estimated largest number of cases worldwide (total 10–20 million cases). Definitive hosts are human, dogs, canids and felids, while the second intermediate hosts are freshwater fishes (Scholz & Kuchta, 2016).

5. Pollution

Pollutants negatively and significantly affect ecosystems in different ways leading to biodiversity and habitat loss. In modern life, humans mostly inhabit urban cities, especially the bigger ones, where improperly resource management leads to an unhealthy environment. As air pollution is mostly present in bigger cities or industrial cities, most citizens do not recognize their personal crucial role in nature and ecosystem services protection through spreading green surfaces. This includes introducing green infrastructure as a solution for air polluted cities. Policy makers were searching for effective methods of reducing human exposure to air pollutants in urban cities and nowadays they find the most effective method for air pollution mitigation - green infrastructure: street and park trees, green walls, green roofs (Berardi, GhaffarianHoseini, & Hoseini, 2013) and other ways of expanding the vegetation and green surfaces in urban landscape. Thus, green infrastructure can provide human health, economics and social co-benefits, otherwise improperly built buildings with no green surfaces increase air pollution. Urban trees provide habitats that enhance biodiversity, provide shade and other microclimate services (Salmond et al. 2016). The process of reducing air pollution through trees and forest as ecosystem benefits are presented by particulate accumulation on the leaves surface. It has been estimated that 1 ha of trees removed 9,7 kg of pollution in one year in Chicago. Therefore, the contribution of ecosystem services provided by urban green surfaces and forests allowed healthier quality of life. Positive impact of green spaces enhanced immune functioning and reduced chronic diseases and mental health disorders.

Creating the long-term sustainability perspective from the point of human needs is a long way and makes it almost impossible to achieve. Lack of awareness of the overall issues of ecosystem services importance, avoid the understanding of human significant roles in nature protection which consequently focus their attention on profit and cost. Most urban citizens cannot recognize the value and benefits of urban forests, but can recognize the cost through personal experience or losses. The world's population is expected to increase in urban areas and cities which would lead to extremely high air pollution through industrial production. China, as one of the world's economic developed countries, rapidly expanded industrial development, resulting in increased energy consumption, emissions of air pollutants, and the number of poor air quality days. Thus, this problem has become one of the most significant environmental issues in China. (Chan, 2008). However, China is not the only air polluted city, but all other big cities have the same problem worldwide. Also, urbanization affects biodiversity through light pollution. Furthermore, the artificial light at night (ALAN) has increased worldwide due to urbanization, affecting nocturnally bird migrations where in spring and autumn effects of ALAN are higher than during the breeding period. The intensity of ALAN, as urban light pollution, has an impact on bird migration causing disorientation and resulting collisions and mortality. This is considered as a big problem for bird populations, especially juveniles who are having first migration. Changes in migrations flyways due to ALAN cause changes in migratory behavior of migratory bird populations. Furthermore, ALAN can affect the precision of the magnetoreception system which is a main driver for bird disorientation (La Sorte et al., 2017).

Being aware of the significant importance of water through the high water percentage in the human body and on Earth's surface, water pollutants are highly important because of their negative impact on living organisms. Therefore, water pollutants include pollutants and contaminants of domestic waste, insecticides and herbicides, food processing waste, heavy metals, chemical waste and many others. Lack of improved sanitation seems to be one of the most severe impacts on human health which is related to the lack of safe drinking water. Accessible renewable freshwater is consumptively used for agricultural, industrial and domestic purposes (Schwarzenbach et al., 2006) which activities lead to water contamination containing various synthetic and geogenic natural chemicals. This scenario is one of the major problems for water sanitation worldwide. Moreover, freshwater and oceans provide the most important human benefits, giving the drinking water and food. Considering these benefits, any threat can be harmful to ecosystem services which are linked to the human needs, as well as their health, disturbing the ecosystem balance and causing the loss of food security, livelihoods and good health.

However, one of the greatest environmental global concerns are persistent organic pollutants (POPs) and plastics because of their long-range transport and persistence potential in the environment and ability to biomagnify and bioaccumulate in ecosystems. This also includes that chemical, photochemical and biological transformation processes do not lead to a significant removal of the compound in any environmental compartment which gives the focus of their toxic features to living organisms. The ability of plastic generation, fragmentation and propensity to sorb/release persistent organic pollutants (POPs) are determined by the features of constituted polymers. This can be very harmful to biodiversity, especially as microplastics that can be formed through fragmentation known as secondary microplastics (Anthony, 2007).

As the microplastics issue is correlated to ocean global problems, it is a big concern to marine wildlife, especially in estuaries that provide several valuable ecosystem services such as fixing carbon, recycling nutrients and coastline protection from erosion and wave action (Schaafsma & Turner, 2015). Because of this, it is important to understand the negative impact on ecosystem services, economic value, and environment, human and animal health. Thus, by growing the population in coastline areas the higher potential of valuable ecosystem services would become compromised (Gray, Wertz, Leads, & Weinstein, 2018). This should encourage institutions, policymakers and citizens to become more engaged in water management through education and policy.

For a long time, it has been considered that microplastics had a negative impact on biodiversity and habitat loss affecting the health status through physiological changes. Because of their prevalence, microplastics affect animals creating the changes in feeding and depleting energy stores which link it with negative impacts on fecundity and growth. Thus, the chronic exposure to microplastic is rarely lethal (Galloway, 2017).

Zettler, Mincer, and Amaral-Zettler (2013) analyzed the microplastics debris collected from the sea surface and discovered a diverse microbial community of colonizing bacteria. These bacteria form biofilms on the surface allowing access to nutritious matter and dispersal enhancing. Thus, the microplastics biofilms are much different than other marine substrata (Wotton, 2004) where *Vibrio* is most frequently reported bacteria on plastic surface linking the changes in organisms such as the pathogen infection in oysters caused by *Vibrio crassostrea* (De Tender et al., 2015). Last years, most studies have focused on the individual effects of microplastics ingestion on cellular and subcellular level where it has been recorded the sub-cellular oxidative stress as a response on polyester (2–6 μm) ingestion in mussels (Paul-Pont et al., 2016). Furthermore, the reproductive output is one of the affected parts under microplastic pressure especially fecundity and fertility. he Pacifiy oysters *Crassostrea gigas* it has been recorded the significant changes in sperm motility, oocyte number (fecundity) and size (Cole & Galloway, 2015). This is a great example of how microplastics can be used as vectors for harmful microorganisms that can have negative ecological impacts. Furthermore, the microplastics can be a good substrate for various invasive species allowing their colonization worldwide (Debeljak et al., 2017). Through this process, microplastics as a substrate of wide range species can cause the introduction of invasive and increase some populations by transporting them to remote regions desregulating ecosystems.

As water pollution is a serious problem globally causing various diseases and changes in organisms that directly affect the human population, there are also other diseases such as zoonoses which can be more related to improper animal waste management and water sanitation. El-Tras, Tayel, Eltholth, and Guitian (2009) report cases of Nile catfish infected by *Brucella* in Nila Delta region, Egypt. This region is well known as an area where people throw animal waste. Although there are very rare cases regarding the records of infected fish species, this can have a significant role in the epidemiology of brucellosis which should be investigated in future studies. Furthermore, there are studies indicating disease infection by consuming raw fish in the U.S. and New Zealand. These indicate that raw fish products can cause brucellosis. However, animal waste should be considered more carefully which also needs better waste management as well as water sanitation. Therefore, brucellosis is considered as a great example of how polluted waters and water sanitation problems can cause serious problems globally.

Furthermore, a slurry pit used by farmers for throwing animal waste together with other unusable organic matter, causes decomposition and produces deadly gases. Vidal, Lopez, Santoalla, and Valles (1999) analyzed physical chemical parameters where they recorded saline and organometallic contamination in rivers due to slurries negative effects. The saline contamination was recorded in wells and conduits. In a farm slurry pit, the conditions of temperature or environment suitable for *B. abortus* is 12°C.

Expansion of animal industries and urbanization consequently caused the lack of hygienic measures in animal husbandry as well as the contaminated food or water leading to public health hazard. Due to urbanization, population growth and caused problems in life pollution sources (Teng, Yang, Zuo, & Wang, 2011) which could be considered as a trigger for disease. Furthermore, increasing the international travel and importation of products can transport contaminated food contributing to increased disease spreading (Corbel et al., 2006). *Brucella spp.* can survive for a long period on different surfaces such as dust, dung, water, soil, meat and dairy products which depend on the type of substrate, as well as the environmental conditions such as temperature or pH. It is intracellular gram-negative cocco-bacilli, non-spore-forming and non-capsulated (Seleem et al., 2009). There are more species recognized as important for brucellosis disease development, including terrestrial animals: *B. abortus, Brucella melitensis, B. suis, B. ovis, B. canis, B. neotomae,* and *Brucella microti* i (Scholz et al., 2008; Verger, Grimont, Grimont, & Grayon, 1987) and two that affect marine mammals: *Brucella ceti* and *Brucella pinnipedialis* (Foster, Osterman, Godfroid, Jacques, & Cloeckaert, 2007). However, hygiene practices have one of the main roles in disease prevention which are a great challenge for poor and developing countries.

Soils as a natural resource are also a renewable resource, but the rates of soil formation or restoration could not match the rates of soil degradation making the degradation process much faster than the formation or restoration. Moreover, soils have high biodiversity where species interactions are necessary for soil processes and ecosystem functions which include nutrient cycling, organic matter decomposition, soil structure formation, pest regulation and bioremediation of contaminants (Pulleman et al., 2012). All mentioned processes are related to ecosystem services such as food production, climate regulation or provision of clean water. Loss of biodiversity is in correlation with loss of ecosystems which are due to expansions and intensification of agriculture across Europe, but also other related pressures such as soil erosion, organic matter decline, compaction, contamination, salinization and climate changes. Therefore, it is important to understand the soil biodiversity, their distribution, interaction and, functions in ecosystems, and how they transform and provide ecosystem services. Property soil and land management can provide a healthy environment by allowing property implementation of regulations.

Medical drugs found in soil ecosystems can have a negative impact on biodiversity loss containing the harmful components. Cuthbert et al. (2007) reported negative impact of non-steroidal anti-inflammatory drugs (NSAID) on Asia population of Gyps vultures. Carprofen, diclofenac, flunixin, and ibuprofen toxicity caused negative effects or mortality on raptors, storks, cranes and owls too, affecting the wide range of different species. However, Gyps vultures are scavengers so they have an important role in ecosystems by consuming dead animals. Decrease in scavengers populations leaves carcasses in the wild, which can also contribute to disease spreading.

Pesticides are very harmful and they enter into the food chain through plant roots absorption. Therefore, they contaminated food causing disease such as itai-itai disease in Japan caused through ingestion of cadmium contaminated rice (Pan, Plant, Voulvoulis, Oates, & Ihlenfeld, 2010; Volpe et al., 2009). Negative impacts of pesticide on soil organisms and functions are shown through different examples, such as the possibility of some organochlorine pesticide that suppress symbiotic nitrogen fixation resulting in lower crop yields (Fabra, 1997). Furthermore, if the content of nitrogen is higher than plants limit, the nitrification of microbial activity will lead to accumulation of nitrates (NO_3). When soil nutrient availability increased microbial biomass it caused the imbalance in the nutrient cycle (Lu & Tian, 2017).

However, the main problem are harmful pesticides which are available in markets, but only in Low and Middle Income Countries, because High Income Countries take them off from the market due to negative effects on human health. Furthermore, Low and Middle Income Countries are often facing these problems because the assessment and monitoring of products' ecotoxicity are nonexistent due to low regulation capacity (Brodesser et al., 2016). Uncontrolled use of high pesticides concentration to increase productivity can induce soil degradation processes where productivity can be affected too. Also, intestinal pathogens could enter the soil through manure or faces and survive and stay there for several months or years. Pathogens mostly presented in soil are *Salmonella, Campylobacter* and *Echerichia coli* viruses.

Furthermore, pesticides have effects in in-crop areas where some species such as pest predators or natural pollinators colonize crop fields. Species exposed in areas with high pesticides percentage needs recovery. No matter what substance is used, pesticides or herbicides, both of them affect ecosystem services through habitat modification. Most invasive species can be found in areas that are under pesticide and herbicide pressure, affecting plant populations and making it suitable for invasion species. High concentration of these chemicals results in better adaptation and spreading of invasive species which affect native species and deregulating ecosystems.

Considering facts above, the soil ecosystem services refer to the capacity of natural processes and components to provide goods and services that satisfy human needs, directly or indirectly. Therefore, researchers were trying to figure out how to put price tags meaningfully on "nature's services" through better understanding various functions and benefits of soils.

6. Invasive species

Every minute thousands of species of microbes, plants, fungi and animals are being transported, intentionally or unintentionally, to new environments across the planet. Nowadays, globalization process has a high impact on biodiversity, environment and society. Although it provides benefits for human well-being, there are many challenges which could be harmful for environment and biodiversity, and one of the most significant is the introduction of alien species. Although the majority of these species fails to establish populations and spread in new habitats, some manage to adapt, reproduce and spread beyond the place of introduction. Some of them present a threat to native species or cause changes in the structure and composition of ecosystems, often affecting the ecosystem services valuable for mankind. CBD (2008, pp. 284–299) defines native or indigenous species as those species that naturally inhabit given area, while invasive alien species (IAS) are defined as species which are intentionally or unintentionally introduced by human activities to an ecosystem in which they do not live naturally and present as threats for habitats, ecosystems or native species.

Invasive alien species are one of the main causes of biodiversity loss and one of the most powerful means by which humans alter the planet, due to their ability to disperse and cause negative effects on native species and the environment (Simberloff et al., 2012). They belong to all major taxonomic groups, from viruses to mammals. In some of those groups, e.g. vascular plants, as many as 10% out of 300,000 known species, have the potential to become invasive (Rejmánek, 2000). The number of species introduced to new regions by human activities is continuously increasing (Seebens et al., 2017).

The major influences of invasive alien species include: a) environmental and ecological impacts on native species and ecosystems; b) economic and social impacts and c) impacts on human health (Mazza et al., 2013). Populations of native species can be directly affected through predation, herbivory and disease (Simberloff, 2013), e.g. the brown tree snake *Boiga irregularis* that caused the extinction of nine bird species on Guam Island (Simberloff & Rejmánek, 2011). Indirectly, invasive species may cause native species declines due to resource competition and habitat alteration (Davis, 2009), e.g. the salt cedar *Tamarix spp.* that can make the soil unsuitable to native species due to the large amounts of salt deposited into the surrounding soil (Bell et al., 2002). Economic and social impacts that can represent direct effects of a species on property values, agricultural productivity, public utility operations, native fisheries, tourism, and outdoor recreation, or costs associated with invasive species control actions. According to the study from the United States it estimated that the economic costs associated with invasive species in 2005 reached $120 billion per year (FWS, 2012). They affect human health, whether as vectors of diseases or by inflicting bites, stings or causing allergies and poisoning. The Asian tiger mosquito *A. albopictus* that is considered as the most invasive mosquito in the world, represents a vector for several diseases, including West Nile Virus and Dengue fever (Benedict et al., 2007).

According to the Millennium Ecosystem Assessment (2005), the impact of invasive alien species on biodiversity is either steady or increasing. It is the highest on island biodiversity, followed by coastal and inland waters and Mediterranean ecosystems. The very rapid increase of impact is seen in several biomes, including temperate and tropical forests, Mediterranean and tropical grasslands and inland waters.

Invasive pathways represent the dispersal mechanisms used by alien species to enter certain territory. It is a series of stages that include transport, introduction, establishment and spread of alien species (Cassey, Garcia-Diaz, Lockwood, & Blackburn, 2018; Tsiamis, Cardoso, & Gervasini, 2017). The introduction can be intentional or unintentional. Intentional introduction refers to the deliberate movement and/or release by man of an alien species outside its natural range, while unintentional introduction represents all other introductions. As defined by the European Strategy on invasive alien species (Genovesi & Shine, 2004) the pathways represent: a) The geographic route by which a species moves outside its natural range (past or present); b) The corridor of introduction (e.g. road, canal, tunnel) and c) The human activity that gives rise to an intentional or unintentional introduction. The invasion pathway starts with species transportation to introduced area. When a species arrives into a "non-native" area due to human intervention, it becomes an introduced or alien species. By arriving in the new area, these species are not self-sustainable. Therefore, they need to pass the pathway of becoming self-sustainable, through further changes and adaptation, when the species survives, escapes and reproduces without human direct help. After achieving all of these criteria, a species becomes established. When this species is dispersing and spreads widely with negative consequences to other species, habitats or human wellbeing, it becomes an invasive species (e.g. Keller, Geist, Jeschke, & Kuhn, 2011).

Intentional introduction is a result of human activities, e.g. through international trade, while unintentional introduction could be a result of unintentional transport from one to another region. This also include the stowaways on ships where smaller animals or plants are found on the commodity or in its packaging (Cassey et al., 2018). Rats and mice are good examples of a worldwide spread of a species by accidental transport on ships. Generally, introduction of these species often has devastating influence on island biodiversity (Drake & Hunt, 2009; Harper & Bunbury, 2015). These animals have been accidently introduced by ships worldwide and had a strong impact on bird's diversity, particularly on islands. In the 16 United Kingdom Overseas Territories (UKOT) with the world's great seabird colonies, including several endemic and threatened species, significant changes and reductions of populations and extinctions of colonies (among which are four critically endangered endemics on Gough Island, St Helena and Montserrat) were caused by invasive mammals, feral cats, rats and mice (Hilton & Cuthbert, 2010).

Intensification of global trade after the First World War resulted with many new invasive species of plants introduced to Europe. One of the most notable among them is common ragweed *Ambrosia artemisiifolia* (Fig. 9.2), a species infamous for its adverse effects as an agricultural weed and specially as a source of allergenic pollen (Comtois, 1998). Common ragweed is a highly invasive species native to North America, whose introduction and spread in Eastern Europe in recent decades resulted with large environmental and economic costs in agriculture and public health.

The slider terrapin *Trachemys scripta* (Fig. 9.3), native to the eastern US and northeast Mexico, is probably the most commonly traded reptile worldwide that has been very popular pet because it is cheap and easy to keep animal (more than 52 million individuals exported from the US between 1989 and 1997) (Telecky, 2001). They are usually sold as very young and only a few centimeters in size, but they grow quickly and owners are rarely willing to keep large adults so they often release them into natural or semi-natural habitats (Teillac-Deschamps et al., 2009). Non-native populations are

FIG. 9.2 Common ragweed *Ambrosia artemisiifolia* L.

FIG. 9.3 The slider terrapin *Trachemys scripta* (Thunberg in Schoepff, 1792) (right) basking on the sun with European pond terrapin *Emys orbicularis* (L.) (left) on its back.

established worldwide, including Mediterranean region, Europe, Asia, non-native range in North America, Australia and New Zealand (Lever, 2003; Cadi et al., 2004; Ramsay et al., 2007; Ficetola et al., 2009; Kikillus et al., 2010). These terrapins are large and aggressive, and have considerable impact on native species, including reptiles, amphibians, fish and invertebrates. They outcompete native terrapins for food, nesting and basking places (Lindeman, 1999; Cadi & Joly, 2003; Spinks et al., 2003).

In addition to threats to native species and ecosystems, the spread of invasive species can be very costly, causing significant damages in agriculture, forestry, fisheries, and other human activities. Therefore, quantification of invasive alien species potential impact on ecosystem services is necessary for implementing effective protection and management measures. They can cause huge outflow of global resources (Early et al., 2016), e.g. in UK at least £1.7 billion annually and at least €12.5 billion annually in EU (Pimentel, Lach, Zuniga, & Morrison, 2000; Williams et al., 2010, pp. 1–99).

Invasive species represent a global problem that asks for international cooperation. Numerous global and regional policies are already addressing this issue. The Convention on Biological Diversity (CBD, 2010) in Article eighth calls on its parties to "prevent the introduction of, control or eradicate those alien species which threaten ecosystems, habitats, or species" and the Aichi Biodiversity Target 9 from Strategic Plan for Biodiversity 2011–20 urges that "By 2020, invasive alien species and pathways are identified and prioritized, priority species are controlled or eradicated, and measures are in place to manage pathways to prevent their introduction and establishment" (Decision X/2 of the Conference of the Parties to the Convention on Biological Diversity, annex.)

In Europe, the fast growing number of new invasive species becomes an increasing and serious problem that requires substantial joint engagement of all stakeholders, development and implementation of proper IAS regulations. As the problem of IAS is growing pretty fast, all regulations should be updated regularly based on the newest situation and knowledge in Europe and worldwide. To tackle the problem of IAS, European Parliament adopted the EU Regulation 1143/2014 that incorporates the list of Invasive Alien Species of Union concern that is updated regularly (Pagad, Genovesi, Carnevali, Scalera, & Clout, 2015).

Giving the significant importance to IAS as a great threat to the environment, the International Union for Conservation of Nature (IUCN) established Invasive Species Specialist Group (ISSG) that is working with others to mitigate biodiversity loss from IAS. Furthermore, The Intergovernmental Science-Policy Platform on Biodiversity and Ecosystem Services (IPBES) Regional Assessments and Global Assessment Report identified IAS as one of the main causes of biodiversity loss worldwide and launched assessment of invasive alien species and their control.

Many IAS could act as the potential vectors of human and animal disease. Some of these species are very popular as pets globally, which is also a serious problem as it represents an organized and intentional introduction of potentially harmful species throughout the world. One of them is a small squirrel-like animal, the Siberian Chipmunk *Tamias sibricus*, that is very popular among pet owners. However, this species could be infested by the castor bean tick *Ixodes ricinus* and consequently infected by *B. burgdorferi* that causes the Lyme borreliosis, increasing the risk of spreading this disease to humans (Marsot et al., 2013). Another example is raccoon dog *N. procyonoides*, the canid indigenous to East Asia that is today widespread in Europe. This carnivore is an ideal host and vector for a variety of pathogens, like leishmaniosis, trichinellosis and *E. multilocularis*. Due to its good migration ability, it has large potential for distribution of pathogens between different regions in Europe and its establishment increases health risks for livestock, wildlife, and humans (Sutor et al., 2014). There are many other examples of various zoonotic diseases caused by IAS and further increase in introductions of alien species in non-native areas will certainly become a growing problem leading to expansion of animal and human diseases globally.

For centuries introduced pathogens have devastated human populations. Sea travels to the New World and distant islands brought pathogens, like smallpox, influenza, measles and other diseases to natives with devastating effect. It is estimated that up to 80% of all native Americans died of introduced diseases by the end of the 17th century. Probably the best-known example of introduced human disease are the pandemics of plague in 14th and 19th centuries that killed millions in Europe. Most likely they were caused by infected fleas from Asia carried by rats on ships. In a similar way around 1830 cholera reached Europe. Several influenza epidemics in the past and recent times, like Spanish and Asian flu in the 20th century or avian influenza and swine flu in the 21st century are a warning examples of the public health consequences of introduction of new pathogens (Simberloff, 2013). Expansion of new trade routes between countries and regions as well as modern transportation systems have increased both the frequency and magnitude of invasions and disease outbreaks worldwide (Aide & Grau, 2004; Crowl et al., 2008).

The impacts of invasive alien species on human health are various, from psychological, discomfort, nuisance and phobias to skin irritations, allergies, poisoning, disease and death. Some species are of particular importance to human health, like strongly allergenic plant species, such as common ragweed, *A. artemisiifolia* (Chapman et al., 2016) or mosquitoes that are of high medical significance being the vectors of human diseases (Martinet et al. 2019; Romi et al., 2018). These impacts will further increase with climate change. As it is highlighted by Millennium Ecosystem Assessment, major drivers of environmental change will interact together. Climate change and invasive alien species are described as a 'deadly duo' at the Nagoya Biodiversity Summit in 2010, with abundant evidences on the growing effect of climate change on the already immense impact of invasive species. Roy et al. (2009) emphasized the increasing impact of several invasive species, including mosquitoes, invasive garden ant *Lasius neglectus* and Argentine ant *Linepithema humile*.

Global climate change effects include changes in species distributions and abundance, changes in abiotic factors, changed opportunities for reproduction and recruitment and altered interactions among species (Karieva et al., 1993; Sutherst, 2000). Climate change alters the conditions for the establishment and spread of invasive species, as well as the suitability of local climates for native species and the nature of interactions among native communities. These changes can reduce the ability of native species to resist invaders and make them more sensitive to pathogens and predators. The extreme climatic events can disturb ecosystems and make them vulnerable to invasions, e.g. invasive species can spread more easily after droughts, flooding's of fires that destroy native vegetation.

Disease outbreaks caused by arboviruses that were previously confined to tropical or subtropical regions are increasing in Europe (Martinet et al., 2019). With the global changes and increase in travel and transport arboviruses are more easily expanded outside of their historical range. Establishment of major disease vectors, mosquitoes and ticks, facilitates transmission of new pathogens. The mosquitoes Culicidae are the main taxonomic group of medical importance within the Arthropods. In addition of being biting nuisance, these blood-sucking insects are vectors of numerous diseases, such as malaria, yellow fever or dengue (Beaty & Marquardi, 1996). They are among the most successful invaders and at the top of threat to human life globally due to adaptations of many species to live with humans (particularly *Aedes*, *Anopheles* and *Culex* mosquitoes) and the ability to spread easily (Toto et al., 2003;

Almeida et al., 2007; Bullivant & Martinou, 2017; Lounibos, 2002; Romi et al., 2018). Arboviruses like Zika, dengue or chikungunya are already causing autochthonous human infections in Europe in last two decades and the number of countries in Europe where exotic invasive mosquitoes that represent disease vectors are being established is growing (e.g. Martinet et al., 2019; Medlock et al., 2012). Humans have played a major role in the worldwide dissemination of mosquito vectors. The global transportation of used tires is responsible for importation of invasive aedine mosquito species to many areas. Their drought-resistant eggs laid in these tires can survive long time journeys between continents (Medlock et al., 2012). In Europe, establishment of invasive *Aedes* mosquitoes, particularly *A. albopictus* and *A. aegypti* raises concern about the future spread of dengue or chikungunya arboviruses and the possibility that they will become adapted to local mosquito species (Almeida et al., 2007). Predictions on climate change in Europe suggests further spreading of invasive *Aedes* mosquitoes, important vectors of yellow fever, dengue, Zika and chikungunya viruses, beyond its current range and its adaptation to colder climates (Paupy et al., 2009; ECDC, 2009; 2012).

In some cases, invasions also had significant cultural impacts, like the chestnut blight disease caused by pathogenic fungus that actually eliminated dominant chestnut trees in eastern North America, eliminating with it a major economy based on this tree, its wood, and chestnuts associated cuisine and traditions. Another example is the introduction of phylloxera, a small aphid-like insect, to Europe at the end of 19th century. It devastated grape vines throughout the continent, destroying 75% of all vineyards, and as a consequence many traditions associated with grape culture and harvest were lost when numerous rural settlements dependent on wine production vanished (Simberloff, 2013).

7. Conclusion

The concern about the ecosystems health is one of the leading environmental challenges today, where the links between biodiversity loss, habitat loss and human health play a significant role in human livelihoods, where humans are trying to find solutions for sustainable natural resources use, putting the price tag on ecosystem services. As natural resources are being overexploited, their value increase. Each of four major ecosystem services categories (provisioning, regulating, cultural, and supporting services) are fundamental for human well-being, health, and livelihoods because they maintain biodiversity and the production of ecosystem goods, including water, food, nutrient cycling, waste management, cultural, and recreational activities, which are main ecosystems services linked to public health. As there are many challenges to ecosystem services which are mostly caused by increased human activities, all of them are leading to biodiversity and habitat loss. Biodiversity loss is caused by climate change, habitat loss, pollution, trade, illegal hunting, and spreading of invasive species, which are linked to public health through potential zoonotic disease. One of the main causes of the increase of zoonotic diseases is rapid urbanization, enabling faster transmission of zoonoses because of the high human population density, allowing easier infection through human-animals contact. No matter what are the human needs, solving the ecosystem service challenges need to satisfy both, social-economic and environmental demands, which include the balance of economic growth, sustainable environment, and public health. Otherwise, increased biodiversity loss will disbalance ecosystems and lead to loss of ecosystems services which are fundamental for human well-being, and consequently result with increase of habitat degradation, climate change, invasive species spread or pollution, which are main factors accelerating the spread of zoonotic diseases.

References

Aide, T. M., & Grau, H. R. (2004). Globalization, migration, and Latin American ecosystems. *Science, 305*, 1915–1916.

Allan, B. F., Keesing, F., & Ostfeld, R. S. (2003). Effect of forest fragmentation on Lyme disease risk. *Conservation Biology, 17*, 267–272.

Almeida, A. P., Gonçalves, Y. M., Novo, M. T., Sousa, C. A., Melim, M., & Gracio, A. J. (2007). Vector monitoring of Aedes aegypti in the Autonomous Region of Madeira, Portugal. *Euro Surveill., 12*(11), E071115.6.

Almond, R. E., Grooten, M., & Peterson, R. (2020). *Living Planet Report 2020-Bending the curve of biodiversity loss*. World Wildlife FUnd.

Anthony, L. A. (2007). The plastic in microplastics: A review. *Marine Pollution Bulletin, 119*(1), 12–22.

Antonenko, Y. N., Khailova, L. S., Knorre, D. A., Markova, O. V., Rokitskaya, T. I., Ilyasova, T. M., ... Skulachev, V. P. (2013). Penetrating cations enhance uncoupling activity of anionic protonophores in mitochondria. *PLoS One, 8*(4), e61902. https://doi.org/10.1371/journal.pone.0061902

Baker, M. L., Schountz, T., & Wang, L. F. (2013). Antiviral immune responses of bats: A review. *Zoonoses Public Health, 60*, 104–116. https://doi.org/10.1111/j.1863-2378.2012.01528.x

Bausch, D. G., & Schwarz, L. (2014). Outbreak of Ebola virus disease in Guinea: where ecology meets economy. *PLoS Negl Trop Dis, 8*(7), e3056.

Beaty, B. J., & Marquardi, W. C. (1996). *The Biology of Disease Vectors*. Denver, Colorado: University Press of Colorado.

Bell, D., Roberton, S., & Hunter, P. R. (2004). Animal origins of SARS coronavirus: Possible links with the international trade in small carnivores. Philosophical Transactions of the Royal Society of London. *Series B: Biological Sciences, 359*, 1107–1114.

Bell, C. E., Neill, B., DiTimaso, J. M., Lovich, J., DeGouvenain, R., Chavez, A., ... Barrows, C. (2002). *Saltcedar: a non-native invasive plant in the Western U.S.* California: University of California.

Benedict, M. Q., Levine, R. S., Hawley, W. A., & Lounibos, L. P. (2007). Spread of the tiger: global risk of invasion by the mosquito Aedes albopictus. *Vector-Borne and Zoonotic Diseases, 7*(1), 76–85.

Benvenuto, D., Giovanetti, M., Ciccozzi, A., Spoto, S., Angeletti, S., & Ciccozzi, M. (2020). The 2019-new coronavirus epidemic: Evidence for virus evolution. *Journal of Medical Virology, 92*(4), 455–459.

Berardi, U., GhaffarianHoseini, A. H., & Hoseini, G. (2013). State-ofthe-art analysis of the environmental benefits of green roofs. *Journal of Applied Energy, 115*, 411–428.

Berthová, L., Slobodník, V., Slobodník, R., Olekšák, M., Sekeyová, Z., Svitálková, Z., ... Špitalská, E. (2016). The natural infection of birds and ticks feeding on birds with Rickettsia spp. and Coxiella burnetii in Slovakia. *Experimental and Applied Acarology, 68*, 299–314.

Biagini, P., Thèves, C., Balaresque, P., Geraut, A., Cannet, C., Keyser, C., ... Orlando, L. (2012). Variola virus in a 300-year-old Siberian mummy. *New England Journal of Medicine, 367*, 2057–2059.

Bidle, K. D., Lee, S., Marchant, D. R., & Falkowski, P. G. (2007). Fossil genes and microbes in the oldest ice on Earth. *Proceedings of the National Academy of Sciences, 104*, 13455–13460.

Bogoch, I. I., Watts, A., Thomas-Bachli, A., Huber, C., Kraemer, M. U., & Khan, K. (2020). Potential for global spread of a novel coronavirus from China. *Journal of Travel Medicine, 27*(2), taaa011.

Bonizzoni, M., Gasperi, G., Chen, X., & James, A. A. (2013). The invasive mosquito species Aedes albopictus: current knowledge and future perspectives. *Trends in Parasitology, 29*, 460–468.

Brevik, E. C., & Burgess, L. C. (2015). Soil: Influence on human health. In S. V. Jorgensen (Ed.), *Encyclopedia of environmental management* (pp. 1–13). Boca Raton, FL: CRC Press.

Brodesser, J., Byron, D. H., Cannavan, A., Ferris, I. G., Gross-Helmert, K., Hendrichs, J., ... Zapata, F. (2006). *Pesticides in developing countries and the international code of conduct on the distribution and the use of pesticides*. Consultant, FAO: FAO/IAEA Joint Programme.

Brownstein, J. S., Skelly, D. K., Holford, T. R., & Fish, D. (2005). Forest fragmentation predicts local scale heterogeneity of Lyme disease risk. *Oecologia, 146*(3), 469–475.

Bruggen, A. H. C., Goss, E. M., Havelaar, A., Diepeningend, A. D., Finckh, M. R., & Glenn Morris, J. (2019). One Health - Cycling of diverse microbial communities as a connecting force for soil, plant, animal, human and ecosystem health. *Science of The Total Environment, 664*, 927–937.

Bullivant, G., & Martinou, A. F. (2017). Ascension Island: a survey to assess the presence of Zika virus vectors. *J R Army Med Corps, 163*(5), 347–354.

Cadi, A., Delmas, V., Prévot-Julliard, A. C., Joly, P., & Girondot, M. (2004). Successful reproduction of the introduced slider turtle (Trachemys scripta elegans) in the south of France. *Aquatic Conservation: Marine and Freshwater Ecosystems, 14*, 237–246.

Cadi, A., & Joly, P. (2003). Competition for basking places between the endangered European pond turtle (*Emys orbicularis galloitalica*) and the introduced red-eared slider (Trachemys scripta elegans). *Canadian Journal of Zoology, 84*, 1392–1398.

Cassey, P., Garcia-Diaz, P., Lockwood, J. L., & Blackburn, T. M. (2018). Invsion biology: Searching for prediction and prevention, and avoiding lost causes (pp. 3-13). In J. M. Jeschke, & Tina Heger (Eds.), *Invasion biology: Hypotheses and evidence*. Wallingford: CAB International. https://doi.org/10.1079/9781780647647.0003

Castañeda-Moya, E, Twilley, R, & Rivera-Monroy, V. (2013). Allocation of biomass and net primary productivity of mangrove forests. *Forest Ecology and Management, 307*, 226–241.

CBD – Convention of Biological Diversity. (2008). *Biodiversity glossary*. Available at https://www.cbd.int/cepa/toolkit/2008/doc/CBD-Toolkit-Glossaries.pdf. (Accessed 7 July 2020).

CBD – Convention of Biological Diversity. (2010). *X/2.Strategic plan for biodiversity 2011-2020*. Available at https://www.cbd.int/decision/cop/?id=12268 (Accessed 07 August 2020).

Chan, C.,K., & Yao, X. (2008). Air pollution in mega cities in China. *Atmospheric Environment, 42*(1), 1–42.

Chapman, C. A., Gillespie, T. R., & Goldberg, T. L. (2005). Primates and the ecology of their infectious diseases: How will anthropogenic change affect host-parasite interactions? *Evolutionary Anthropology: Issues, News, and Reviews, 14*(4), 134–144.

Chapman, D. S., Makra, L., Albertini, R., Bonini, M., Páldy, A., Radinkova, V., ... Bullock, J. M. (2016). Modelling the introduction and spread of non-native species: international trade and climate change drive ragweed invasion. *Glob Chang Biol, 22*(9), 3067–3079.

Chomel, B. B., Belotto, A., & Meslin, F. X. (2007). Wildlife, exotic pets, and emerging zoonoses. *Emerging Infectious Diseases, 13*(1), 6–11. https://doi.org/10.3201/eid1301.060480

Clarke, S., Milner-Gulland, E. J., & Cemare, T. B. (2007). Perspectives: Social, economic and regulatory drivers of the shark fin trade. *Marine Resources Economics, 22*, 305–327.

Cole, M., & Galloway, T. (2015). Ingestion of nanoplastics and microplastics by Pacific oyster larvae. *Environmental Science and Technology, 49*, 14625–14632.

Comtois, P. (1998). Ragweed (Ambrosia spp.): the phoenix of allergophytes. *Spieksma, F.T. (ed.) Ragweed in Europe* (pp. 3–5). Denmark: Alk–Abelló A/S, Horsholm.

Corbel, M. J., & Food and Agriculture Organization of the United Nations, World Health Organization & World Organisation for Animal Health. (2006). *Brucellosis in humans and animals*. World Health Organization. https://apps.who.int/iris/handle/10665/43597.

Costanza, R., & Folke, C. (1997). Societal dependence on natural ecosystems. In G. Daily (Ed.), *Nature's services* (pp. 49–70). Washington DC: Island.

Coutts, C., & Hahn, M. (2015). Green infrastructure, ecosystem services, and human health. *International Journal of Environmental Research and Public Health, 12*(8), 9768–9798.

CRS (U.S. Congressional Research Service). (2008). In L. S. Wyler, & P. A. Sheikh (Eds.), *International illegal trade in wildlife: Threats and U.S. policy*. Washington, DC: Library of Congress.

Crowl, T. A., Crist, R. O., Parmenter, R. R., Belovsky, G., & Lugo, A. E. (2008). The spread of invasive species and infectious disease as drivers of ecosystem change. *Front. Ecol. Environ, 6*(5), 238–246.

Cuthbert, R., Parry-Jones, J., Green, R. E., & Pain, D. J. (2007). NSAIDs and scavenging birds: potential impacts beyond Asia's critically endangered vultures. *Biology letters, 3*(1), 91–94.

Davis, M. A (2009). *Invasion Biology*. Oxford: Oxford University Press.

Daszak, P., Plowright, R., Epstein, J. H., Pulliam, J., Abdul Rahman, S., Field, H. E., ... Cunningham, A. A., & the Henipavirus Ecology Research Group (HERG). (2006). The emergence of Nipah and Hendra virus: Pathogen dynamics across a wildlife-livestock-human continuum. *Disease Ecology: Community Structure and Pathogen Dynamics*, 186–201.

de Rocha Taranto, M. F., Pessanha, J. E. M., dos Santos, M., dos Santos Peraira, A. C., Camargos, V. N., Alves, S. N., & Ferreira, J. M. S. (2015). Dengue outbreaks in Divinopolis, south-eastern Brazil and the geographic and climatic distribution of *Aedes albopictus* and *Aedes aegypti* in 2011–2012. *Tropical Medicine & International Health, 20*(1), 77–88.

De Tender, C. A., Devriese, L. I., Haegeman, A., Maes, S., Ruttink, T., & Dawyndt, P. (2015). Bacterial community profiling of plastic litter in the Belgian part of the North Sea. *Environmental Science and Technology, 49*, 9629–9638.

Debeljak, P., Pinto, M., Proietti, M., Reisser, J., Ferrari, F. F., Abbas, B., & Herndl, G. J. (2017). Extracting DNA from ocean microplastics: a method comparison study. *Analytical Methods, 9*(9), 1521–1526.

Degallier, N., Servain, J., Lucio, P. S., Hannart, A., Durand, B., De Souza, R. N., & Ribeiro, Z. M. (2012). The influence of local environment on the aging and mortality ofAedes aegypti (L.): Case study in Fortaleza-CE, Brazil. *Journal of Vector Ecology, 37*(2), 428–441.

Dobson, A. P., Pimm, S. L., Hannah, L., Kaufman, L., Ahumada, J. A., Ando, A. A., ... Vale, M. M. (2020). Ecology and economics for pandemic prevention. *Science, 369*(6502), 379–381.

Dodd, C. K., & Smith, L. L. (2003). Habitat destruction and alteration. Historical trends and future prospects for amphibians. In R. D. Semlitsch (Ed.), *Amphibian conservation* (pp. 94–112). Washington D.C: Smithsonian Institution.

Donnelly, C., Ghani, A., Leung, G., Hedley, A., Fraser, C., Riley, S., ... Chau, P. (2003). Epidemiological determinants of spread of causal agent of severe acute respiratory syndrome in Hong Kong. *Lancet, 361*, 1761–1766.

Drake, D. R., & Hunt, T. L. (2009). Invasive rodents on islands: integrating historical and contemporary ecology. *Biol Invasions, 11*, 1483–1487.

Duarte, C. M. (2000). Marine biodiversity and ecosystem services: An elusive link. *Journal of Experimental Marine Biology and Ecology, 250*, 117–131.

Early, R., Bradley, B. A., Dukes, J. S., Lawler, J. J., Olden, J. D., Blumenthal, D. M., ... Tatem, A. J. (2016). Global threats from invasive alien species in the twenty-first century and national response capacities. *Nature Communications, 7*, 12485.

Eisen, R. J., & Eisen, L. (2018). The blacklegged tick, Ixodes scapularis: An increasing public health concern. *Trends Parasitol, 34*, 295–309.

Eisen, R. J., Kugeler, K. J., Eisen, L., Beard, C. B., & Paddock, C. D. (2017). Tick-borne zoonoses in the United States: Persistent and emerging threats to human health. *ILAR Journal, 58*(3), 319–335.

El-Tras, W. F., Tayel, A. A., Eltholth, M. M., & Guitian, J. (2009). Brucella infection in fresh water fish: Evidence for natural infection of Nile catfish, *Clarias gariepinus*, with *Brucella melitensis*. *Veterinary Microbiology, 141*(3–4), 321–325.

Epstein, J. H., Field, H. E., Luby, S., Pulliam, J. R. C., & Daszak, P. (2006). Nipah virus: Impact, origins, and causes of emergence. *Current Infectious Disease Reports, 8*(1), 59–65.

Eriksen, M., Lebreton, L. C. M., Carson, H. S., Thiel, M., Moore, C. J., Borerro, J. C., ... Reisser, J. (2014). Plastic pollution in the world's oceans: More than 5 trillion plastic pieces weighing over 250,000 tons afloat at sea. *PLoS One, 9*, 12.

Evander, M., & Ahlm, C. (2009). Milder winters in northern Scandinavia may contribute to larger outbreaks of haemorrhagic fever virus. *Global Health Action*. https://doi.org/10.3402/gha.v2i0.2020

Fabra, A. (1997). Toxicity of 2,4-dichlorophenoxyacetic Acid to *Rhizobium* sp in pure culture. *Bulletin of Environmental Contamination and Toxicology, 59*(4), 645–652.

Fahrig, L. (2003). Effects of habitat fragmentation on biodiversity. *Annual Review of Ecology, Evolution, and Systematics, 34*(1), 487–515.

Fatmi, S. S., Zehra, R., & Carpenter, D. O. (2017). Powassan virus—a new reemerging tick-borne disease. *Front Public Health, 5*(342), 1–12.

Ficetola, G. F., Thuiller, W., & Padoa-Schioppa, E. (2009). From introduction to the establishment of alien species: Bioclimatic differences between presence and reproduction localities in the slider turtle. *Diversity and Distributions, 15*, 108–116.

Foster, G., Osterman, B. S., Godfroid, J., Jacques, I., & Cloeckaert, A. (2007). *Brucella ceti* sp. nov. and *Brucella pinnipedialis* sp. nov. for *Brucella* strains with cetaceans and seals as their preferred hosts. *Int. International Journal of Systematic and Evolutionary Microbiology, 57*, 2688–2693.

Fredriksson-Ahomaa. (2019). Wild boar: a reservoir of foodborne zoonoses. *Foodborne pathogens and disease, 16*(3), 153–165.

Friant, S., Ayambem, W. A., Alobi, A. O., Ifebueme, N. M., Otukpa, O. M., Ogar, D. A., ... Rothman, J. M. (2020). Eating bushmeat improves food security in a biodiversity and infectious disease "Hotspot". *Ecohealth*, 1–14.

Galloway, T. S., Cole, M., & Lewis, C. (2017). Interactions of microplastic debris throughout the marine ecosystem. *Nature Ecology & Evolution, 116*(1), 1–8.

Genovesi, P., & Shine, C. (2004). *European strategy on invasive alien species: Convention on the conservation of european wildlife and habitats (Bern convention). Nature and environment 137*. Strasbourg: Council of Europe Publishing.

Ghazali, G. D., Guericolas, M., Thys, F., Sarasin, F., Arcos Gonzalez, P., & Casalino, E. (2018). Climate change impacts on disaster and emergency medicine focusing on mitigation disruptive effects: An international perspective. *International Journal of Environmental Research and Public Health, 15*(7), 1379.

Gibb, R, Moses, L. M., Redding, D. W., & Jones, K. E. (2017). Understanding the cryptic nature of Lassa fever in West Africa. *Pathogens and global health, 111*(6), 276–288.

Gilchrist, M. J., Greko, C., Wallinga, D. B., Beran, G. W., Riley, D. G., & Thorne, P. S. (2007). The potential role of concentrated animal feeding operations in infectious disease epidemics and antibiotic resistance. *Environmental Health Perspectives, 115*(2), 313–316.

Githeko, A. K., Lindsay, S. W., Confalonieri, U. E., & Patz, J. A. (2000). Climate change and vector-borne diseases: A regional analysis. *Bulletin of the World Health Organization, 78*(9), 1136–1147.

Glass, G. E., Schwartz, B. S., Morgan, J. M., Johson, D. T., Noy, P. M., & Israel, E. (1995). Environmental risk factors for Lyme disease identified with geographic information systems. *American Journal of Public Health, 85*(7), 944–948.

Godfroid, J., Cloeckaert, A., Liautard, J. P., Kohler, S., Fretin, D., Walravens, K., & Letesson, J. J. (2005). From the discovery of the Malta fever's agent to the discovery of a marine mammal reservoir, brucellosis has continuously been a re-emerging zoonosis. *Veterinary Research, 36*(3), 313–326.

González, C., Wang, O., Strutz, S. E., González-Salazar, C., Sánchez-Cordero, V., & Sarkar, S. (2010). Climate change and risk of Leishmaniasis in North America: Predictions from ecological niche models of vector and reservoir species. *PLOS Neglected Tropical Diseases, 4*(1), e585.

Gould, E. A., & Higgs, S. (2009). Impact of climate change and other factors on emerging arbovirus diseases. *Trans Royal Society of Tropical Medicine, 103*, 109–121.

Graham, J. P., Leibler, J. H., Price, L. B., Otte, J. M., Pfeiffer, D. U., Tiensin, T., & Silbergeld, E. K. (2008). The animal-human interface and infectious disease in industrial food animal production: Rethinking biosecurity and biocontainment. *Public Health Reports, 123*(3), 282–299.

Gray, A. D., Wertz, H., Leads, R. R., & Weinstein, J. E. (2018). Microplastic in two South Carolina Estuaries: Occirrence distribution and composition. *Marine Polllution Bulletin, 128*, 223–233.

Guan, Y., Zheng, B. J., He, Y. Q., Liu, X. L., Zhuang, Z. X., Cheung, C. L., ... Poon, L. L. (2003). Isolation and characterization of viruses related to the SARS coronavirus from animals in southern China. *Science, 30*(2), 276–278.

Gummow, B. (2010). Challenges posed by new and re-emerging infectious diseases in livestock production, wildlife and humans. *Livestock Science, 130*(1/3), 41–46.

Guo, T., Fan, Y., Chen, M., Wu, X., Zhang, L., He, T., ... Lu, Z. (2020). Cardiovascular implications of fatal outcomes of patients with coronavirus disease 2019 (COVID-19). *JAMA Cardiology, 5*(7), 1–8. https://doi.org/10.1001/jamacardio.2020.1017. Advance online publication.

Gylys, L., Chomel, B. B., & Gardner, I. A. (1998). Epidemiological surveillance of rabies in Lithuania from 1986 to 1996. *Revue Scientifique et Technique-Office International des Épizooties, 17*, 691–698.

Hahn, B. H., Shaw, G. M., De Cock, K. M., & Sharp, P. M. (2000). AIDS as a zoonosis: Scientific and public health implications. *Science, 287*(5453), 607–614.

Hamilton, S. E., & Casey, D. S. (2016). Creation of a high spatiotemporal resolution global database of continuous mangrove forest ocover for the 21st Century (CGMFC-21). *A Journal of Macroecology. (Global Ecology and Biogeogprahy), 25*, 229–738.

Han, J. W., Choi, H., Jeon, Y. H., Yoon, C. H., Woo, J. M., & Kim, W. (2016). The effects of forest therapy on coping with chronic widespread pain: Physiological and psychological differences between participants in a forest therapy program and a control group. *International Journal of Environmental Research and Public Health, 13*, 255.

Harper, G. A., & Bunbury, N. (2015). Invasive rats on tropical islands: Their population biology and impacts on native species. *Global Ecology and Conservation, 3*, 607–627.

Hilton, G. M., & Cuthbert, R. J. (2010). The catastrophic impact of invasive mammalian predators on birds of the UK Overseas terrtiroies: A review and synthesis. *The International Journal of Avian Science, 152*, 443–458.

Himsworth, C. G., Bidulka, J., Parsons, K. L., Feng, A. Y. T., Tang, P., Jardine, C. M., ... Patrick, D. M. (2013). Ecology of *Leptospira interrogans* in Norway rats (*Rattus norvegicus*) in an inner-city neighborhood of Vancouver, Canada. *PLOS Neglected Tropical Diseases, 7*(6), e2270.

Horwitz, P., & Finlayson, C. M. (2011). Wetlands as settings for human health: incorporating ecosystem services and health impact assessment into water resource management. *BioScience, 9*, 678–688.

Huizinga, H., & McLaughlin, G. (1990). Thermal ecology of Naegleria fowleri from a power plant cooling reservoir. *Applied and Environmental Microbiology, 56*, 2200–2205.

IPBES. (2019). *Glossary, Habitat degradation*. Available at https://ipbes.net/sites/default/files/ipbes_global_assessment_glossary_unedited_31may.pdf. (Accessed 10 July 2020).

Jimenez-Clavero, M. A. (2012). Animal viral diseases and global change: Bluetongue and West Nile fever as paradigms. *Frontiers in Genetics, 3*(105), 3–15.

Johnson, C. K., Hitchens, P. L., Evans, T. S., Goldstein, T., Thomas, K., Clements, A., ... Mazet, J. K. (2015). Spillover and pandemic properties of zoonotic viruses with high host plasticity. *Scientific Reports, 5*, 14830.

Jones, B. A., Grace, D., Kock, R., Alonso, S., Rushton, J., Said, M. Y., ... Pfeiffer, D. U. (2013). Zoonosis emergence linked to agricultural intensification and environmental change. *Proceedings of the National Academy of Sciences, 110*(21), 8399–8404.

Jones, K. E., Patel, N. G., Levy, M. A., Storeygard, A., Balk, D., Gittleman, J. L., & Daszak, P. (2008). Global trends in emerging infectious diseases. *Nature, 451*. https://doi.org/10.1038/nature06536

Karieva, P. M., Kingsolver, J. G., & Huey, R. B. (1993). *Biotic Interactions and Global Change*. Sunderland, Massachusetts, USA: Sinauer Associates Inc.

Katayama, T., Tanaka, M., Moriizumi, J., Nakamura, T., Brouchkov, A., Douglas, T. A., ... Asano, K. (2007). Phylogenetic analysis of bacteria preserved in a permafrost ice wedge for 25,000 years. *Applied and Environmental Microbiology, 73*, 2360–2363.

Keesing, F., Holt, R. D., & Ostfeld, R. S. (2006). Effects of species diversity on disease risk. *Ecology Letters, 9*(4), 485–498.

Keitha, A. M., Schmidt, O., & McMahon, B. J. (2016). Soil stewardship as a nexus between Ecosystem Services and One Health. *Ecosystem Services, 17*, 40–42.

Keller, R. P., Geist, J., Jeschke, J. M., & Kuhn, I. (2011). Invasive species in Europe, Ecology, status and policy. *Environmental Sciences Europe, 23*(23), 1–17.

Kernif, T., Leulmi, H., Raoult, D., & Parola, P. (2016). Emerging tick-borne bacterial pathogens. *Emerging Infections, 10*, 295–310.

Keusch, G. T., Pappaioanou, M., Gonzalez, M. C., Scott, K. A., & Tsai, P. (2009). *Committee on achieving sustainable global capacity for surveillance and response to emerging diseases of zoonotic origin, National research council. Sustaining global surveillance and response to emerging zoonotic diseases.*

Kikillus, K. H., Hare, K. H., & Hartley, S. (2010). Minimizing false-negatives when predicting the potential distribution of an invasive species: A bioclimatic envelope for the red-eared slider at global and regional scales. *Animal Conservation, 15*, 5–15.

Kitron, U., & Kazmierczak, J. J. (1997). Spatial analysis of the distribution of Lyme disease in Wisconsin. *American journal of Epidemiology, 145*(6), 558–586.

Koch, L. K., Kochmann, J., Klimpel, S., & Cunze, S. (2017). Modeling the climatic suitability of leishmaniasis vector species in Europe. *Scientific Reports, 7*.

Kondolf, G. M., Podolak, K., & Grantham, T. E. (2012). Restoring mediterranean-climate rivers. *Hydrobiologia Springer, 719*(1), 527–545.

Kovats, R. S., Bouma, M. J., Hajat, S., Worrall, E., & Houries, A. (2003). El Niño and health. *Lancet, 362*, 1481–1489.

La sorte, F. A., Fink, D., Buler, J. J., Farnsworth, A., & Cabrera-Cruz, S. A. (2017). Seasonal associations with urban light pollution for nocturnally migrating bird populations. *Global Change Biology, 23*(11), 4609–4619.

Lafferty, K. D. (2009). The ecology of climate change and infectious diseases. *Ecology, 90*(4), 888–900.

Lambert, K. A., Bowatte, G., Tham, R., Lodge, C., Prendergast, L., Heinrich, J., … Erbas, B. (2017). Residential greenness and allergic respiratory diseases in children and adolescents – a systematic review and meta-analysis. *Environmental Research, 159*, 212–221.

Leroy, E. M., Rouquet, P., Formenty, P., Souquière, S., Kilbourne, A., Froment, J. M., … Rollin, P. E. (2004). Multiple Ebola virus transmission events and rapid decline of central African wildlife. *Science, 303*(5656), 387–390.

Lever, C (2003). *Naturalized Amphibians and Reptiles of the World.* New Work: Oxford University Press.

Li, W., Shi, Z., Yu, M., Ren, W., Smith, C., Epstein, J. H., … Wang, L. (2005). Bats are natural reservoirs of SARS-like coronaviruses. *Science, 310*, 676–679.

Lindeman, P. V. (1999). Aggressive interactions during basking among four species of Emydid turtles. *Journal of Herpetology, 33*, 214–219.

Loh, E. H., Zambrana-Torrelio, C., Olival, K. J., Bogich, T. L., Johnson, C. K., Mazet, J. A., … Daszak, P. (2015). Targeting transmission pathways for emerging zoonotic disease surveillance and control. *Vector-Borne and Zoonotic Diseases, 15*(7), 432–437.

Lounibos, L. P. (2002). Invasions by insect vectors of human disease. *Annual Review of Entomology, 47*, 233–266.

Lu, J., Gu, J., Li, K., Xu, C., Su, W., Lai, Z., … Yang, Z. (2020). COVID-19 outbreak associated with air conditioning in restaurant, Guangzhou, China, 2020. *Emerging Infectious Diseases, 26*(7), 1628.

Lu, C., & Tian, H. (2017). Global nitrogen and phosphorus fertilizer use for agriculture production in the past half century: Shifted hot spots and nutrient imbalance. *Earth System Science Data, 9*(1), 181–192. https://doi.org/10.5194/essd-9-181-2017

Lu, P., Zhou, Y., Yu, Y., Cao, J., Zhang, H., Gong, H., ... Zhou, J. (2016). RNA interference and the vaccine effect of a subolesin homolog from the tick Rhipicephalus haemaphysaloides. *Experimental and Applied Acarology, 68*, 113–126.

Lueck, & Jeffrey. (2003). Preemptive Habitat Destruction under the EndangeredSpecies Act. *Journal of Law & Economics, 46*, 27–60.

Mac Kenzie, W. R., Hoxie, N. J., Proctor, M. E., Gradus, M. S., Blair, K. A., Peterson, D. E., ... Rose, J. B. (1994). A massive outbreak in Milwaukee of cryptosporidium infection transmitted through the public water supply. *New England Journal of Medicine, 331*(3), 161–167.

MacArthur, R. H., & Wilson, E. O. (1967). *The theory of island biogeography*. Princeton, NJ: Princeton University Press.

Mackey, T. K., Liang, B. A., Cuomo, R., Hafen, R., Brouwer, K. C., & Lee, D. E. (2014). Emerging and reemerging neglected tropical diseases: A review of key characteristics, risk factors, and the policy and innovation environment. *Clinical Microbiology Reviews, 27*, 949–979.

Macpherson, C. N. L. (2005). Human behaviour and the epidemiology of parasitic zoonoses. *International Journal for Parasitology, 35*(11–12), 1319–1331.

Maguire, H. C., & Heymann, D. L. (2016). Yellow fever in Africa. *BMJ*, i3764.

Marsot, M., Chapuis, J.-L., Gasqui, P., Dozières, A., Masséglia, S., Pisanu, B., ... Vourc'h, G. (2013). *Introduced Siberian chipmunks (*Tamias sibiricus *barberi) contribute more to Lyme borreliosis risk than native reservoir rodent.*

Martin, V., Chevalier, V., Ceccato, P., Anyamba, A., De Simone, L., Lubroth, S., ... Domenech, J. (2008). The impact of climate change on the epidemiology and control of Rift Valley Fever. *Rev Scientific Technical Off Int Epiz, 27*(2), 413–426.

Martinet, J. P., Ferte, H., Failloux, A. B., Scgaffner, F., & Depaquit, J. (2019). Mosquitoes of North-Western Europe as Potential Vectors of Arboviruses: A Review. *Viruses, 11*(11), 1059.

Mas-Coma, S., Valero, M. A., & Bargues, M. D. (2009). Climate change effects on trematodiases, with emphasis on zoonotic fascioliasis and schistosomiasis. *Veterinary Parasitology, 163*(4), 264–280.

Mazza, G., Tricario, E., Genovesi, P., & Gherardi, F. (2013). Biological invaders are threats to human health: an overview. *Ethology Ecology & Evolution, 26*, 112–129.

McMullan, L. K., Folk, S. M., Kelly, A. J., MacNeil, A., Goldsmith, C. S., Metcalfe, M. G., ... Nichol, S. T. (2012). A new Phlebovirus associated with severe Febrile illness in Missouri. *New England Journal of Medicine, 367*(9), 834–841.

McPhaden, M. J., Zebiak, S. E., & Glantz, M. H. (2006). ENSO as an integrated concept in earth science. *Science, 314*(5806), 1740–1745.

Medlock, J. M., Hansford, K. M., Schaffner, F., Versteirt, V., Hendrickx, G., Zeller, H., & Van Bortel, W. (2012). A review of the invasive mosquitoes in Europe: ecology, public health risks, and control options. *Vectorborne and Zoonotic Diseases, 12*, 435–447.

Merhej, V., Angelakis, E., Socolovschi, C., & Raoult, D. (2014). Genotyping, evolution and epidemiological findings of rickettsia species. *Infection Genetics and Evolution, 25*, 122–137.

Millennium Ecosystem Assessment. (2003). *Ecosystems and human well-being a framework for assessment*. Washington DC: Island Press.

Mitchell, C. E., Tilman, D., & Groth, J. V. (2002). Effects of grassland plant species diversity, abundance, and composition on foliar fungal disease. *Ecology, 83*, 1713–1726.

Mollentze, N., & Streicker, D. G. (2020). Viral zoonotic risk is homogenous among taxonomic orders of mammalian and avian reservoir hosts. *Proceedings of the National Academy of Sciences, 117*(17), 9423–9430.

Monath, T. P. (2001). Yellow fever: an update. *The Lancet infectious diseases, 1*(1), 11–20.

Moore, S., Shrestha, S., Tomlinson, K. W., & Vuong, H. (2011). Predicting the effect of climate change on African trypanosomiasis: Integrating epidemiology with parasite and vector biology. *Journal of the Royal Society Interface, 9*, 817–830.

Morse, S. S., Mazet, J. A., Woolhouse, M., Parrish, C. R., Carroll, D., Karesh, W. B., … Daszak, P. (2012). Prediction and prevention of the next pandemic zoonosis. *The Lancet, 380*(9857), 1956–1965. https://doi.org/10.1016/S0140-6736(12)61684-5

MS-BR. (2017). *Centro de Operações de emergências em Saúde Pública sobre Febre Amarela – No 43/2017 [Internet]*. Ministério da Saúde do Bras (pp. 1–7). Available at http://portalarquivos.saude.gov.br/images/pdf/2017/junho/02/COES-FEBRE-AMARELA%2D%2D-INFORME-43%2D%2D-Atualiza%2D%2D%2D%2Do-em-31maio2017.pdf.

MS-BR. (2018). *Monitoramento do Período Sazonal da Febre Amarela Brasil – 2017/2018 - informe no 27 | 2017/2018 [Internet]*. Ministério da Saúde do Bras (pp. 1–24). Available at http://portalarquivos2.saude.gov.br/images/pdf/2018/outubro/08/Informe-FA.pdf.

MS-BR. (2019). *Monitoramento de Febre Amarela Brasil 2019 - informe no 18 | 9 de Junho 2019 [Internet]*. Ministério da Saúde do Bras (pp. 1–8). Available at https://portalarquivos2.saude.gov.br/images/pdf/2019/junho/13/Informe-de-Monitoramento-de-Febre-Amarela-Brasil%2D%2Dn-18.pdf.

Newbold, R., Hudson, L. N., Kill, S. L., Contu, S., Lysenko, I, Senior, R. A., & Day, J. (2015). Global effects of land use on local terrestrial biodiversity. *Nature, 520*(7545), 45–50.

Ogden, N. H., & Lindsay, L. R. (2016). Effects of climate and climate change on vectors and vector-borne diseases: Ticks are different. *Trends in Parasitology, 32*, 646–656.

Pacheco, R. C., Echaide, I. E., Alves, R. N., Beletti, M. E., Nava, S., & Labruna, M. B. (2013). *Coxiella burnetii* in ticks, Argentina. *Emerging Infectious Diseases, 19*, 344–346.

Paddock, C. D., & Childs, J. E. (2003). *Ehrlichia chaffeensis*: A prototypical emerging pathogen. *Clinical Microbiology Review, 16*, 37–64.

Paddock, C. D., & Goddard, J. (2015). The evolving medical and veterinary importance of the Gulf coast tick (Acari: Ixodidae). *Journal of Medical Entomology, 52*, 230–252.

Pagad, S. N., Genovesi, P., Carnevali, V., Scalera, R., & Clout, M. (2015). IUCN SSC invasive species specialist group: Invasive alien species information management supporting practitioners, policy makers and decision takers. *Management of Biological Invasions, 6*(2015), 127–135.

Pan, J., Plant, J. A., Voulvoulis, N., Oates, C. J., & Ihlenfeld, C. (2010). Cadmium levels in Europe: Implications for human health. *Environmental Geochemistry and Health, 32*(1), 1–12. https://doi.org/10.1007/s10653-009-9273-2

Parola, P., Paddock, C. D., & Raoult, D. (2005). Tick-borne rickettsioses around the world: Emerging diseases challenging old concepts. *Clinical Microbiology Review, 18*, 719–756.

Paul-Pont, I., Lacroix, C., Fernandezm, C. G., Hegaret, H., Lambert, C., Goic, N. L., … Soudant, P. (2016). Exposure of marine mussels *Mytilus* spp. to polystyrene microplastics: Toxicity and influence on fluoranthene bioaccumulation. *Environmental Pollution, 216*, 724–737.

Paupy, C., Delatte, H., Baagny, L., Corbel, V., & Fontenille, D. (2009). Aedes albopictus, an arbovirus vector: from the darkness to the light. *Microbes & Infection, 11*(14–15), 1177–1185.

Peeters, M., Courgnaud, V., Abela, B., Auzel, P., Pourrut, X., Bibollet-Ruche, F., … Delaporte, E. (2002). Risk to human health from a plethora of simian immunodeficiency viruses in primate bushmeat. *Emerging Infectious Diseases, 8*(5), 451.

Peiris, J. S. M., Lai, S. T., Poon, L. L. M., Guan, Y., Yam, L. Y. C., Lim, W., … Yuen, K. Y. (2003). Coronavirus as a possible cause of severe acute respiratory syndrome. *Lancet, 361*, 1319–1325.

Peters, A. (2014). *Global trade impacts on biodiversity and ecosystem services in ecosystem services (191-219)*. Elsevier Inc.

Pimentel, D., Lach, L., Zuniga, R., & Morrison, D. (2000). Environmental and economic costs of non-indigenous species in the United States. *BioScience, 50*, 53–65.

Plumer, L., Davison, J., & Saarma, U. (2014). Rapid urbanization of red foxes in Estonia: distribution, behaviour, attacks on domestic animals, and health-risks related to zoonotic diseases. *PLoS One, 9*(12), 115–124.

Primavera, J. H. (1997). Socio-economic impacts of shrimp culture. *Aquaculture Research, 28*, 815–827.

Pulleman, M., Creamer, R., Hamer, U., Helder, J., Pelosi, J., Peres, G., & Rutgers, M. (2012). Soil biodiversity, biological indicators and soil ecosystem services-an overview of European approaches. *Current Opinion in Environmental Sustainability, 4*, 529–538.

Rabitsch, W., Essl, F., & Schindler, S. (2017). The rise of non-native vectors and reservoirs of human diseases. In Impact of biological invasions on ecosystem services. *Springer, Cham.*, 263–275.

Ramsay, N. F., Ng, P. K. A., O'Riordan, R. M., & Chou, L. M. (2007). The read-eared slider (Trachemys scripta elegans) in Asia: A review. *F. Gherardi (ed), Biological Invaders in Inland Waters: Profiles, Distribution, and Threats* (pp. 161–174). Dordrecht: Springer.

Randolph, S. E. (2004). Evidence that climate change has caused'emergence'of tick-borne diseases in Europe? *International Journal of Medical Microbiology Supplements, 293*, 5–15.

Ready, P. (2010). Leishmaniasis emergence in Europe. *Eurosurveillance, 15*(10), 19505.

Redshaw, C. H., Stahl-Timmins, W. M., Fleming, L. E., Davidson, I., & Depledge, M. H. (2013). Potential changes in disease patterns and pharmaceutical use in response to climate change. *Journal of Toxicology and Environmental Health, 16*, 285–320.

Reid, W. V., & Mooney, H. A. (2016). The Millennium Ecosystem Assessment: testing the limits of interdisciplinary and multi-scale science. *Current Opinion in Environmental Sustainability, 19*, 40–46.

Reiter, P. (2008). Climate change and mosquito-borne disease: Knowing the horse before hitching the cart. *Revue scientifique et technique, 27*(2), 383–398.

Reithinger, R., Dujardin, J. C., Louzir, H., Pirmez, C., Alexander, B., & Brooker, S. (2007). Cutaneous leishmaniasis. *The Lancet infectious diseases, 7*(9), 581–596.

Rejmánek, M (2007). Invasive plants: approaches and predictions. *Austral Ecol, 25*, 497–506.

Reynolds, J. D., & Peres, C. A. (2006). Overexploitation. In M. J. Groom, G. K. Meffe, & C. R. Carroll (Eds.), *Principles of conservation* (pp. 253–277). Sunderland, Massachusetts: Sinauer Associates, Inc.

Rezza, G., Nicoletti, L., Angelini, R., Romi, R., Finarelli, A., Panning, M., … Magurano, F. (2007). Infection with chikungunya virus in Italy: An outbreak in a temperate region. *Lancet, 370*, 1840–1846.

Rioux, J. A., Jarry, D. M., Lanotte, G., Maazoun, R., & Killickkendrick, R (1984). Ecology of leishmaniasis in Southern France. 18. Enzymatic identification of Leishmania infantum Nicolle, 1908, isolated from Phlebotomus ariasi Tonnoir, 1921, spontaneously infected in the Cévennes. *Annales de Parasitologie Humaine et Compareé, 59*(4), 331–333.

Rizzoli, A., Hauffe, H. C., Carpi, G., Vourc'h, G., Neteler, M., & Rosa, R. (2011). Lyme borreliosis in Europe. *Eurosurveillance, 16*(27), 19906.

Rogalski, M. A., Gowler, C. D., Shaw, C. L., Hufbauer, R. A., & Duffy, M. A. (2017). Human drivers of ecological and evolutionary dynamics in emerging and disappearing infectious disease systems. *Philosophical Transactions of the Royal Society B: Biological Sciences, 372*(1712), 20160043.

Romi, R., Boccolini, D., DI Luca, M., Medlock, J. M., Schaffner, F., Severini, F., & Toma, L (2018). Invasive Mosquitoes of Medical Importance. *G. Mazza and E. Tricarico (eds): Invasive Species and Human Health* (pp. 76–90). CABI invasives series.

Rothan, H. A., & Byrareddy, S. N. (2020). The epidemiology and pathogenesis of coronavirus disease (COVID-19) outbreak. *Journal of Autoimmunity*. https://doi.org/10.1016/j.jaut.2020.102433 (in press).

Russell, V. S. (1974). Pollution: Concept and definition. *Biological Conservation, 6*(3), 157–161.

Salmond, J. A., Tadaki, M., Vardoulakis, S., Arbuthnott, K., Coutts, A., Demuzere, M., ... Wheeler, B. W. (2016). Health and climate related ecosystem services provided by street trees in the urban environment. *Environmental Health, 15*, 36.

Sanderman, J., Hengl, T., Fiske, G., Solvik, K., Adame, M. F., Benson, L., & Duncan, C. (2018). A global map of mangrove forest soil carbon at 30 m spatial resolution. *Environmental Research Letters, 13*, 5.

Schaafsma, M., & Turner, R. K. (2015). *Valuation of coastal and marine ecosystem services: A literature Review in coastal zones Eccosystem services*. Springer International Publishing Switzerland. https://doi.org/10.1007/978-3-319-17214-9_6

Schmidt, J. P., Park, A. W., Kramer, A. M., Han, B. A., Alexander, L. W., & Drake, J. M. (2017). Spatiotemporal fluctuations and triggers of Ebola virus spillover. *Emerging Infectious Diseases, 23*, 415–422.

Scholz, H. C., Hubalek, Z., Sedlacek, I., Vergnaud, G., Tomaso, H., Al Dahouk, S., ... Nockler, K. (2008). *Brucella microti* sp. nov., isolated from the common vole *Microtus arvalis*. *International Journal of Systematic and Evolutionary Microbiology, 58*, 375–382.

Scholz, T., & Kuchta, R. (2016). Fish-borne, zoonotic cestodes (diphyllobothrium and relatives) in cold climates: A never-ending story of neglected and (re)-emergent parasites food waterb. *Parasitology, 4*, 23–38.

Schwarzenbach, R. P., Escher, B. I., Fenner, K., Hofstetter, T. B., Johnson, C. A., Von Gunten, U., & Wehrli, B. (2006). The challenge of micropollutants in aquatic systems. *Science, 313*, 1072–1077.

Seebens, H., Blackburn, T. M., Dyer, E. E., Genovesi, P., Hulme, P. E., Jeschke, J. M., ... Essl, F. (2017). No saturation in the accumulation of alien species worldwide. *Nature Communications, 8*, 14435.

Seleem, M. N., Boyle, S. M., & Sriranganathan, N. (2009). Brucellosis: A re-emerging zoonosis. *Veterinary Microbiology, 140*, 392–398.

Shi, P., Dong, Y., Yan, H., Li, X., Zhao, C., & Liu, W. (2020). *The impact of temperature and absolute humidity on the coronavirus disease 2019 (COVID-19) outbreak-Evidence from China*. medRxiv.

Simberloff, D (2013). *Invasive Species: What Everyone Needs to Know*. New York: Oxford University Perss.

Simberloff, D., & Rejmánek, M (2011). *Encyclopedia of Biological Invasions. Berkeley*. California: University of California Press.

Simberloff, D., Martin, J. L., Genovesi, P., Maris, V., Wardle, D. A., Aronson, J., ... Vilà, M. (2012). Impacts of biological invasions: what's what and the way forward. *Trends in Ecology & Evolution, 28*(1), 58–66.

Solomon, S., Qin, D., Manning, M., Chen, Z., Marquis, M., Averyt, K., ... Miller, H. (2007). *Climate change 2007: The physical science basis. Contribution of working group I to the fourth assessment report of the intergovernmental panel on climate change*. New York, USA: Cambridge University press.

Sonenshine, D. E. (2018). Range expansion of tick disease vectors in north America: Implications for spread of tick-borne disease. *International Journal of Environmental Research and Public Health, 15*(3), 478.

Spinks, P. Q., Pauly, G. B., Crayon, J. J., & Shaffer, H. B. (2003). Survival of the western pond turtle (Emys marmorata) in an urban California environment. *Biological Conservation, 113*, 257–267.

Steere, A. C., Gross, D., Meyer, A. L., & Huber, B. T. (2001). Autoimmune mechanisms in antibiotic treatment-resistant Lyme arthritis. *Journal of Autoimmunity, 16*(3), 263–268. https://doi.org/10.1016/j.jaut.2020.102433

Sutherst, R. W. (2000). Climate change and invasive species: A conceptual framework. *Mooney, H. A., & Hobbs R.J. (eds.). Invasive Species in a Changing World* (pp. 211–240). Washington D.C.: Island Press.

Sutor, A., Schwarz, S., & Conraths, F. J (2014). The biological potential of the raccoon dog (Nyctereutes procyonoides, Gray 1834) as an invasive species in Europe—new risks for disease spread? *Acta Theriol, 59*, 49–59.

Suzán, G., Esponda, F., Carrasco-Hernandez, R., & Aguirre, A. (2012). Habitat fragmentation and infectious disease ecology. In A. A. Aguirre, R. S. Ostfeld, & P. Daszak (Eds.), *New directions in conservation medicine: Applied cases of ecological health.* Oxford University Press.

Suzán, G., Marcé, E., Giermakowski, J. T., Armién, B., Pascale, J., Mills, J., … Yates, T. (2008). The effect of habitat fragmentation and species diversity loss on hantavirus prevalence in Panama. *Annals of the New York Academy of Sciences, 1149*(1), 80–83.

Sykora, J., Keleti, G., & Martinez, A. J. (1983). Occurrence and pathogenicity of *Naegleria fowleri* in artificially heated waters. *Applied and Environmental Microbiology, 45*, 974–979.

Tan, J., Mu, L., Huang, J., Yu, S., Chen, B., & Yin, J. (2005). An initial investigation of the association between the SARS outbreak and weather: With the view of the environmental temperature and its variation. *Journal of Epidemiology and Community Health, 59*, 186–192.

Teillac-Deschamps, P., Lorrilliere, R., Servais, V., Delmas, V., Cadi, A., & Prévot-Julliard, A. C. (2009). Management strategies in urban green spaces: Models based on an introduced exotic pet turtle. *Biological Conservation, 142*, 2258–2269.

Telecky, T. M. (2001). US import and export of live turtles and tortoises. *Turtle and Tortoise Newsletter, 4*, 8–13.

Teng, Y., Yang, J., Zuo, R., & Wang, J. S. (2011). Impact of urbanization and industralization upon surface water quality: A pilot study of Panzhihua mining town. *Journal of Earth Science, 22*(5), 658–668.

Toto, J. C., Abaga, S., Carnevale, P., & Simard, F. (2003). First report of the oriental mosquito Aedes albopictus on the West African island of Bioko, Equatorial Guinea. *Medical and Veterinary Entomology, 17*(3), 343–346.

Thomas, R. J., Dumler, J. S., & Carlyon, J. A. (2009). Current management of human granulocytic anaplasmosis, human monocytic ehrlichiosis and *Ehrlichia ewingii* ehrlichiosis. *Expert Review of Anti-infective Therapy, 7*, 709–722.

Thompson, R. C. A. (2013). Parasite zoonoses and wildlife: One health, spillover and human activity. *International Journal for Parasitology, 43*, 1079–1088.

Thompson, A. A., Matamale, L., & Kharidza, S. D. (2012). Impact of climate change on children's health in Limpopo Province, South Africa. *International Journal of Environmental Research and Public Health, 9*, 831–854.

Torres-Guerrero, E., Quintanilla-Cedillo, M. R., Ruiz-Esmenjaud, J., & Arenas, R. (2017). Leishmaniasis: a review. *F1000Research, 6*, 750.

Tosepua, R., Gunawanb, J., Effendyc, D. S., Ahmadd, L. O. A. I., Lestarie, H., Baharf, H., & Asfiang, P. (2020). Correlation between weather and covid-19 pandemic in Jakarta, Indonesia. *Science of the Total Environment, 725*, 1–4.

Tse, C. C. K., Bullard, J., Rusk, R., Douma, D., & Plourse, P. J. (2019). Surveillance of Echinococcus tapeworm in coyotes and domestic dogs of Winnipeg, Manitoba. *Canada Communicable Disease Report, 45*(7/8), 171. https://doi.org/10.4745/ccdr.v45i78a01

Tsiamis, K., Cardoso, A. C., & Gervasini, E. (2017). The European alien species information network on the convention on biological diversity pathways categorization. *NeoBiota, 32*, 21–29. https://doi.org/10.3897/neobiota.32.9429

Tu, C., Crameri, G., Kong, X., Chen, J., Sun, Y., Yu, M., … Wang, L. F. (2004). Antibodies to SARS coronavirus in civets. *Emerging Infectious Diseases, 10*, 2244–2248.

United Nations. (2007). *World population prospects: The 2006 revision, medium variant.* New York: United Nations: Department of Economics and Social Affairs, Population Division.

Van Vliet, N., Moreno Calderón, J. L., Gomez, J., Zhou, W., Fa, J. E., Golden, C., … Nasi, R. (2017). Bushmeat and human health: Assessing the evidence in tropical and sub-tropical forests. *Ethnobiology and Conservation, 6*(3), 1–45.

Vandini, S., Corvaglia, L., Alessandroni, R., Aquilano, G., Marsico, C., & Spinelli, M. (2013). Respiratory syncytial virus infection in infants and correlation with meteorological factors and air pollutants. *Italian Journal of Pediatrics, 39*(1), 1.

Vannier, E., & Krause, P. J. (2020). *Babesiosis, Hunter's tropical medicine and emerging infectious diseases* (pp. 799–802). Elsevier.

Vasconcelos, M., Datta, K., Oliva, N., Khalekuzzaman, M., Torrizo, L., Krishnan, S., & Datta, S. K. (2003). Enhanced iron and zinc accumulation in transgenic rice with the ferritin gene. *Plant Science, 164*(3), 371–378.

Verger, J. M., Grimont, F., Grimont, P. A., & Grayon, M. (1987). Taxonomy of the genus. *Brucella. Ann. Inst. Pasteur Microbiol., 138*, 235–238.

Vidal, M., Lopez, A., Santoalla, M. C., & Valles, V. (1999). Factor analysis for the study of water resources contamination due to the use of livestock slurries as fertilizer. *Agricultural Water Managment, 45*(2000), 1–15.

Vittor, A. Y., Gilman, R. H., Tielsch, J., Glass, G., Shields, T. I. M., Lozano, W. S., … Patz, J. A. (2006). The effect of deforestation on the human-biting rate of Anopheles darlingi, the primary vector of falciparum malaria in the Peruvian Amazon. *The American Journal of Tropical Medicine and Hygiene, 74*(1), 3–11.

Volpe, M. G., La Cara, F., Volpe, F., De Mattia, A., Serino, V., Petitto, F., … Di Stasio, M. (2009). Heavy metal uptake in the enological food chain. *Food Chemistry, 117*(3), 553–560. https://doi.org/10.1016/j.foodchem.2009.04.033

Walkowiak, J., Wiener, J. A., Fastabend, A., Heinzow, B., Kramer, U., Schmidt, E., … Winneke, G. (2001). Environmental exposure to polychlorinated biphenyls and quality of the home environment: Effects on psychodevelopment in early childhood. *Lancet, 358*, 1602.

Wallace, R. G., Bergmann, L., Kock, R., Gilbert, M., Hogerwerf, L., Wallace, R., & Holmberg, M. (2015). The dawn of structural one health: a new science tracking disease emergence along circuits of capital. *Social Science & Medicine, 129*, 68–77.

Walsh, M., Willem De Smalen, A., & Mor, S. (2018). Climatic influence on anthrax suitability in warming northern latitudes. *Scientific Reports, 8*(1), 1–9.

Watson, R. T., Albritton, D. L., Barker, T., Bashmakov, I. A., Canziani, O., Christ, R., … Zhou, D. (2001). *Climate change 2001: Synthesis report. Contribution of working groups I, II, and III to the third assessment report of the intergovernmental panel on climate change.* Cambridge, United Kingdom: The press syndicate of the Univeristy of Cambridge.

Wilcove, D. S., McLellan, C. H., & Dobson, A. P. (1986). Habitat fragmentation in the temperate zone. *Conservation Biology, 6*, 237–256.

Williams, F., Eschen, R., Harris, A., Djeddour, D., Pratt, C., Shaw, R., … Murphy, S. T. (2010). *The economic cost of invasive nonnative species on Great Britain. CABI Proj No VM10066.*

Williamson, M. (1993). Invaders, weeds and the risk from genetically manipulated organisms. *Cellular and Molecular Life Sciences, 49*, 219–224.

Wilson, P. R. (2002). Advances in health and welfare of farmed deer in New Zealand. *New Zealand Veterinary Journal, 50*(3), 105–109.

Wotton, R. S. (2004). The essential role of exopolymers (EPS) in aquatic systems. *Oceanography and Marine Biology, 42*, 57–94.

Yasuoka, J., & Levins, R. (2007). Impact of deforestation and agricultural development on Anopheline ecology and malaria epidemiology. *The American Society of Tropical Medicine and Hygiene, 76*, 450–460.

Yuan, J., Yun, H., Lan, W., Wang, W., Sullivan, S. G., & Jia, S. (2006). A climatologic investigation of the SARS-CoV outbreak in Beijing, China. *American Journal of Infection Control, 34*, 234–236.

Zettler, E. R., Mincer, T. J., & Amaral-Zettler, L. A. (2013). Life in the 'plastisphere': Microbial communities on plastic marine debris. *Environmental Science & Technology, 47*, 7137–7146.

Zhang, D. (2004). Endangered species and timber harvesting: The Case of Red-Cockaded Woodpeckers. *Economic Inquiry, 42*(1), 150–165.

Zhang, T., Wu, Q., & Zhang, Z. (2020). *bioRxiv Preprint*. https://doi.org/10.1101/2020.02.19.950253

Van Langevelde, F., & Rivera Mendoza, H. R. (2021). *The link between biodiversity loss and the increasing spread of zoonotic diseases*. (Accessed 5 July 2021).

Chapter 10

Challenges and future perspectives for the application of One Health

Bolajoko Muhammad-Bashir and Balogun A. Halimah
National Veterinary Research Institute, Vom, Plateau State, Nigeria

1. Introduction

With the increasing risk of emerging and re-emerging infectious diseases and its spread, the risk of pandemics has become more critical (Zinsstag et al., 2018). These diseases are medically, socially, economically, and environmentally expensive, with their wide consequences requiring interdisciplinary solutions. The "One Health" approach can proffer such solutions, because it is a growing global strategy that health organizations and policy makers must adopt in response to global needs (Barrett, Bouley, Stoertz, & Stoertz, 2010).

Although, the concept of One Health is a modern and growing field of research, it can be traced back to over two centuries ago. In the 19th century, Dr. Rudolf Virchow formally accepted the connection between animal and human medicine. The concept was promoted by his followers, but for practical purposes, the concept lost its values, despite the revelations about the similarities in the physiology and pathology of both humans and animals (Hristovski, Cvetkovik, Cvetkovik, & Dukoska, 2010).

Before it was proclaimed One Health, late in the 1990s and the early 2000s, ecologists and wildlife conservationists began to resuscitate this ancient concept. It was embraced by the veterinary profession and the human medical profession following their awareness of it, as a result of outbreaks of the Highly Pathogenic Avian Influenza - HPAI, the Bovine Spongiform Encephalopathy −BSE, the Severe Acute Respiratory Syndrome − SARS, the Nipah Virus, and the Ebola Virus Disease (EVD).

The need for increased One Health awareness rose following the realization that; human and animal health is impacted by climates, *for example, in vector-borne diseases;* food and water safety depend heavily on climate stability; antimicrobial resistance has an environmental component; animal,

One Health. https://doi.org/10.1016/B978-0-12-822794-7.00007-1

ecological and environmental health are important to human health; and that human and animal health professionals can acquire more knowledge by comparative medicine and translational research (APHA, 2017).

In August 2009, the One Health Commission was established, to enhance close professional interactions, collaborations among all health disciplines. The commission was endorsed by the 2006 G8 summit, the Food and Agricultural Organization (FAO), the World Organization for Animal Health (OIE), the World Health Organization (WHO), United Nations Children Fund, the Wildlife Conservation Society, and the Centers for Disease Control and Prevention. It is believed that One Health will proffer solutions to global health problems and address the challenges of conservation and development (Barrett et al., 2010).

Defining the boundaries of One Health may appear difficult. However, One Health is known to promote health through interdisciplinary studies and action, across all species of animals (Gibbs, 2014). The One Health Initiative Taskforce Report from the American Veterinary Medical Association (AVMA) defines One Health as "the collaborative effort of multiple disciplines working locally, nationally, and globally to attain optimal health for people, animals and our environment". However, the external action arm of the European Union (EU) defined it as 'The improvement of health and well-being through (i) the prevention of risks and the mitigation of effects of crizes that originate at the interface between humans, animals and their various environments, and (ii) promoting a cross-sectoral, collaborative, *whole of society* approach to health hazards, as a systemic change of perspective in the management of risks' (Okello, Gibbs, Vandersmissen, & Welburn, 2011). The FAO's definition of One Health is in line with the approach adopted by the EU. The definitions of the WHO and the OIE focus on zoonotic threats (Okello et al., 2011). This was why in 2004, there was a global concern that the zoonotic disease, Highly Pathogenic Avian Influenza H5N1(HPAI H5N1), could cause a pandemic and possibly have more mortalities than the estimated 50 million caused by the 1918 Spanish Flu pandemic (Gibbs, 2005) (Fig. 10.1).

2. Benefits of One Health

The introduction of the One Health Initiative provided international organizations (FAO, OIE, WHO and the World Bank) with a vehicle for interinstitutional and interdisciplinary approach to address the threat of the emerging and re-emerging zoonotic diseases. It further enabled these international bodies and national authorities to search holistically for solutions to disease outbreaks that pose threats in the livestock-human-environmental interphase (Gibbs, 2014).

The immediate threat of human pandemic caused by HPAI H5N1 had decreased, however, the global coordination established under the principles of One Health remained in place. This coordination was activated when the

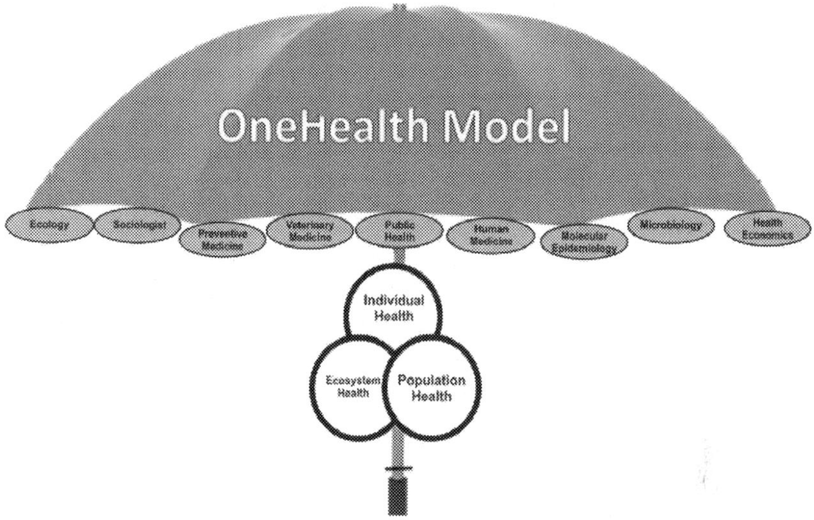

FIG. 10.1 Illustration of different disciplines coming together to constitute the One Health Model.

H1N1 pandemic emerged in 2009. The surveillance systems that were established during the HPAI H5N1 pandemic were able to detect new strains of Avian Influenza, that had the potential to result in widespread disease in humans. As a result of the response to the pandemic influenza, the world is better protected via One Health in action (Gibbs, 2014).

Control programs of other diseases that do not have high public profiles like the Influenza have also gained, by receiving increased attention and financial support. An example is the control and prevention of endemic zoonoses such as Anthrax, Bovine tuberculosis, Brucellosis, Cysticercosis, Echinococcosis, Rabies, and zoonotic Trypanosomiasis, in developing countries.

The Integrated Control of Neglected Zoonoses in Africa (ICONZ), a five-year research project, funded by the European Union and coordinated by the University of Edinburgh, researched on many integrated animal interventions for the control of neglected zoonoses, with innovation and public engagement as the biggest strengths of the project.

ICONZ involved 21 African and European Universities and Research institutes focusing on case studies of zoonotic diseases clusters in seven African countries; Morocco, Mali, Mozambique, Nigeria, Zambia, Uganda, and Tanzania (Gibbs, 2014). This huge collaborative project was aimed at the neglected zoonotic diseases and assisted in linking important interdisciplinary and interinstitutional knowledge gaps on the burden of neglected zoonoses. It also provides a strong evidence base to promote policy decisions at all levels – global, regional, and national levels in developing countries (Okello et al., 2011).

Another good example of the One Health concept that is attracting renewed interest is the global control of rabies. In the year 2011, a memorandum of understanding was signed to encourage increased One Health collaboration between the World Small Animal Veterinary Association (WSAVA) and the OIE. Partners held a symposium, in November 2013, titled *One Health: Rabies and Other Disease Risks from Free-Roaming Dogs*, where the OIE and WSAVA promoted public-private partnerships to implement the appropriate prevention and control methods for rabies. Working closely with donor organizations, the OIE has founded regional vaccine banks to improve the fight against rabies. One of the primary focus of the WSAVA's One Health Committee is Rabies. A dog collar and wristband campaign were launched in 2014, as part of a continuous program in Africa (Gibbs, 2014).

In Chad, a simultaneous evaluation of the health statuses of animals and humans in nomadic pastoralists was carried out. Outcome showed that no nomadic child or woman was fully immunized while a large population of cattle was vaccinated because of compulsory vaccination campaigns (Bechir et al., 2004). This happened because preventive veterinary measures were designed to reach out to the moving population, but the public health services were restricted to immobile health centers. In 2002, a pilot campaign to have a joint human and animal vaccinations was developed, involving the national authorities and a target population. This was aimed at providing simultaneous health service delivery to the people and animals of two provinces in Chad. The first assessment of this joint campaign showed the technical and organizational feasibility of a simultaneous vaccination program in human and animals, and a 15% reduction in costs, from sharing transportation cost and efficiency of public health workers, when compared to the separate campaigns (Bechir et al., 2004).

One Health approach has also proven to improve resource efficiency. An epidemiological study of Rift Valley Fever, *Understanding Rift Valley Fever in Republic of South Africa,* funded by the United States Defense Threat Reduction Agency, was carried out to evaluate the weather and/or climate, vegetation, soil, vector populations, ruminant host exposures and human risk factors of Rift Valley Fever. The cost of transportation between October 2014 and January 2016 for this epidemiological study was evaluated - a total of 333 trips were recorded by the project vehicles during the study period, however, if vehicles were not shared, 483 trips would have been made, showing a 31% decrease in number of trips made. The projected cost of making individual trips was $18,177.08, but with the One Health approach, sharing vehicles reduced the cost to $11,744.41. The savings realized from this could be used in supporting other activities within the project (Rostal et al., 2018). All workers on the project, animal, and health workers, were trained together, on ethics, consent, and confidentiality. This helped to maximize the efficiency of the workforce, achieving the most with a minimum number of staff (Rostal et al., 2018). The decrease in cost seen during this project, as an integrated One

Health program, is in line with estimates of efficiency gain reported by World Bank (Rostal et al., 2018).

There are also economic benefits of addressing public health issues with the One Health approach. For example, in the developing countries, dairy animals are the major source of transmission of brucellosis from animal to man, via the consumption of milk, which is a common source of protein and other nutrients. Even though brucellosis is not fatal in humans, it can lead to chronic and debilitating disease. Vaccinating the animals against brucellosis will break the chain of transmission to human, reducing the population burdened by disease, improve animal productivity, hence ensuring food security for the humans that depend on them. This in turn reduces medical costs via disease prevention and increased productivity (World Bank Report, 2014).

If laboratories are designed to cater for human and animal disease diagnosis, economies of scale will be achieved. Proximity to diagnostic facilities will help to reduce the time it takes to transport test samples to the testing laboratory and in turn reduce the delay of getting results (World Bank Report, 2014). An excellent example is the use of the National Veterinary Research Institute Laboratory in the North Central Zone of Nigeria, as a testing laboratory in the 2020 outbreak of COVID-19.

A better understanding of the epidemiology of noncommunicable diseases can also be achieved using the One Health approach. Comparative medicine has been beneficial to human health through the development of drugs and other scientific discoveries using animal models. However, with improved technology and controversies about animal use and welfare in research, there is a need to improve human and animal health through translational research. One Health issues may be found by monitoring surveillance aspects of comparative medicine, especially assessing multiple species within specific ecosystems for signs that imply new infections, or environmental toxicity. For example, genetic, lifestyle, and environmental risk factors for cancers in humans are highly resource and time intensive to explain, carrying out parallel studies in canine subjects, may provide timely information about cancer prevention and targeting in humans (APHA, 2017).

The United Nations Sustainable Development Goal (SDG) tagged "*Transforming our World; the 2030 Agenda for Sustainable Development*" is focused on alleviating poverty and enhancing sustainable development. The SDGs take an integrated approach and are pertinent to all countries. Health is an important factor in achieving all 17 goals of the United Nations SDGs. An interdisciplinary One Health approach for zoonotic diseases that addresses the connection between health and its social and economic determinants aligns with the SDG framework and implementing health projects with this approach will contribute to the success of SDGs (FAO, OIE, & WHO, 2019). Understanding the effect of emerging and re-emerging zoonoses, and the successful implementation of control measures can reduce poverty and enhance food security through a strengthened livestock management system (Vandermissen & Welburn, 2014).

3. Challenges to the success of One Health approach

One of the major challenges faced by One Health, denying the successful integration of the environment, human and animal health is the ability to define the state of health of the ecosystems (Destoumieux-Garzón et al., 2018). The documents and publications on which the One Health concept and the strategic framework are built around have mainly focused on the fight against emerging zoonoses of domestic or wildlife origin and/or their interactions, without putting the role of inclusive ecosystem into consideration (Destoumieux-Garzón et al., 2018).

The concept also did not have answers to the question of chronic non-infectious diseases (Destoumieux-Garzón et al., 2018), which accounts for about 71% of all deaths globally. However, One Health intends to extend to other fields such as antimicrobial resistance, ecotoxicology, and health in urban environment. (Destoumieux-Garzón et al., 2018).

Data on the burden of zoonotic diseases are not readily available, especially when it is not a re-emerging disease, or posing immediate threat. This makes it difficult to convince policy makers on the benefit of planning and investing in animal surveillance for public health use (Munyua et al., 2019).

One Health is impeded by the dysfunction in the governance of global health and inability to adequately articulate the One Health agenda (Lee & Brumme, 2012). This is because the various definitions of One Health convey a general idea of collaboration, without engaging the specifics of professional roles (Gibbs, 2014).

The role of education in achieving One Health is particularly important, however, the paucity of collaborative programs for students, inadequate environmental training for health professionals and absence of institutional support slows down progress (Barrett et al., 2010).

Following a research conducted in the UK, among scientists, preliminary data showed that many see One Health as an avenue to obtain large research grants and to collaborate with other researchers from other fields. Some researchers saw it as a useful way to rebrand or advertise ongoing works by them, they did not regard One Health as a concept that will promote new research ideas or better collaborations (Gibbs, 2014). The diversity of the One Health definition is a major weakness, and an operational definition is required to achieve the goals of One Health (Lee & Brumme, 2012). As can be deduced from Gibbs (2005, 2014) and Okello et al. (2011), the operational definition of One Health can be said to be a dynamic system that comprises of all ethical and transparent activities that will promote and ensure optimal health for humans and animals in time and space via collaborative effort of the general public and multiple disciplines across the globe.

Many in the medical profession are said to see One Health as a Veterinary *land grab.* According to Atlas (2013), some people in the field of human medicine that see One Health as being championed by veterinarians believe that they have ulterior motives. The reactions from health professionals are divided, as many in the human health sector have not been active in One Health, whereas professionals in animals and environmental health sector have shown huge interest in the concept (Gibbs, 2014). This is so for the professionals in the animal and environmental health sector because their work gives them more opportunities to appreciate the role environment play in the dynamics of zoonosis/infectious diseases (that is, the causation, transmission, and persistence of diseases among susceptible human and animal population). As a matter of fact, Veterinary medicine also includes human health protection, as evident in the science of food safety. Moreover, to extinguish all misgivings among different disciplines about One Health, there is need to introduce curricular about One Health from the post-secondary education level with a view to making it evident early enough in life, the need for different disciplines to come together to contribute to strategies that will ensure sustainable optimal health for humans, animals, and their shared environment.

Research is important to identifying effective ways of promoting health. However, many research funding is usually aimed at specific diseases, making it difficult for interdisciplinary research, as seen in One Health, to get adequate financial support. The EU, US Agency for International Development (USAID) and the Department for International Development (DFID) in the United Kingdom (UK) have however proven that One Health can be funded.

The EU has provided financial support for the ICONZ project. The USAID has also established programs to address emerging pandemic threats. The USAID program is made up of four complementary projects; PREDICT – which focuses on detection of zoonotic diseases at the wildlife interface, PREVENT, IDENTIFY and RESPOND. These programs have technical assistance from the Centers for Disease Control and Prevention and utilizes professionals from both animal and human health sectors, to build local, regional and global One Health capacities, for early disease detection, laboratory-based diagnosis, rapid response and containment and risk reduction (Gibbs, 2014). Effectively, funding programs such as those mentioned above could benefit a sustainable global implementation of One Health.

Many countries, especially developing countries, do not have adequate structures for effective administrative and technical collaboration among animal health, public health, and environmental sectors (WHO, 2019). During zoonotic disease outbreaks or emergencies, insufficient joint preparation and existing structure for collaboration can result in delayed responses which can lead to poorer health outcomes. In the case of endemic zoonotic disease threats, the absence of organized planning, information dissemination, evaluation and control activities across relevant sectors can delay and complicate implementation of effective disease control programs (WHO, 2019). Corruption, at all levels of government is another bane to the success of One Health. Funds allocated to research and/or other One Health projects are diverted to other areas of interest of the persons at the helm of affairs.

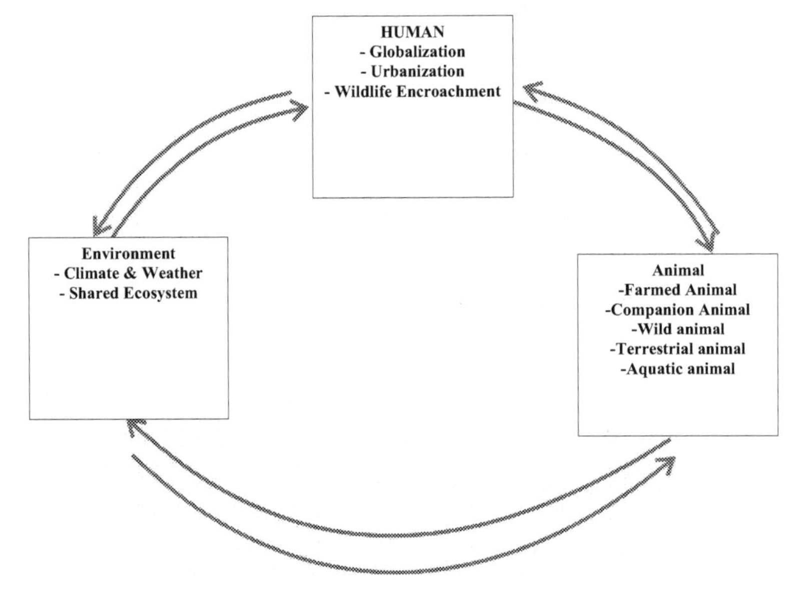

FIG. 10.2 Illustrating how the human and animal share a common ecosystem and how human activities and their increased interaction with animals can impact on their shared environment.

With the progress made so far in the One Health, sustainability of the efforts and success made so far cannot be ascertained, because many of the organizations funding the programs cannot be relied on. They might see a need to divert funds to another agenda presumed more pertinent. Most governments also have many interests including endemic diseases outbreaks and non-communicable diseases within the health sector and focusing more on export of animal and animal products, which appears more important (Munyua et al., 2019) (Fig. 10.2).

4. Practical solutions to challenges of One Health

It is important to involve policy analysts, institutional advocates, and the public to reduce the burden of neglected diseases. Many countries do not regard zoonotic diseases as important because capacity to diagnose them rarely exist (Zinsstag & Tanner, 2008). In rural areas of developing countries, where the risk factors for zoonoses are high, there is the likelihood to underreport diseases because of unavailability or inadequate access to health services (Okello et al., 2011).

Public health departments at all levels of government should communicate and collaborate with agricultural, veterinary, and environmental authorities and other stakeholders in routine and emergency situations. This could be through the use of data-sharing systems, routine meetings, working groups, and communication channels via relevant networks, to build capacity and encourage interdisciplinary coordination and sharing of expertize, including in joint planning, surveillance, and analysis exercises to prevent, detect, respond to, and recover from outbreaks and other health emergencies (APHA, 2017).

Animal and human health professionals working together is complicated. Despite calls for unity (Van der Zeijst, 2008) results have shown slow progress, especially in developing countries where collaboration is almost non-existent (Zinsstag, Schelling, Wyss, & Mahamat, 2005). Better interdisciplinary collaboration can be enhanced through zoonotic disease research centers or joint response teams, as seen in the model founded for the pastoralist-livestock vaccination programs in Chad in 2002 (Zinsstag et al., 2005). To encourage and sustain interest in comparative research, a change in the way research is funded has been recommended. Joint funding from human, animal and environmental health sectors can promote integrated public health programs (Okello et al., 2011). Prioritizing research grants to groups involved in professional collaboration and interdisciplinary research, will promote One Health (Kahn, 2006).

Obtaining the total value of One Health approach calls for the support of and consultation with sectors and industries that are stakeholders in health governance (FAO et al., 2019). The aim of existing One Health partnerships is to reinforce and enhance zoonosis surveillance and control practices to improve livelihoods, ecosystems management and human and animal health.

Surveillance and disease control, at national level, calls for political readiness and an appreciation of the links between human and animal health, the environment, people's livelihoods, and policy processes. This will inform policy recommendations, supported by interdisciplinary approaches that integrate epidemiological, socio-economic, and sociocultural research methodologies (Vandermissen & Welburn, 2014).

Local key stakeholders should encourage discussion and exchanges between local government workers in all involved sectors, to boost understanding and contribute to shared learning among local community members, local associations and group and public services. These will contribute to sustaining One Health (Vandermissen & Welburn, 2014).

One Health will be embedded at local, district and national levels, if communication is maintained among livestock owners, members of the community and authorities, with the intent of identifying acceptable and affordable interventions. Participation from all levels of the society will ensure that vulnerable individuals and communities are made more resilient. Interventions that integrate the appreciation of local knowledge, cultural practices, risk perceptions and gender issues are more inclined to being accepted in local communities (Vandermissen & Welburn, 2014).

Among the numerous challenges humanity is faced with is the emergence and re-emergence of infectious diseases that spread irrespective of local, national and international boarders (Musoke, Ndejjo, Atusingwize, & Halage,

2016; WHO, 2008). Most if not all these diseases are due to unsustainable alterations of the environment and ecosystems by human activities, increasing interactions among humans and animals in the wild and globalization of trade in animals and their products (Musoke et al., 2016; WHO, 2008, 2011). These challenges underscore the existence of dynamic links which have always existed between environment and health of the animal and human population in time and space (Graham et al., 2008; WHO, 2005; WHO, 2011). These linkages explain why from generation to generation, humanity have always adopted and adapted the existing ecosystem to support healthy living of animals and humans by provision of wholesome food, safe water, and medicines, pollution remediation and pathogen regulation (AVMA, 2008; WHO, 2008). Therefore, the environment sector has always been and remains a major contributing sector in public health, however, it is still yet to receive that recognition as a KEY partner for health security in most countries, especially in developing nations of the world. "One Health" concept from inception recognizes the existed connections between human, animal, and the ecosystem/environment, which will ensure that the concept reaches its full potential in practice at all levels: local, national, and global levels. Incorporation of the environmental sector in One Health will provide data, expertize, and management approaches in the environment and natural resources disciplines for comprehensive understanding of the root causes of diseases, and how complexity of environmental factors and natural resources can be tweaked and/or adapted to benefit and maintain a healthy animal and human population in time and space (AVMA, 2008; Graham et al., 2008).

As such, the need to link environment and health can never be overemphasized particularly as we are witnessing increasing awareness in the need for global adoption of One Health for a safe and healthy world. Without doubt, healthy environment is key contributor to optimum health and well-being of any given animal and human population in time and space. A healthy environment will always provide early indices that will allow for timely and adequate preparedness against endemic, epidemic, and pandemic threats (AVMA, 2008).

Educational processes should be initiated at all levels of the society - post-secondary academic levels, all levels of government and among professionals. For examples, an established curriculum across medical, veterinary, public health, engineering, architecture, agriculture, land management, economic, social science, political science, and conservation/wildlife science, etc. faculties can be introduced in the post-secondary academic levels. This will, in the long term, help the society to accept necessary changes in the health model and how society tackles contemporary health challenges. The goal can be a common One Health curriculum for all relevant disciplines and maybe culminated to a continuing professional development and government training programs (Queenan, 2017). Essentially, more focus should be placed on One Health educational programs and effective training should be incorporated early in the process. With the increased institutional support that One Health has gathered over the years, it is important to establish National Centers of One Health Excellence (COHE) for these educational programs to succeed. This will enlighten and strengthen the global health workforce (Barrett et al., 2010).

The concept of participatory epidemiology should be explored as a practical solution to the challenges of One Health. Participatory Epidemiology (PE) is an evolving arm of veterinary epidemiology that utilizes a practitioner's communication skills and participatory methods to enhance the inclusion of animal keepers in the analysis of animal diseases, the design, implementation and evaluation of disease control programs and policies. It is a systematic use of participatory approaches and methods to better understand diseases and options for animal disease control (Catley, Alders, & Wood, 2012). PE stems from Participatory Rural Appraisal (PRA), a multidisciplinary approach that was adopted in the early 1970s, as an alternative system of inquiry during data collection, with emphasis on local analysis and action with communities (Chambers, 1994). A multidisciplinary team used PE and conventional structured questionnaires to investigate the effects of anthrax on human lives and related perceptions of conservation, public health, and veterinary health efforts in the Queen Elizabeth National Park area of Western Uganda. The combination of Participatory methods and One Health resulted in a rich and complex view of the disease and context (Coffin, Monje, Asiimwe-Karimu, Amuguni, & Odoch, 2015).

The growing use of PE demonstrates the versatility of the participatory methods and importance of the method in settings where resources are low. PE appears to play a vital role in aiding researchers and government epidemiologists to relate better with livestock farmers and have a better understanding of diseases from the local point of view (Catley et al., 2012).

5. Conclusion

It is important to simultaneously study zoonoses in people and animals, as this will enhance the One Health approach to making the world healthier. The trend of emerging and re-emerging zoonoses appears to have increased in the last two decades. Many high-income countries can contain them, but many low-income countries struggle to give an adequate response to emerging and even existing zoonoses (Klempner & Shapiro, 2004).

For example, outbreaks of Rift Valley fever in Mauritania were misdiagnosed as Yellow Fever, until contact with the livestock services revealed abortions resulting from Rift Valley fever (Digoutte, 1999). The only case of a human infected with the Ebola Virus Disease in Cote d'Ivoire would probably have been avoided if the behavioral scientists had respected the veterinary directive of not opening infected Chimpanzee cases (Zinsstag et al., 2005). The importance of Q-fever and brucellosis and their animal hosts identified among the Chadian nomadic pastoralist by simultaneous epidemiological studies in human and animal health provided sustainable approaches to assess the importance and epidemiological links of many other zoonoses (Schelling et al., 2003). The relationship between Bovine Spongiform Encephalopathy (BSE) and its human variant Creutzfeldt-Jakob disease also show the need to promote appreciation of how domestic animal systems and related policy impact the human health (Zinsstag & Weiss, 2001).

Furthermore, the present COVID-19 pandemic across the globe, clearly reiterates and underscores how embracing the One Health model will fill-in many gaps in understanding the epidemiology and transmission dynamics of the COVID-19 virus and how to prevent future outbreaks of such diseases (UNEP & ILRI, 2020). A cursory analysis of the present COVID-19 outbreak, informs clearly that One Health can be used to raise awareness at all levels of society about drivers of zoonoses/pandemics and to determine what information is out there in the public on the social, economic and ecological aspects of zoonotic diseases (UNEP & ILRI, 2020). The present global pandemic of COVID-19 revealed that One Health can be used to monitor and regulate systems linked to zoonosis, by so doing we are able to determine key incentives and drivers that are critical for sustainable food systems to avoid drivers of outbreaks from both emerging and re-emerging zoonoses. Therefore, the efficient adoption, adaption and application of One Health Model in this ever-increasing globalized world is long overdue so that across the globe, biosecurity and sustainable control in entrenched in livestock agriculture to avoid the drivers of any form of future pandemics and zoonotic diseases (UNEP & ILRI, 2020). What is now left is how we can globally push for governance and one health approach to sustainably control and prevent zoonoses and pandemics across the globe. To achieve this, there is urgent need to build sustainable capabilities for all relevant stakeholders in the human, animal, and environmental health sectors toward operationalizing the One Health approach to avoid the drivers of pandemic zoonoses.

As stated in the Manhattan principle of 2004, "It is clear that no one discipline, or sector of society has enough knowledge and resources to prevent the emergence or resurgence of diseases in today's globalized world. No one nation can reverse the patterns of habitat loss and extinction that can and do undermine the health of people and animals (Council on Foreign Relations, 2004). Only by breaking down the barriers among agencies, disciplines, institutions, individuals, specialties and sectors can we unleash the innovation and expertize needed to meet the many serious challenges to the health of people, domestic animals, and wildlife and to the integrity of ecosystems (Council on Foreign Relations, 2004). Solving today's threats and tomorrow's problems cannot be accomplished with yesterday's approaches. We are in an era of "One World, One Health" and we must devise adaptive, forward-looking, sustainable and multidisciplinary solutions to the challenges that undoubtedly lie ahead".

References

American Veterinary Medical Association. (2008). *One health: A new professional imperative, One Health initiative task force final report 2008.*

APHA. (2017). *Advancing a "One Health" approach to promote health at the human-animal-environment interface. 2017 12 (Nov.07, 2017).* Retrieved June 29, 2020, from https://www.apha.org/policies-and-advocacy/public-health-policy-statements/policy-database/2018/01/18/advancing-a-one-health-approach.

Atlas, R. M. (2013). One Health: Its origins and future. *Current Topics in Microbiology and Immunology, 365*, 1−13. https://doi.org/10.1007/82_2012_223. PMID: 22527177.

Barrett, M. A., Bouley, T. A., Stoertz, A. H., & Stoertz, R. W. (2010). Integrating a One Health approach in education to address global health and sustainability challenges. *Frontiers in Ecology and the Environment, 9*(4), 239−245. https://doi.org/10.1890/090159

Bechir, M., Schelling, E., Wyss, K., Daugla, D. M., Daoud, S., Tanner, M., & Zinsstag, J. (2004). An innovative approach combining human and animal vaccination campaigns in nomadic settings of Chad: Experiences and costs. *Medecine Tropicale, 64*, 497−502.

Catley, A., Alders, R. G., & Wood, J. L. (2012). Participatory epidemiology: Approaches, methods, experiences. *The Veterinary Journal, 191*(2), 151−160. https://doi.org/10.1016/j.tvjl.2011.03.010

Chambers, R. (1994). The origins and practice of participatory rural appraisal. *World Development, 22*(7), 953−969. https://doi.org/10.1016/0305-750x(94)90141-4

Coffin, J. L., Monje, F., Asiimwe-Karimu, G., Amuguni, H. J., & Odoch, T. (2015). A One Health, participatory epidemiology assessment of anthrax (*Bacillus anthracis*) management in Western Uganda. *Social Science & Medicine, 129*, 44−50. https://doi.org/10.1016/j.socscimed.2014.07.037

Council on Foreign Relations. (2004). *Manhattan principles on "one world, one health"*. Published September 29, 2004 https://www.cfr.org/world/manhattan-principles-one-world-one-health/p22091. (Accessed 10 October 2020).

Destoumieux-Garzón, D., Mavingui, P., Boetsch, G., Boissier, J., Darriet, F., Duboz, P., … Voituron, Y. (2018). The One Health concept: 10 years old and a long road ahead. *Frontiers in Veterinary Science, 5*. https://doi.org/10.3389/fvets.2018.00014

Digoutte, J. P. (1999). Present status of an arbovirus infection: Yellow fever, its natural history of hemorrhagic fever, Rift Valley fever. *Bulletin de la Société de Pathologie Exotique, 92*, 343−348.

FAO, OIE, WHO. (2019). *Taking a multisectoral, One Health approach: A tripartite guide to addressing zoonotic disease in countries*. Retrieved June 30,2020 from https://www.oie.int/fileadmin/Home/eng/Media_Center/docs/EN_TripartiteZoonosesGuide_webversion.pdf.

Gibbs, E. P. (2005). Emerging zoonotic epidemics in the interconnected global community. *The Veterinary Record, 157*(22), 673−679. https://doi.org/10.1136/vr.157.22.673

Gibbs, E. P. (2014). The evolution of one health: A decade of progress and challenges for the future. *The Veterinary Record, 174*(4), 85−91. https://doi.org/10.1136/vr.g143

Graham, J. P., Leibler, J. H., Price, L. B., Otte, J. M., Pfeiffer, D. U., Tiensinet, T., & Silbergeld, K. (2008). The animal-human interface and infectious disease in industrial food animal production: Rethinking biosecurity and biocontainment. *Public Health Reports, 123*, 282−299.

Hristovski, M., Cvetkovik, A., Cvetkovik, I., & Dukoska, V. (2010). Concept of one health - a new professional imperative. *Macedonian Journal of Medical Sciences, 3*(3), 229−232. https://doi.org/10.3889/mjms.1857-5773.2010.0131

Kahn, L. H. (2006). Confronting zoonoses, linking human and veterinary medicine. *Emerging Infectious Diseases, 12*, 556–561.

Klempner, M. S., & Shapiro, D. S. (2004). Crossing the species barrier—one small step to man, one giant leap to mankind. *New England Journal of Medicine, 350*, 1171–1172.

Lee, K., & Brumme, Z. L. (2012). Operationalizing the One Health approach: The global governance challenges. *Health Policy and Planning, 28*(7), 778–785. https://doi.org/10.1093/heapol/czs127

Munyua, P. M., Njenga, M. K., Osoro, E. M., Onyango, C. O., Bitek, A. O., Mwatondo, A., … Widdowson, M. (2019). Successes and challenges of the One Health approach in Kenya over the last decade. *BMC Public Health, 19*(S3). https://doi.org/10.1186/s12889-019-6772-7

Musoke, D., Ndejjo, R., Atusingwize, E., & Halage, A. A. (2016). The role of environmental health in one health: A Uganda perspective. *One Health*. https://doi.org/10.1016/j.onehlt.2016.10.003

Okello, A. L., Gibbs, E. P., Vandersmissen, A., & Welburn, S. C. (2011). One Health and the neglected zoonoses: Turning rhetoric into reality. *The Veterinary Record, 169*(11), 281–285. https://doi.org/10.1136/vr.d5378

Queenan, K. (2017). Roadmap to a one health agenda 2030. *CAB Reviews: Perspectives in Agriculture, Veterinary Science, Nutrition and Natural Resources, 12*(014). https://doi.org/10.1079/pavsnnr201712014

Rostal, M. K., Ross, N., Machalaba, C., Cordel, C., Paweska, J. T., & Karesh, W. B. (2018). Benefits of a One Health approach: An example using Rift Valley fever. *One Health, 5*, 34–36. https://doi.org/10.1016/j.onehlt.2018.01.001

Schelling, E., Diguimbaye, C., Daoud, S., Nicolet, J., Boerlin, P., Tanner, M., & Zinsstag, J. (2003). Brucellosis and Q-fever seroprevalences of nomadic pastoralists and their livestock in Chad. *Preventive Veterinary Medicine, 61*, 279–293.

United Nations Environment Programme and International Livestock Research Institute. (2020). *Preventing the Next Pandemic: Zoonotic diseases and how to break the chain of transmission. Nairobi, Kenya*, ISBN 978-92-807-3792-9. Job No: DEW/2290/NA.

Vandermissen, A., & Welburn, S. (2014). Current initiatives in one health: Consolidating the One Health global network. *Revue Scientifique Et Technique De L'OIE, 33*(2), 421–432. https://doi.org/10.20506/rst.33.2.2297

Van der Zeijst, B. A. (2008). Infectious diseases know no borders: A plea for more collaboration between researchers in human and veterinary vaccines. *Veterinary Journal (London, England 1997), 178*(1), 1–2. https://doi.org/10.1016/j.tvjl.2008.02.023

World Bank Report. (2014). *Implementation completion and results Report on the European commission Avian and Human influenza Trust fund (EC-AHI) – TF012273 in the amount of US $10.00 million equivalent to the Government of Nepal for a zoonoses control. Project (P130089), Report No: ICR00003260 September 15, 2014.*

World Health Organization. (2005). *Combating emerging infectious diseases in the South East Asia region*. http://www.searo.who.int/entity/emerging_diseases/documents/SEA_CD_139/en/index.html. (Accessed 10 January 2020).

World Health Organization. (2008). *Contributing to One World, One Health* A strategic framework for reducing risks of infectious diseases at the animal –Human–Ecosystems interface*.

World Health Organization. (2011). *High-level technical meeting to address health risks at the human-animal-ecosystems interfaces Mexico City, Mexico*. The Food and Agriculture Organization of The United Nations and The World Organisation for Animal Health, 2011.

Zinsstag, J., Crump, L., Schelling, E., Hattendorf, J., Maidane, Y. O., Ali, K. O., ... Cissé, G. (2018). Climate change and one health. *FEMS Microbiology Letters, 365*(11). https://doi.org/10.1093/femsle/fny085. fny085.

Zinsstag, J., Schelling, E., Wyss, K., & Mahamat, M. B. (2005). Potential of cooperation between human and animal health to strengthen health systems. *The Lancet, 366*(9503), 2142–2145. https://doi.org/10.1016/s0140-6736(05)67731-8

Zinsstag, J., & Tanner, M. (2008). One Health: The potential of closer cooperation between human and animal health in Africa (Bovine tuberculosis in Ethiopia). *The Ethiopian Journal of Health Development, 22*, 105–108.

Zinsstag, J., & Weiss, M. G. (2001). Livestock diseases and human health. *Science (New York, N.Y.), 294*(5542), 477. https://doi.org/10.1126/science.294.5542.477

Index

Note: 'Page numbers followed by "f" indicate figures and "t" indicate tables.'

D

E

P

R

S

Printed in the United States
by Baker & Taylor Publisher Services